Tributes
Volume 5

Approaching Truth
Essays in Honour of
Ilkka Niiniluoto

Volume 1
We Will Show Them! Essays in Honour of Dov Gabbay, Volume 1
S. Artemov, H. Barringer, A. d'Avila Garcez, L. Lamb and J. Woods, eds.

Volume 2
We Will Show Them! Essays in Honour of Dov Gabbay, Volume 2
S. Artemov, H. Barringer, A. d'Avila Garcez, L. Lamb and J. Woods, eds.

Volume 3
Probability and Inference: Essays in Honour of Henry E. Kyburg
Bill Harper and Greg Wheeler, eds.

Volume 4
The Way Through Science and Philosophy: Essays in Honour of Stig Andur Pedersen
H. B. Andersen, F. V. Christiansen, K. F. Jørgensen, and V. F. Hendricks, eds.

Volume 5
Approaching Truth: Essays in Honour of Ilkka Niiniluoto
Sami Pihlström, Panu Raatikainen and Matti Sintonen, eds.

Tributes Series Editor
Dov Gabbay   dov.gabbay@kcl.ac.uk

Approaching Truth
Essays in Honour of Ilkka Niiniluoto

edited by
Sami Pihlström,
Panu Raatikainen,
Matti Sintonen

© Individual author and College Publications 2007. All rights reserved.

ISBN 978-1-904987-37-6

College Publications
Scientific Director: Dov Gabbay
Managing Director: Jane Spurr
Department of Computer Science
King's College London, Strand, London WC2R 2LS, UK

http://www.collegepublications.co.uk

Original cover design by orchid creative   www.orchidcreative.co.uk
Printed by Lightning Source, Milton Keynes, UK

All rights reserved. No part of this publication may be reproduced, stored in a retrieval system or transmitted in any form, or by any means, electronic, mechanical, photocopying, recording or otherwise without prior permission, in writing, from the publisher.

Table of Contents

Preface vii

I. Logic, Language, and Mathematics

Tuomo Aho: Seen from a Perspective 3
Leila Haaparanta: The Third Realm Revisited 19
Jaakko Hintikka: Hilbert Was an Axiomatist, Not a Formalist 33
Marja-Liisa Kakkuri-Knuuttila & Simo Knuuttila: Meaning Finitism –
   An Antirealist Position? 49
Martin Kusch: Niiniluoto on Meaning Finitism 65
Jan von Plato: Dialogue on Foundations of Geometry 83
Panu Raatikainen: Truth, Correspondence, Models, and Tarski 99
Veikko Rantala: Perception in Context and Its Logic 113
Gabriel Sandu: Read on the Liar 129

II. Induction, Truthlikeness, and Scientific Progress

Roberto Festa: Verisimilitude, Qualitative Theories, and Statistical
   Inferences 143
Theo A. F. Kuipers: The Hypothetico-Probabilistic (HP)-Method as a
   Concretization of the HD-Method 179
Isaac Levi: Is a Miss as Good as a Mile? 209
Graham Oddie: Truthlikeness and Value 225
Jan Woleński: Induction and Metalogic 241

III. Epistemology, Culture, and Religion

Juha Manninen: Beginning the Logical Construction of Cognition 253
Uskali Mäki: Putnam's Realisms: A View from the Social Sciences 295
Sami Pihlström: Values as World 3 Entities 307
Juha Sihvola: Religion within the Limits of Rational Acceptability 345
Matti Sintonen: The Social and Cultural Roots of the Scientific Method 357
Raimo Tuomela: On the Ontological Nature of Social Groups 381

Ilkka Niiniluoto: Replies 399

Preface

Our friend and colleague Professor Ilkka Niiniluoto celebrated his 60th birthday on March 12, 2006. As his colleagues and former students, we decided to edit a Festschrift consisting of contributions that either directly deal with his work or address issues close to his philosophical interests. It is our great pleasure to present him this collection of papers in honor of his remarkable achievements.

Ilkka Niiniluoto started his career in the tradition arising from the "Finnish school of inductive logic" and received his doctoral degree in theoretical philosophy from the University of Helsinki in 1973, with a thesis on conceptual enrichment and the "hypothetico-inductive" method. Partly in collaboration with Raimo Tuomela (one of the authors of this volume), he developed a realist view of scientific progress. Since the late 1970s, Niiniluoto's work focused on the concept of truthlikeness (verisimilitude), with the aim of making the realist theory of scientific progress mathematically exact. His main technical works on these themes are included in his books, *Is Science Progressive?* (D. Reidel, Dordrecht, 1984) and *Truthlikeness* (D. Reidel, Dordrecht, 1987), but his defense of scientific realism culminated in his more recent monograph, *Critical Scientific Realism* (Oxford University Press, Oxford and New York, 1999). In addition to the books and articles intended for an international expert audience, Niiniluoto has contributed hundreds of articles and other smaller pieces to Finnish journals and anthologies. He has also written several books in Finnish, some of which have been widely used as university-level textbooks in the philosophy of science.

Niiniluoto is, since 1981, Professor of Theoretical Philosophy at the University of Helsinki. As a well-known intellectual who was in 2003 chosen as the Rector of the University, he has constantly defended science and reason in public debates. He has for a long time been one of the most prominent public intellectuals in Finland.

This Festschrift is organized in three main parts. The first part, *Logic, Language, and Mathematics*, contains essays on various issues in the philosophy of logic, the philosophy of language, and the philosophy of mathematics. These include the Fregean "third realm", meaning finitism, and the foundations of geometry. The second part, *Induction, Truthlikeness, and Scientific Progress*, continues with articles on a number of key topics in the kind of "technical" philosophy of science Niiniluoto has significantly contributed to, such as the hypothetico-inductive method and verisimilitude. Finally, the third part, *Epistemology, Culture, and Religion*, collects essays on somewhat broader themes, including the early history of Carnapian logical empiricism, social ontology, and the dispute between theism and atheism.

We should like to thank all our contributors, as well as Ms. Auli Kaipainen for excellent editorial help. With this selection of original articles, we warmly congratulate our teacher, friend, and colleague Ilkka Niiniluoto.

Helsinki, March 2007

The Editors

I
Logic, Language, and Mathematics

Tuomo Aho

Seen from a Perspective

Our modern logic and basic philosophy of language has its well-established conceptual framework, which also conveys some standard ways of interpreting natural sentences formally. In this approach, maximal economy is sought for and achieved, by building upon few syntactic categories and using an apparatus of extensions, intensions, and compositionality. Furthermore, in philosophical study the propositions of language are often seen as abstract or objectively existent entities, in the same manner in which mathematical entities are often viewed. The resulting vision of language is of course rather abstract and condensed.

The amazing success of the logic operating with such a pattern can easily lead one to think that this pattern also gives an adequate picture of what the full logic of language really is: that in order to describe the logical relations in real occurrences of linguistic phenomena we only need to apply the same apparatus ingeniously enough. But such a thought would be misconceived, because language in the primary sense of the word is not a set-theoretic thing but a social and psychological one. We may well start with the abstract basis of logic, but in order to examine the functions of language we need an ascent to the concrete. This modest paper is intended to illustrate the necessity and possibilities of this ascent by a case study, but in the extremely restricted space I can, unfortunately, only give a few theses, not arguments.

In reality, each declarative sentence is a statement *of somebody* or *by somebody*. By using a certain linguistic form, the speaker makes one statement, and propositions really exist only as such phenomena. Thus, the ultimate concern of our philosophy of language must be for sentence-tokens, which are unique and occur in special historical contexts and have particular subjects who produce them. A classic exposition of all these points can be found in Buridan's works, like *Summulae de Dialectica*. As he makes clear, the field of reality that logical theory is about consists of propositional acts. Obviously, it cannot restrict itself to the sentences that have actually occurred, but it disregards such empirical matters and speaks instead about all arbitrary statement-makers and all statements that *might* be uttered.

Here Buridan agreed with the majority of late medieval logicians, who further always divided propositions onto three levels: mental, spoken, and written propositions. We need not care about writing now, but the mental propositions are essential. For, if we want to speak about propositional attitudes, we must speak about that kind of mental states which have an irreducible *propositional* intentional correlate, in other words, about mental propositions, which are mental acts. This has gradually been generally understood when the grasp of intentions has improved, though it ought not be imagined that all attitudes could be reduced to propositional ones. The "attitude contents" can thus be seen as intentions of such mental propositions.

Every attitude is someone's attitude: let us call that someone an attitude-subject. (Generally speaking, this thing need not even be a person: a social group has its attitudes, or a theory has its theses or beliefs.) And every statement is someone's statement: let us call that someone a statement-subject. Let all these subjects of mental and spoken sentences be called *subjects*. (It is hard to find a good term here, "agent" and "person" would be even less satisfactory.) Each occurrence of a sentence-content that p has a subject a; the act of a's judging that p is a proposition that can be referred to as

a: p.

Reflections like this can sound entirely trivial – but they obtain an especial importance in the propositions that are about propositions. That is the case with the sentences describing propositional attitudes. Such sentences display the functioning of subjects and their propositions, and therefore the theory of propositional attitudes has considerable general significance not only as a part of the philosophy of mind but as a field for logical observations as well. Let us first note that, since the proposition may be mental, it is possible that it is qualified by various mental modes, not only by belief, which is conceptually connected to judgment and spoken statements, but also by such modes as perceptual awareness and its varieties, memory, or emotional modes.[1] These

---

[1] The class and criteria of propositional attitudes is, however, somewhat vague. Some attitudes have lexical names, others do not, like various grades and shades of belief. Other examples might be imagination and fancy, wish and will. But the most

modes are expressed with attitude verbs, and in the symbolic notation, with attitude operators like B for belief, S for seeing, and so on. In this way, we come to the theory of propositional attitudes, where each attitude sentence $Att_a p$ reports the fact that a undergoes the proposition that p, qualified by the attitudinal mode Att. (Moreover, it is clear without saying that in speaking about propositional attitudes we must take care of the customary distinction between express attitudes and mental states that could be represented by such propositions, that is, between expressed and unexpressed beliefs, etc. The theory of attitudes mainly operates with the wider interpretation.)

Simple propositions, as we said, could be represented as a: p. But when a makes such a proposition he only says p; saying that p, a commits the propositional act a: p. As the speaker speaks *in propria persona* and presents p as true, the subject need not be mentioned; moreover, for normal speakers and circumstances, the conversational implicature of the utterance surely is that $B_a p$, but we must not write

   a: $B_a p$

since a does not make the judgment that $B_a p$. (The difficult case of first-person attitude sentences cannot be examined now.)

The proposition that a makes – and that expresses a's propositional attitude – can be about the propositions or propositional attitudes that somebody has. Suppose that this somebody is b, and the attitude of b's that is judged about is belief. In this case, a's propositional act has the form

   a: $B_b p$.

Logically, the other subject and attitude-mode can be arbitrary. It is also clear that this procedure can be repeated, producing propositions about attitudes about attitudes about attitudes. All sentences where such propositions are represented are called attitude-sentences. If we fix a certain set of attitude operators, the set of attitude-sentences is definable quite formally, as the closure with respect to these simple operations.

---

famous example of all, knowledge, is an attitude only in a derivative sense, since the occurrence of a knowledge is not the subject's real property but depends on something completely independent of his state, that is, the fact that indeed p.

It is natural to ask next about formal semantics of these sentences. Undoubtedly the most popular approach here has so far been the classical possible worlds semantics of modal logic. As we know, in this semantics a model includes a set W of worlds and for each modal operator O a function O: $W \to P(W)$, which gives for each world w the set of its O-alternatives. If O_a is an attitudinal operator it is customary to interpret $O_a(w)$ as the set of the worlds compatible with a's all O-attitudes in w, that is, the worlds in accordance with everything that a believes, perceives, imagines.

Then it is usually continued by laying down the *truth-definition* for modal sentences. In the case of attitudinal operators, it would read

(1) $w \vDash O_a A$ iff $w' \vDash A$ for all $w' \in O_a(w)$.

My opinion is that this line of thought is completely and fundamentally mistaken. The suggested definition immediately leads to logical omniscience in the well-known way.[2] Moreover, logical omniscience is just a special case of the general problem that arises when one tries to transfer the logical properties and relations of attitude-contents to concern the attitude-sentences themselves. After all, it ought to be no news anymore that there are no logical relations between two contingent natural events. Thus, if the sentences $O_a A$ and $O_a B$ do refer to different attitudinal states (really distinct qualities of a subject) then they cannot be logically interdependent.[3]

But we can preserve one half of the proposed definition, the necessary condition

(2)   if $w \vDash O_a A$ then $w' \vDash A$ for all $w' \in O_a(w)$.

This much seems incontestable, and it gives at least some minimal restrictions for the structure of the alternativeness functions O_a.

---

[2] It is true, of course, that various *ad hoc* attempts have been made for overcoming this problem. However, none of them has been sufficient; and, more importantly, they have been without the slightest philosophical basis.

[3] However, there still remains the possibility that the sentences refer to the *same* real event – the distinction is merely "rational", as the medievals would have said. In that case there might be conceptually necessary or analytic interconnections between them, and it is arguable that some of them could be called logical. These issues have been badly neglected. I hope to be able to return to them in some connection.

So far we have kept to propositional logic. The next remarkable step comes with quantification. Here the distinction between *de dicto* and *de re* sentences is essential. A *de dicto* sentence,

a Os that for some x, $\varphi(x)$,

produces no special problems in principle. It can be understood as a simple sentence $O_a(\exists x)\varphi(x)$ and handled without greater worry like any O_a-sentence. But a *de re* sentence,

concerning some x, a Os that $\varphi(x)$,

is much more problematic. As every textbook of modal logic will tell us, *de re* contexts are always more obscure than *de dicto* contexts, since they apply to objects such predicates (or sentence-functions) that could not even be expressed in a non-modal language of actual world. This has led to innumerable debates. And in the attitudinal case there appears a *further* complication. If we try to represent our sentence straightforwardly as $(\exists x)O_a\varphi(x)$ we run to difficulties since it is possible that a's attitude is about some entity that only exists according to *his* attitudes. ("Concerning the person a believes / imagines to be on the left, he believes / imagines ...") Here it is hard to find examples that are absolutely unambiguous and that no one can ever dispute. Yet the phenomenon itself seems undeniable: people refer, as to well-defined objects, to something that does not really exist.

Thus it is not clear in what way the something that the attitude is about is something. Let us look more closely at this problem in order to find an accurate formulation for it. One illuminating clue may possibly be in the very notion of quantification. Indeed, it has been suggested that it would be essential to distinguish between *physical* and *perspectival* quantification. I would think that this is the most important initiative that survives the collapse of the so-called "attitude logics". It was propounded for the first time perhaps in Hintikka (1969, "On the Logic of Perception"), and utilized in several articles collected in Hintikka (1975).[4] The distinction seems to be very helpful in the translation of natural attitude sentences to logical idiom. (Nevertheless, this task can be surprisingly difficult even with quite elementary sentences.)

---

[4] See especially the instructive paper Hintikka (1970). Similar ideas were also expressed in Kaplan (1969).

Hintikka's original remarks were somewhat brief and hasty, and the functioning of the two ways of quantification has never been sufficiently clarified. As far as I know, the most careful example is still Niiniluoto's study of visual perception, "Knowing that One Sees" from 1979. In fact, later the whole idea has apparently been largely forgotten.[5] The nature of the two quantifiers certainly remains a bit obscure, and hence I would like to propose a reinterpretation in order to acquire a better grasp of them.

De re quantifiers have semantically been construed by means of *world-lines*: functions selecting objects from a given class of worlds. The values of world-lines are said to be "identical", since they are the same *res*. Originally, *physical* quantification was connected to physical world-lines which select objects individuated according to suitable physical criteria, like similarity and spatiotemporal continuity. Hintikka and his followers have assumed that the concrete details of these criteria can be left unexplained, as a task that does not belong to logics at all (e.g. 1969, pp. 169–170). (Somebody might wonder how this identity suits non-physical objects: existent or not, they are spoken about.) If we symbolize the usual physical quantifier as (Ex), this view has read a sentence like

$$(Ex)B_a \varphi(x)$$

as: there is a *physical* object x concerning which a believes that $\varphi(x)$. This does not necessarily entail the actual existence of x but the physical x must be present in the B_a-alternatives, that is, in all the worlds in $B_a(w)$, the worlds compatible with a's beliefs. Its presence is shown by a "physical world-line" that picks objects from the domains of the worlds in this world-class.

The other quantifier is called *perspectival* because the object-selecting world-lines are different for it: they are given in respect with an attitude subject. The objects so identified are subject-bound, but nothing more is said about their identity. Remarkably, no criteria are mentioned here, unlike in the case of physical quantification; in other words, it is not explained what constitutes this perspectival identity. Anyway, the objects so identified belong to a's field of thought. The perspectival quantifier can be symbolized $(\exists x)$, and then a sentence like

---

[5] As an overview of the two quantifiers, please allow me to refer to my old sketch Aho (1994), ch. IV.3.

$(\exists x)B_a\varphi(x)$

will say: there is a *perspectival* object x concerning which a believes that $\varphi(x)$.[6] Neither does this entail anything about the actual existence of x, that is, about its presence in the "actual world" w.

Now that we have introduced the two ways of quantification, we must ask if it is quite obvious what their sense is and what they quantify upon. Unfortunately, the case still is the same as in 1994: "Formulations of the theory of perspective identities have so far been rather brief and unspecific" (Aho 1994, p. 259). In the very first texts the whole question was assumed to be immediately clear. Ilkka Niiniluoto tried to give at least a summary definition of the distinction of physical and perceptual cross-identification between individuals in different worlds (1979; 1982). As he said, speaking about visual perception, "while the former relies on the physical properties of individuals, the latter identifies those individuals which play the same role in the visual field of the percipient" (1982, p. 118; analogously 1979, p. 261). This is an explanation of Hintikka's "public figures" and "acquaintances". The paper Niiniluoto (1985) contains a number of ingenuous points but the crucial characterization remains the same: in the case of imagination, physical world-lines "pick out from each imaginary world the same, 'physically cross-identified' individual", whereas perspectival world-lines cross-identify the things that "play the same role in [the subject's] field of imagination" (pp. 192–193).[7]

Perspectival world-lines go via perspectivally indistinguishable entities. However, the Finnish advocates of perspectival quantification made an extra supposition: that the perspectival world-line can even be extended to the actual world by connecting the possible-world entities to the actual object that is causally responsible for the perspectival impression. One reason for this thought may have been that they often were speaking merely about perception and not about other attitudes. But anyway, such a conclusion appears to be

---

[6] Note that the notation has often been exactly the opposite (and something else too); here we reserve the most usual quantifier symbols for physical quantifiers, and also bypass universal sentences for brevity.

[7] The appeal to "same role" seems to be the best explanation that has been achieved so far. The same jargon was uncritically used in Aho (1994). Twenty years after the introduction of the two quantifiers, in 1987, there still seemed to be more quarrel than understanding about them in the anthology edited by Radu J. Bogdan, *Jaakko Hintikka*.

mistaken. Perspectival objects are identified only in regard to the subject's perspective, regardless of any actual causal factors. (Of course the actual world can belong to the class of alternatives, and perspectival world-lines can go through it, but this depends on the content of the experience and not on additional causal matters.)[8]

After removing this complication, we can consider what it means that something is a perspectival entity for a subject – remembering that all attitudes must be counted among perspectives, not only perception but belief, imagination, will, fear, etc. Then it begins to seem that a perspectival entity in the perspective of a is simply an *entity for a*. Now, what can this mean? Does it mean that we must suppose that every subject has his own world with his own private entities?

This question is by no means futile or preposterous. In fact, Walter Edelberg, who has studied the so-called "Hob–Nob sentences" perhaps more passionately than anyone else, favours such a conclusion. According to him, the solutions of difficult cases can be divided to two camps: the "realist" views, which suppose an objective ontology, and the "perspectivalist" views, which suppose that terms in attitude ascriptions have their references only within a subject's private ontology.[9] And he thinks that only the perspective views can be wholly successful in interpreting attitude sentences. As van Rooy points out (2000, p. 157), what this amounts to is that "what is expressed by a sentence is not said to be true or false in a world, but true or false relative to a theory or belief state". This may suffice to show that the matter is not just grammatical sophistry but remarkable philosophical issues are involved.

The possible-worlds approach does not require such a radical perspectivalism. Though some *world-lines* are called perspectival, that does not necessarily mean that there exist any perspectival *objects* in the real world or in any given world, that is, objects depending on the attitude-contents of any subjects. This, apparently, has not been clear to all readers, though it was rather carefully put forth already in the 1970's by Hintikka, Saarinen and Niiniluoto.[10]

---

[8] See Aho (1994), p. 266.

[9] For an especially concise picture see Edelberg (1992), § 12.

[10] Edelberg's sense for the word "perspective" is quite different from that of Hintikka's followers, and this has given rise to confusion. That seems to be the case e.g. in Leijting-Paulekaite (2001).

Then there still is one important ambiguity in the distinction between physical and perspectival identification. "The former method is called physical or descriptive cross-identification, the latter perspectival or demonstrative cross-identification" (Niiniluoto 1979, p. 261). But the dichotomy, if it is a dichotomy, between descriptive and demonstrative has nothing to do with physical and perspective. It is perfectly possible to identify an object in a purely demonstrative way and then describe a subject's attitudes about it. Likewise, it is possible that a subject identifies an entity with a definite description in the framework of his perspective and then refers to it anaphorically. So let us abandon even this additional characterization of the two ways of identification.

After the preceding eliminations, we can ask: what remains of the distinction between the two quantifiers? And the answer seems to be, nothing. They only quantify in different domains. It is easiest to think of domains here as classes of possible substituents for variables; the members of domains are individuals but not necessarily existent. A completely simple proposition

a: $(Ex)A(x)$

involves referential relations in the sense that it is about the individuals identified by a; those are the individuals in a's repertoire. If $A(x)$ further includes a sentence about b's attitudes, it can happen that the sentence also claims something about b's individuals. But the person b herself individuates her individuals in the same way as a, "physically", as something that she takes to be real constituents of the world according to physical criteria. You can't think of something *de re* unless you think that what you are thinking of is a thing, *res*, that has its identity independently of your present thinking act. And therefore the entities in the class of worlds compatible with a subject's attitude are cross-identified with some physical, or perhaps better *natural*, identity if they are objects of thought.

The overall ideas of world-lines can be applied to special cases and formalized with full definitions. In fact, this work was started in the 1970's and only smaller adjustments are needed. This paper is not a place for technical details, so let me only give a simple example. For a sentence $(Ex)B_a S_b A(x)$ about believing (B) and seeing (S), we get: if

$$w \vDash (Ex)B_a S_b A(x)$$

then there is a's world-line f through $(B_a S_b)(w) = S_b(B_a(w))$ so that whenever $w' \in B_a(w)$, then for all $w'' \in S_b(w')$,

$$w'' \vDash A(f(w)) \ .$$

But if there are only natural objects in every possible world, and if each subject refers to them as natural objects, it seems that we are going to have only the ordinary quantifier of dear old existence. What has happened? Indeed, the *perspective* is not in the quantifier, but in the alternativeness functions. The word "perspectival" may be a little misleading. Dictionaries give *perspectived* as a synonym for it, and that might really be more illuminating. The heart of the matter is that objects are seen and referred to from different points of view which determine their special classes of relevant worlds. We must assume that there is an enormous class of possible worlds – what they are, I am not willing to say, but probably it is reasonable to interpret them linguistically; at least no Lewisian ultra-realism must be accepted. And then there is a family of alternativeness functions, which ensure that some references are situated within perspectives.

In "Über Sinn und Bedeutung" Frege, after defining the two basic notions, proceeds to enquire such contexts where the first main distinction does not seem to work. They are the referentially opaque contexts, and the case he studies is just that of attitude sentences. With his usual genius, he calls them *indirect* contexts, thus pointing straight to the grammatical concepts of direct and indirect discourse. His further analysis of the situation builds upon the abstract notions of *Sinn* and *Bedeutung*, postulating a hierarchy of them; thus the intension of a phrase becomes its extension when it occurs as an attitude content, and it gets a second-order intension, which can again turn into an extension in a new embedding. Thus he analyses the reference shifts in the spirit of logical realism, using an unlimited apparatus of abstract objects.

The same reference shift has, however, been analysed in a less realist manner. For instance Brentano, after Frege, took his clue from the very notions of traditional grammar, those of description *modo recto* and *modo obliquo*. He applied these terms to attitudinal contexts with the idea that they are indirect or oblique in a completely concrete sense, as reports of another subject's mental propositions and its references. Bound to the subject's state of mind, these references can be different from ours.

In other words, since the proposition reported in indirect speech is a certain speaker's or thinker's proposition, even the references there are determined with respect to the entities in the universe of that subject. This does not mean that there are private entities in the world; it only means that for each attitudinal *de re* reference, there must exist an entity in the subject's attitudinal class of alternatives such that the subject makes his judgment concerning that entity. Obviously, all attitudes are not "veridical", i.e., they do not fulfill the success condition $O_aA \rightarrow A$. All attitudes do not even fulfill the "irresistibleness condition" $O_aA \rightarrow B_aA$. So there can be objects of attitudes which are not even believed to exist; it is enough that they are accepted as sufficiently individuated, as entities figuring in the contents of the attitudes in question. One criterion or consequence of sufficient individuation is that the subject can refer to the *same* object again and use anaphora for it. (This is the matter that has been especially difficult to represent logically.)

Let us attempt to apply these considerations to our semantic problem. Perspectival world-lines were supposed to connect perspectivally cross-identified objects, and objects are perspectivally cross-identified if they play the same role in some perspective. Now, what does "the same role" mean? What kind of identities should we draw between *roles*? Role-identity lines, perhaps? Apparently, the whole point behind defining perspectival quantification was to refer to individuals which cannot be "physically" individuated. Thus Niiniluoto explains that if a sees an object but is not able to distinguish whether it is m or n, then m and n play the same role in a's perceptual field (1982, pp. 118-119). It follows that the perceptual field contains an individual that is individuated perceptually but not physically; and therefore a "weaker" (perceptual) quantifier is needed.

However, the whole contrast appears questionable. Attitude contents can indeed be compared, in other words, the different perspectives can be compared. And the speaker or thinker can compare the perspectives of other subjects, or his own other attitudinal perspectives, to the perspective of his beliefs, which contains the individuals he individuates as naturally existent in the actual world. This perspective is then honoured by giving it the name 'physical'. But what is so special in this comparison in itself? If we make such a comparison, we are again using *our* perspective, and the existential claims we make report the state in our implicit attitudes. As far as I can see, this does not entail any relativism. It is possible that such judgments are not true of the actual world (although I believe that a vast amount of our common-sense opinions are literally true). Then it of course would be recommendable to improve our judgments and make them more like truth. But *this* matter has nothing to do with physical and perspectival reference. To put it briefly, the "roles" seem to be nothing but perspectived objects.

If we suppose that classes of alternatives have perspectived objects, an objection is bound to arise very soon: how can the same object appear in two classes? The problem becomes more urgent if it is supposed that the classes correspond to the attitudes of two distinct subjects, in other words, that two subjects have their propositional attitudes, but a same object is present in the perspectives of both. And, to make the issue even more obvious, we can suppose that the object that the attitudes are about is not really existent. How can this be possible? For it surely is possible. The puzzle arose from the 'Hob-Nob sentences' introduced by Geach in his classic paper (Geach 1967):

> Hob believes that a witch blighted Bob's mare, and Nob believes that she killed Cob's sow.

This type of sentences has later been developed to ever more complicated forms. However, if we consider them as connected to a wider philosophical framework, their problem can perhaps be cut in two. *First*, such cases can be seen as problematic for the theory of propositional attitudes themselves. When it has been declared that the entities referred to *de re* in attitudinal contexts are perspectived, and that each perspective corresponds to a certain subject's one attitudinal mode, it seems to demand an explanation how it can happen that one and the same perspectived entity figures in two perspectives. *Second*, from a logico-linguistic point, the Hob-Nob sentences ought to be given a satisfactory grammatical or structural analysis, as any sentences; indeed, they are particularly exciting and challenging sentences for a logical grammarian, and they require sophisticated methods. It is easier to get a grasp of the situation if these two issues are kept apart. The second issue is technically very difficult but it is not necessarily an obstacle for understanding the first issue. I try to say a few words about both questions.

(A) This is in fact an instance of a very fundamental matter. According to a view that seems very convincing to me, the very notions of propositional attitudes are social in the sense that it is essential to have attitudes about other subjects' attitudes. The way we think and speak about attitudes is deeply connected to and dependent on the uses of expressing attitudes that are present in the social community; an attitude could be another subject's attitude as well. (Thus iteration of attitude operators is easy to understand in practice and conceptually necessary in theory, while in the case of genuine modalities the iteration has given rise to much debate.) Followed faithfully, this would have the conclusion that one can also have attitudes about the things about which another subject has attitudes. How is that possible, when the objects of the other subject are perspectived? If we exclude innate ideas, it can be possible only if something has caused the same thing to appear in both perspectives.

In an ordinary case, the reason for this resemblance is that there really exists a natural entity that both subjects are thinking of. It has, in some way, caused their ability to refer to it. But the Hob-Nob case was planned to deny this, because there supposedly are no witches. Therefore it is necessary to suppose that some other causal factor has made the same perspectived object appear in both perspectives. There seem to be two possibilities.[11] Either the joint community of Hob and Nob has caused the two subjects to have their similar witch beliefs. (Perhaps there reigns a common witch mania in the neighbourhood.) In this case the mob causes the beliefs of Hob and Nob. Or it is possible that Hob tells about the first misfortune to Nob, who then draws his own conclusion. In this case Hob causes the shared reference. The main point is that the sentence would be impossible in a situation where Hob and Nob have no connection: there must be a common causal background.[12]

From a completely general point of view, there ought to be nothing astonishing in these situations. They simply mean that a same perspectived entity is present in two perspectives. There is nothing logically repugnant in this. But the only way to *realize* this in fact is to *cause* it by means of a connection, which is normally communication.

---

[11] Strangely, both possibilities are discussed in literature but hardly ever together. Edelberg (1985) is an exception. See Aho (2000).

[12] In one famous example, the "Edelberg Astronomers", American astronomers think that a galaxy cluster has distance x, Soviet astronomers *independently* think that it has distance y. In fact the cluster does not exist at all, the observations result from independent factors. Now, I doubt if this is a Hob-Nob case at all, or rather a case of two simultaneous errors.

(B) As I mentioned, the logical and formal analysis of Hob-Nob sentences has been very difficult. Such a sentence,

> Hob believes that a witch blighted Bob's mare, and Nob believes that she killed Cob's sow,

includes a logically troublesome anaphora (of *she*). As a problem-type, this case is clearly related to the so-called donkey sentences:

> Every philosopher who owns a donkey discusses with it.

Donkey sentences have provoked an enormous amount of literature. The difficulty in them is that the anaphora of *it* refers over a quantifier scope. But Hob-Nob sentences are in some respects even more obscure: here the anaphora acts over the scope of an intensional operator (attitude verb). It is not surprising that grammatically minded philosophers have worked with these sentences,[13] but even genuine linguists have been interested in them.

The observations under (A) are rather well in accord with the assumptions made in some linguistic analyses of the sentence-type. See, for example, Roberts (1996). In Aho (2000) I tried

$$(Ex)(B_H\varphi(x) \,\&\, (Ey)B_N(x = y \,\&\, \psi(y)))$$

as a tentative rendering for the simplest Hob-Nob clause. Edelberg rejects all explanations like that because in his theory all "ideas" are strictly private (Edelberg 1986). It has already become clear that I disagree on that point because of the ontological nature of the objects of thought, but still I do not venture to claim that my reading must be correct. And more complicated sentences surely require much more refined readings and techniques. In any case, the suggested formula is not a sufficient grammatical analysis of the sentence of natural language. We would need accurate and systematic syntactic and semantic representations, which would show also relations between sentences, among other things. Here philosophers can probably do nothing

---

[13] For an outline of the history of the question, see van Rooy (2000), Leijting-Paulekaite (2001).

better than humbly wait for the results of linguists, who are steadfastly working with the secrets of anaphora.[14]

University of Helsinki

Bibliography

Aho, T. 1994: *On the Philosophy of Attitude Logic*, Acta Philosophica Fennica 57, Helsinki.
Aho, T. 2000: "Perspectival Representation", in O. Majer (ed.), *Topics in Conceptual Analysis and Modelling*, The Institute of Philosophy, Academy of Sciences of the Czech Republic, Prague.
Bogdan, R. J. (Ed.) 1987: *Jaakko Hintikka*, Reidel, Dordrecht.
Edelberg, W. 1986: "A New Puzzle about Intentional Identity", *Journal of Philosophical Logic* 15, 1–25.
Edelberg, W. 1992: "Intentional Identity and the Attitudes", *Linguistics and Philosophy* 15, 561–596.
Geach, P. T. 1967: "Intentional Identity", *The Journal of Philosophy* 64, 627–632.
Hintikka, J. 1969: "On the Logic of Perception", in J. Hintikka, *Models for Modalities*, Reidel, Dordrecht.
Hintikka, J. 1970: "The Semantics of Modal Notions and the Indeterminacy of Ontology", in Hintikka (1975).
Hintikka, J. 1975: *The Intentions of Intentionality*, Reidel, Dordrecht.
Kaplan, D. 1969: "Quantifying In", in D. Davidson and J. Hintikka (eds.), *Words and Objections: Essays on the Work of W. V. Quine*, Reidel, Dordrecht.
Leijting-Paulekaite, R. 2001: *Intentional Identity and the Edelberg Asymmetry*, www.illc.uva.nl/Publications/ResearchReports/MoL-2001-05.text.pdf
Niiniluoto, I. 1979: "Knowing that One Sees", in E. Saarinen *et al.* (eds.), *Essays in Honour of Jaakko Hintikka*, Reidel, Dordrecht.
Niiniluoto, I. 1982: "Remarks on the Logic of Perception", in I. Niiniluoto and E. Saarinen (eds.), *Intensional Logic: Theory and Applications*, Acta Philosophica Fennica 35, Helsinki.
Niiniluoto, I. 1985: "Remarks on the Logic of Imagination", in G. Holmström and A. J. I. Jones (eds.), *Action, Logic and Social Theory*, Acta Philosophica Fennica 38, Helsinki.

---

[14] It is not possible to discuss here a further philosophically interesting problem: attitude-sentences have their audience, and therefore the object that the attitude-subject's attitudes are about appears as interpreted in some particular way. This theme was introduced in the work by Bahtin and Vološinov (Vološinov 1929), and nowadays it is investigated in the so-called functional pragmatics in its study of indirect discourse.

Roberts, C. 1996: "Anaphora in Intensional Contexts", in S. Lappin (ed.), *Handbook of Contemporary Semantic Theory*, Blackwell, Oxford.

Van Rooy, R. 2000: "Anaphoric Relations across Attitude Contexts", in K. von Heusinger and U. Egli (eds.), *Reference and Anaphoric Relations*, Kluwer, Dordrecht.

Vološinov, V. N. 1929: *Marxism and the Philosophy of Language*, translated by L. Matejka and I. R. Titunik, Seminar Press, New York, 1973.

Leila Haaparanta

The Third Realm Revisited

1. Introduction

In 1996, Ilkka Niiniluoto's fiftieth birthday was celebrated by a Finnish volume *Tieto, totuus ja todellisuus* (*Knowledge, Truth and Reality*). Anssi Korhonen and I contributed to the volume with a paper titled "Kolmannen valtakunnan vieraina: Huomautuksia loogisten objektien olemassaolosta" ("As visitors to the third realm: Notes on the existence of logical objects"). The title was meant to create a connection between Gottlob Frege and a Finnish author Olavi Paavolainen, who reported his experiences in Hitler's Germany in the 1930s in his book *Kolmannen valtakunnan vieraana*. In that title, the Finnish word 'vieras' means 'a guest', but it may also mean 'strange' or 'a stranger'. Relying on these meanings, we discussed logical objects, which Frege and Edmund Husserl regarded as denizens of the realm of abstract objects. Frege called that realm the third realm, while Husserl called it the ideal realm. We investigated the view put forward by Emil Lask, Frege's and Husserl's contemporary, that logic is homeless; hence, that there is no realm of objects to which logical objects could belong. Lask's thought also influenced Heidegger's philosophy.[1] In the present paper I will take a new look at the problem concerning the existence of propositions and the logical concepts that constitute them. I will revisit the theme of the third realm that Ilkka Niiniluoto has developed in several papers and books, for example, in his recent *Critical Scientific Realism* (1999).

Niiniluoto (1999, 23) stresses the relevance of Karl Popper's distinction between the three worlds to our understanding of how the products of human action exist. In that division, World 1 contains physical things and processes, such as middle-sized ordinary objects, for example, stones and tables, small entities, for example, atoms and cells, large entities, for example, stars and galaxies, and process-like entities, for example, fields of force. World 1 is thus

---

[1] For Lask's influence on Heidegger, see Friedman (1996).

the world of organic and inorganic nature. World 2 is the domain of consciousness, that is, the domain of mental states and processes within individual minds. In Popper's and Niiniluoto's distinction, World 3 consists of artefacts or cultural objects, abstract entities, such as propositions, arguments, theories, problems, numbers, geometric figures, values, and norms, and historical entities and processes, such as events and social institutions (Niiniluoto 1984, 216–217). It is essential to Niiniluoto's realism to first give an answer to the ontological question concerning what there is. However, Niiniluoto is also interested in the epistemological consequences of the ontological difference between Worlds 1 and 3. Particularly, he pays attention to the doctrine of maker's knowledge, which has been an important theme in the tradition of philosophy and geometry. He points out that in this respect the sciences that study the denizens of World 3 have different positions; unlike social scientists, mathematicians have complete maker's knowledge within their reach (*ibid.*, 218–219).

In the present paper, I will discuss the interrelatedness of the being and the being known of the third realm. More specifically, I will pay attention to the relations between the idea of maker's knowledge and the method of analysis in Frege and in ancient geometry.

2. On the History of the Three World Doctrine

The doctrine of the three realms can be found in Hermann Lotze, who was one of Frege's and Bruno Bauch's teachers and whose thought also influenced Heinrich Rickert's philosophy. According to Lotze, abstract objects are not actual, *wirklich*, because their being is not like the being of concrete objects. Instead, they are valid, *geltend*. Lotze takes it to be important to distinguish between what is valid and what is (*was gilt* and *was ist*) (Lotze 1874, 16 and 507). Emil Lask supported that kind of division between the realms (Lask 1923, 6). For Lask, the division was not ontological. Instead, it was a distinction between two different points of view, which we can take towards sensory experience (*ibid.*, 48–49 and 88–91). In everyday experience and in science we are related to psychological, physical and cultural entities, while in the philosophical attitude we are interested in what is valid for those entities. Similarly, from the logical point of view we are interested in the validity, not in the being, of those entities.[2]

Frege presented a doctrine of three realms, by means of which he expressed his view on the being and the being known of logical categories and thoughts that are constituted by those categories. In the first volume of his *Grundgesetze der Arithmetik* (1893) and in his article "Der Gedanke" (1918) Frege makes a distinction between the subjective realm of ideas (*Vorstellungen*), the realm of objective and actual (*wirklich*) objects, and the realm of objects which do not act on our senses but which are objective, that is, the realm of such abstract objects as numbers, truth-values and thoughts (GGA I, XVIII–XXIV; KS, 353). His conceptual notation, which he calls the formula language of pure thought, is meant to mirror parts of the third realm, as it is meant to present the structure of thoughts and the inferential relations between thoughts. Frege's remark in "Der Gedanke", which has to do with what could be called his philosophy of logic, "The third realm must be acknowledged" (KS, 353), receives another meaning from later history. That meaning, even if anachronistic as connected to Frege, is not, after all, very far from Frege's attitudes, as can be seen from his 'political' diary written in 1924.

It is usually assumed that Frege's acknowledgement of the third realm is a Platonic doctrine. Some interpreters have challenged the received view, but others, most notably Tyler Burge (1992), have given strong arguments for the view that Frege held a Platonic ontology; Burge also emphasizes that Frege did not seek to defend his position, except for showing problems in competing views, and that he did not make any effort whatsoever to develop a sophisticated version of his ontology.

In spite of Burge's carefully documented argumentation, other interpretations are still serious candidates if we seek to understand Frege's position. When Frege discusses his third realm in his "Der Gedanke", he remarks that he must use metaphorical language. In other words, such expressions as 'the content of consciousness' and 'grasping the thought' must not be understood literally (KS 359, n. 6). As Frege expresses his worry about the fact that natural language leads us astray as early as in the "Preface" of his *Begriffsschrift*, the interpretation that Frege does not take numbers or thoughts to have being in the proper sense of the word 'being' is at least worth considering. Frege does

---

[2] See Haaparanta and Korhonen (1996).

think that the objects of the third realm are objective, hence, independent of subjective minds. That is not yet an ontological position. On Frege's view, thoughts and their constitutive logical categories are denizens of the third realm but their being is not like the being of the denizens of the objective and actual realm. Thomas Seebohm has also argued that Frege presents a transcendental argument to the effect that the existence of mathematical objects and logical categories is a necessary condition of the meaningfulness of mathematical and logical practice (Seebohm 1989, 348). If that argument holds, Frege's acknowledgement of the third realm would have ensued from his epistemological views.

Husserl argued that we must acknowledge an ideal realm of abstract objects in order to avoid psychologism. He pointed out that there is an unbridgeable difference between the sciences of the real and the sciences of the ideal, as the former are empirical, while the latter are a priori. Husserl realized that if we acknowledge the ideal realm, we must face an epistemological problem concerning our access to this realm. Most of Husserl's logical studies after his *Logische Untersuchungen* are an effort to answer this question by means of the tools of phenomenology. In his last logical works titled *Formale und transzendentale Logik* (1929) and *Erfahrung und Urteil* (1939), which was published posthumously, he seeks to show that we have an access to the denizens of the ideal realm, because we have set the structure of transcendental consciousness to those denizens, hence, we have maker's knowledge of that realm.[3]

Husserl does not think that we could be mistaken about what the correct logical categories are. The problem lies in how to give a justification for what he regards as our true beliefs concerning those categories. In his sixth logical investigation (LU II, 1901, 1921) Husserl studies the components of meaning which determine the form of a proposition, and calls them categorial meaning-forms (*Bedeutungsformen*). In his view, those forms are expressed in natural language in several ways, for example, by definite and indefinite articles, numerals and by such words as 'some', 'many', 'few', 'is', 'and', 'if-then', and 'every' (LU II, A 601/B_2 129; LU II, A 611/ B_2 139). Husserl asks what the origin of logical forms is, when nothing in the realm of real objects seems to correspond to them (LU II, A 611/ B_2 139). He takes it to be a problem how the logical words originally get their meaningfulness, hence, what kind of

---

[3] I have presented Husserl's position in more detail in Haaparanta (1988).

activity of a subject is required in order for the logical words to become meaningful. In his last works on logic Husserl seeks to show that that activity is precisely the activity of transcendental consciousness.

Kant interpreted logical categories as the pure concepts of understanding, which correspond to certain types of judgements and which give form to the objects of experience. In his *Begriffsschrift* Frege, for his part, introduced eight signs as the basic signs of his formula language of pure thought; those signs expressed the basic logical categories and made it also possible for Frege to present most types of judgements listed in Kant's table. As was noted above, in Frege's doctrine of the three realms the logical categories are regarded as constitutive for the denizens of the third realm called thoughts. Two questions arise. First, how did Frege understand his own process of discovering what he considered the correct logical categories? That is, on what grounds did Frege take it to be possible for us to know anything about the third realm? Second, on what grounds did he assume that we have an access to the correct structure of each individual thought? I do not try to give exhaustive answers to these questions; what I will do is to outline Frege's use of the method of analysis. I will focus on Frege's discovery and make a remark on the analysis of individual sentences in the concluding section.

3. Frege and Geometric Analysis

In his work titled *The Ancient Tradition of Geometric Problems* (1986) Wilbur Knorr distinguishes between two ways of understanding the nature of geometry in ancient times. On the one hand, there were the Platonists, the theoreticians; on the other, there were geometers who were close to geometric practice. Theoreticians were interested in theorems; for the practical men the problems mattered more. Solving geometric problems in Euclid's geometry had to do with constructing figures that were described in the given problem. Analysis was the general method for finding the solutions. In geometric analysis, one takes that which is sought *as if* it were admitted and moves from it via its consequences to something that *is* admitted.[4] Taking something as if it were already admitted and constructed normally means drawing a model-figure. The model-figure is an aid but also the object of analysis. The methods of analysis and synthesis were used

---

[4] See Euclid, Book XIII, Prop. 1, and Pappus (1965), *Vol. II*, 634–635.

both in proving propositions and in solving problems. However, Thomas Heath, who comments on the *Elements*, tells us that in ancient geometry analysis was more significant in solving problems than in proving propositions (Heath, "Introduction", 140). Knorr also stresses that the method of analysis was basically meant to offer heuristic power to the ancients in their search for solutions to geometric problems (Knorr 1986, 356). He emphasizes that the activity of investigating problems of construction was prominent in ancient geometry and that the questions of construction primarily concerned problems; what was primarily applicable to problems was then transferred to theorems (*ibid.*, 360 and 368). Knorr remarks that Plato's philosophy gave rise to an opposite view, according to which theorems were eternal and unchanging truths and for that reason proper objects of geometry (*ibid.*, 351). Platonists thought that the being of theorems was absolute; thus, theorems were not constructions. On their view, constructing geometric figures was a lower activity typical of human beings who are close to what is sensible.

In order to be able to draw a certain geometric figure, Euclid relied on the method of analysis. The parts of the required construction were hypothetically assumed or hidden in the model-figure that was drawn in the beginning, but that did not mean that they were really found at that stage. The figure was not known before analysis, even if it was drawn. It was drawn, but not constructed. Analysis was needed for revealing the parts of the required figure. The puzzle we find here is the paradox of analysis. The construction ought to be the same as the result of the process of construction as it is imagined in the beginning; hence, it ought to be the same as the model-figure. If it is not, analysis has not succeeded and we have not constructed what we intended to construct. On the other hand, if it is the same as the model-figure, we must have known in the very beginning how to construct the figure. If this is the case, the process of analysis was not needed at all.

Hence, the peculiar thing in problematic analysis is that even if we draw the model-figure in the beginning, that is, even if we seem to construct the figure, in the real sense of the word we have not constructed it. That is because we do not know how to construct it, which means that we do not have the intuition, the immediate knowledge, which is presupposed by constructive activity. It is the imagined end state, that is, the model-figure, from which we as it were step backwards in analysis; in analysis we reveal matter and the form that the subject gives to the figure in the act of constructing.[5]

Frege's method of analysis can be compared with the method of analysis used in geometric problem-solving. There is very little that Frege has told us about his process of discovering the conceptual notation. This is one of the reasons why interpreters have not discussed this theme in any detail. We cannot even find bold conjectures in Frege scholarship. What I will do in the following is to present a few speculations by using ancient analysis as my model. Even if that comparison were not historically true in the sense that Frege had construed his procedure on the model of geometric analysis, the analogy is still illuminating and reveals certain aspects in Frege's logic of discovery.

Frege states that it is one of the most important differences between his way of thinking and the Boolean way that he does not proceed from concepts to judgments but from judgments to their constituents ("Über den Zweck der Begriffsschrift", 1883, Frege 1964, 101). In fact, Frege starts with assertions formulated in natural language, which express judgments, even if he stresses in the "Preface" of his *Begriffsschrift* that natural language lead us astray. As for model-figures in geometry, the situation is not better; thus, in this respect we can quite well call the assertions of natural language Frege's model-figures. The disanalogy we find here is that, contrary to geometric problem-solving, there is no specific problem in discovering the formula language. We can find a number of remarks in Frege's texts that give us a vague picture of how Frege proceeded. Frege's comments on the philosophical nature of *Begriffsschrift* suggest that in the order of discovery conditionality follows the distinction between arguments and functions.

---

[5] The paradox of analysis can also be found in Husserl's analysis of the objects of the natural attitude or the objects naively conceived. The new objects that a phenomenologist creates via the process of analysis are the objects of the philosophical attitude. It is problematic to argue that the two attitudes deal with the same objects, which we consider from two points of view, because we do not have a third point of view from which we could make such an identity statement. The thesis that the objects are the same is problematic also for the reason that in the process of phenomenological analysis a philosopher leaves the natural attitude behind. Thus, the objects considered from the philosophical attitude cannot be the same as the objects of naive everyday thought and of science. Still, they must be the same objects, if the phenomenologist's analysis is meant to be the analysis of the objects of the natural attitude and not of some other objects. For the model of geometry and Husserl's analysis, see Haaparanta, forthcoming.

For example, when Frege explains why his conceptual notation is better than Boolean logic, he states that it avoids the division of propositions into primary and secondary by construing judgments as prior to concept formation ("Booles rechnende Logik und die Begriffsschrift", 1880/81, NS, 52). Earlier in the same paper he clarifies the difference between Boole's and his own logic in more detail. He states that the real difference is that in logic he avoids a division into two parts, of which the first is dedicated to the relation of concepts, that is, to primary propositions, and the second to the relation of judgments, that is, to secondary propositions (*ibid.*, 14). He continues that, unlike Boole, he reduces his primary propositions to the secondary ones, which can be seen in that he construes the subordination of two concepts as a hypothetical judgment. For example, the subordination of the concept 'square root of 4' to the concept '4th root of 16' is construed as 'if something is a square root of 4, it is a 4th root of 16' (*ibid.*, 17–18). This result comes out when Frege breaks up the judgment which contains subordination. However, before Frege is able to find two thoughts or contents of judgments in one sentence of natural language, he has to realize the distinction between individuals and concepts. This is what he also emphasizes. In the article cited above he remarks that his view does justice to that distinction. That is why the distinction between arguments and functions must be earlier than the concept of conditionality in the order of discovery.

The distinction between individuals and functions, hence also the distinction between the logical functions of being an object and being a function, seems to be a crucial part in Frege's discovery.[6] In the papers where he clarifies his conceptual notation of 1879, Frege stresses the importance of the distinction between individuals and concepts. In "Booles rechnende Logik und die Begriffsschrift" he writes that by turning universal judgments into hypothetical, hence, by reducing what Boole calls primary propositions to secondary ones, we respect the distinction between concept and individual and that this distinction is completely ignored by Boole. In Frege's view, the problem with Boole's notation lies in that Boole's letters never mean individuals but always extensions of concepts. As Frege demands that we must distinguish the case of one concept being subordinate to another from that of a thing falling under a concept, he gives a new form for universal judgments (*ibid.*, 19–20).

---

[6] The priority of the distinction between functions and arguments in the order of discovery has also been stressed by Jean van Heijenoort (1992, 242).

The distinction between individuals and functions, including concepts, seems to ensue from Frege's analysis of judgments. This is because Frege suggests that the names of individuals and the names of functions have crucially different roles in assertions. In the *Grundlagen* Frege connects the distinction with Kant's distinction between intuitions and concepts (GLA, § 27, footnote). In the *Begriffsschrift*, the distinction between individuals and functions is introduced in the form of the distinction between arguments and functions, which Frege construes as parts of expressions (BS, § 9). In his papers in the beginning of the 1890s and in his *Grundgesetze* he considers objects and functions to be two kinds of references (GGA I, § 1 and § 8).

The importance of distinguishing between objects and functions is also stressed in that Frege pays special attention to distinguishing identity from predication in his article "Über Begriff und Gegenstand" (1892). In order to preserve the distinction between objects and concepts Frege considers it necessary to realize that the 'is' of identity differs from the 'is' of predication, and, moreover, that this distinction reflects how things really are ("Über Begriff und Gegenstand", 1892, KS, 168). As was noted above, the distinction between objects and functions appears in the *Begriffsschrift* in the form of the difference between arguments and functions, and it is also mentioned in the paper "Booles rechnende Logik und die Begriffsschrift". The principle according to which objects must be clearly distinguished from concepts manifests itself in that, unlike traditional grammatical analysis, the Fregean analysis of judgments distinguishes the relation between two concepts of the same order from predication, which, for its part, concerns the relation between an individual and a concept or a relation between two concepts of two different orders.[7] The importance of not blurring the distinction between objects and concepts is strongly emphasized by Frege in the *Grundlagen*, where he adds it to the list of his basic principles. Frege also stresses the distinction in the *Grundgesetze* (GGA I, p. X and p. XIV), in "Über die Begriffsschrift des Herrn Peano und meine eigene" (1896, KS, 233) and in "Über die Grundlagen der Geometrie II" (1903, KS, 270). Accordingly, the distinction between the 'is' of predication and the 'is' of class-inclusion, which expresses the relation between two concepts of the same order, ensues from the distinction between individuals and

---

[7] This is emphasized by Frege in "Über die Begriffsschrift den Herrn Peano und meine eigene" (1896), KS, 244.

concepts. Both the conditional form of universal judgments and the logical function of generality are thus consequences of that distinction.

However, when he introduces the concept of conditionality in the *Begriffsschrift*, Frege does not refer to the distinction between individuals and concepts. Nevertheless, it can be seen from his references to Aristotelian inferences in the *Begriffsschrift* that the form of universal judgments is the main motivation for introducing the function called conditionality. In his later paper Frege stresses that a universal sentence is not a sentence about any individuals that we can name; rather, it is a sentence which expresses a condition holding for every object ("Rezension von: E. G. Husserl, Philosophie der Arithmetik I", 1894, KS, 188). It is the distinction between objects and concepts that motivates the difference in the analysis of judgments, as the result of the analysis is determined by the nature of the grammatical subject; the result of analysis depends on whether the subject refers to an object or to a concept. This is what Frege tells us in "Booles rechnende Logik und die Begriffsschrift".

The importance of a specific sign for generality is stressed by Frege in "Booles rechnende Logik und die Begriffsschrift", where he notes that his conceptual notation commands a somewhat wider domain than Boole's formula language, thanks to the notation for generality (NS, 52). For Frege, a universal sentence is not reducible to a conjunction of singular sentences. Instead, Frege extends his universal sentence to concern all objects whether they have a name or not.

Besides showing new developments, Frege's logic also has important sources in the Western tradition of metaphysics. As far as Frege's distinctions between the meanings of the verb 'to be' are concerned, it is illuminating to consider Frege's conceptual notation in the framework of that tradition. However, even if the very structure of the conceptual notation testifies Frege's interest in the traditional question of the meaning of being, it does not follow that much of the content of Frege's works is ontology. The distinction between the 'is' of predication, the 'is' of identity, the 'is' of existence, and the 'is' of class-inclusion is a distinction which Frege makes in the conceptual notation, but those different forms of being are central for him for epistemological reasons; they show some of those important conceptual differences which the conceptual notation must show if it seeks to be a formula language of pure thought.

Hence, Frege stresses the distinction between individuals and functions in various connections and takes it to be one of the basic principles of his thought. As far as Frege's logical discovery is concerned, it is important to realize that after finding the names of objects and the names of functions, Frege can find a number of new things concerning the 'true' structure of the judgment. Those things include the distinction between the 'is' of identity and the 'is' of predication, the distinction between predication and subordination, the hypothetical form of universal judgments, and the concept of generality. This is precisely what we can expect on the basis of the model of geometry, if the primary finding in the process of analysis is essential to the construction. After finding a key element of his model-figure, a geometer is able to take a number of steps in his process of discovery. This is stressed by Euclid and other ancient geometers.[8] Likewise, after finding the distinction between individuals and functions, Frege is able to proceed with his analysis and to reveal a number of features in the structure of judgments.

4. A Concluding Remark

How could we characterize Frege's analytic procedure if we consider individual sentences of natural language and assume that the results of analyses are fully analyzed thoughts? Michael Beaney distinguishes three main modes of analysis, which he calls the regressive, the resolutive or decompositional, and the interpretive or transformative, and argues that these modes may be combined in a variety of ways (Beaney 2002, 54–55). Beaney argues that interpretive or transformative analysis has been important in analytic philosophy (*ibid.*, 67). The interpretive mode of analysis involves 'translating' something into a particular framework, hence also transforming the object of analysis (*ibid.*, 55). In his article, Beaney also compares Frege's, Russell's and Husserl's analyses.

The result of analyzing individual sentences of natural language in the style of Frege is a sentence of the formula language of pure thought, hence, a translation of the sentence of natural language into the formula language of pure thought. However, it is also a new stage in the philosopher's cognitive process. On the one hand, the result of analysis ought to be the same as the original expression that was fixed as the object of analysis; on the other, it must be a different object for the philosopher who knows more after the analytic procedure. The starting-point and the end-point of the process of analysis, the end-point being the construction of the fully analyzed object, can also be recognized in a geometer's analysis of model-

---

[8] See, e.g., Pappus (1965), *Vol. II*, 831–833. Cf. Heath's introduction to the *Elements*, 141–142.

figures, as was noted above. On the one hand, Frege's analysis is remembering something that was there already; hence, it is meant to reveal the hidden knowledge concerning the structure of a denizen of the abstract realm that we had in the very beginning. On the other, it is transformative in the sense that it also creates by making explicit something that was not there for us before the process of analysis.

Department of Mathematics, Statistics and Philosophy, University of Tampere

References

Beaney, M. 2002: "Decompositions and Transformations: Conceptions of Analysis in the Early Analytic and Phenomenological Traditions", *The Southern Journal of Philosophy* 15, 53-99.
Burge, T. 1992: "Frege on Knowing the Third Realm", *Mind* 101, 633-650.
Euclid 1926: *The Thirteen Books of Euclid's Elements, Vol. I-III*, transl., intr. and comm. by T. L. Heath, Cambridge University Press, Cambridge.
Frege, G. 1964: *Begriffsschrift, eine der arithmetischen nachgebildete Formelsprache des reinen Denkens* (1879), in Frege (1964), pp. 1-88.
Frege, G. 1964: "Über den Zweck der Begriffsschrift" (1883), in Frege (1964), pp. 97-106.
Frege, G. 1968: *Die Grundlagen der Arithmetik: eine logisch mathematische Untersuchung über den Begriff der Zahl* (1884), in *The Foundations of Arithmetic/ Die Grundlagen der Arithmetik*, repr. and transl. by J. L. Austin, Basil Blackwell, Oxford. (Referred to as GLA.)
Frege, G. 1893: *Grundgesetze der Arithmetik, begriffsschriftlich abgeleitet, I. Band*, Verlag von H. Pohle, Jena. (Referred to as GGA I.)
Frege, G. 1964: *Begriffsschrift und andere Aufsätze*, hrsg. von I. Angelelli, Georg Olms, Hildesheim.
Frege, G. 1967: *Kleine Schriften*, hrsg. von I. Angelelli, Wissenschaftliche Buchgesellschaft, Darmstadt und Georg Olms, Hildesheim. (Referred to as KS.)
Frege, G. 1969: *Nachgelassene Schriften*, hrsg. von H. Hermes, F. Kambartel, und F. Kaulbach, Felix Meiner Verlag, Hamburg. (Referred to as NS.)
Frege, G. 1994: "Gottlob Freges politisches Tagebuch", Mit Einleitung und Kommentar herausgegeben von G. Gabriel und W. Kienzler, *Deutsche Zeitschrift für Philosophie* 42, 1057-1066.
Friedman, M. 1996: "Overcoming Metaphysics: Carnap and Heidegger", in R. N. Giere and A. W. Richardson (eds.), *Origins of Logical Empiricism, Minnesota Studies in the Philosophy of Science, Vol. XVI*, University of Minnesota Press, Minneapolis, pp. 45-79.
Haaparanta, L. 1988: "Analysis as the Method of Discovery: Some Remarks on Frege and Husserl", *Synthese* 77, 73-97.

Haaparanta, L.: "The Method of Analysis and the Idea of Pure Philosophy in Husserl's Transcendental Phenomenology", forthcoming in M. Beaney (ed.), *The Analytic Turn in Early Twentieth-Century Philosophy*, Routledge, London.

Haaparanta, L. and A. Korhonen 1996: "Kolmannen valtakunnan vieraina – huomautuksia loogisten objektien olemassaolosta", in I. A. Kieseppä, S. Pihlström and P. Raatikainen (eds.), *Tieto, totuus ja todellisuus*, Gaudeamus, Helsinki, pp. 38–44.

van Heijenoort, J. 1992: "Historical Developments of Modern Logic" (1974), *Modern Logic* 2, 242–255.

Husserl, E. 1950: *Logische Untersuchungen I, Husserliana, Band XVIII*, Text der 1. (1900) und der 2. (1913) Auflage, hrsg. von E. Holenstein, Martinus Nijhoff, Den Haag. (Referred to as LU I, A/B.)

Husserl, E. 1984: *Logische Untersuchungen II, Husserliana, Band XIX/1–2*, Text der 1. (1901) und der 2. (1913, 1. Teil; 1921, 2. Teil) Auflage, hrsg. von U. Panzer, Martinus Nijhoff, The Hague/Boston/Lancaster. (Referred to as LU II, A/B$_1$ and LU II, A/B$_2$.)

Husserl, E. 1929: *Formale und transzendentale Logik: Versuch einer Kritik der logischen Vernunft*, Verlag von Max Niemeyer, Halle.

Husserl, E. 1964: *Erfahrung und Urteil: Untersuchungen zur Genealogie der Logik*, red. und hrsg. von L. Landgrebe (1939), Claassen Verlag, Hamburg.

Knorr, W. R. 1986: *The Ancient Tradition of Geometric Problems*, Birkhäuser, Boston.

Lask, E. 1923: "Die Logik der Philosophie und die Kategorienlehre" (1911), in E. Lask, *Gesammelte Schriften II*, hrsg. von Herrigel, J. C. B. Mohr, Tübingen, pp. 1–282. (Referred to as LP.)

Lotze, H. 1874: *System der Philosophie, Erster Teil: Drei Bücher der Logik*, Verlag von G. Hirzel, Leipzig.

Niiniluoto, I. 1984: *Is Science Progressive?*, D. Reidel Publishing Company, Dordrecht/Boston, Lancaster.

Niiniluoto, I. 1999: *Critical Scientific Realism*, Oxford University Press, Oxford.

Pappus 1965: *Collectionis quae supersunt*, 3 vols. (Weidmann, Berlin, 1876–78), Adolf M. Hakkert, Amsterdam.

Seebohm, Th. 1989: "Transcendental Phenomenology", in J. N. Mohanty and W. R. McKenna (eds.), *Husserl's Phenomenology: A Textbook*, Center for Advanced Research in Phenomenology & University Press of America, Washington, D.C., pp. 345–385.

Jaakko Hintikka

Hilbert Was an Axiomatist, not a Formalist

1. Hilbert's Project in Present-day Perspective

In the nineteen-twenties, one of the most important developments in the foundations of mathematics was the project by the prominent German mathematician David Hilbert. This program is often called formalistic, and Hilbert's "formalism" is still routinely listed as one of the three main traditional approaches to the foundations of mathematics, besides logicism and intuitionism.

What Hilbert wanted to do was to free the foundations of mathematics from the doubts and uncertainties that had been surfaced partly as a consequence of set-theoretical paradoxes and partly as a result of the criticisms by Brouwer and Brouwer's followers, notably Hermann Weyl. (See e.g. Brouwer 1921; Weyl 1918 and 1921.) Hilbert proposed to eliminate all these doubts in one fell swoop by proving the consistency of different mathematical theories, in the first place analysis. The success of the graduate process of an arithmetizing of analysis in the nineteenth century suggested that the crucial part of this program was to prove the consistency of arithmetic.

Needless to say, the consistency proof had to be carried out by means that were not subject to the doubts that affected set theory and – at least according to Brouwer – analysis and even logic.

In 1931 Kurt Gödel demonstrated that the formal consistency of as basic a mathematical theory as elementary arithmetic cannot be proved by means of elementary arithmetic itself (Gödel 1931). This proof was a by-product of Gödel's famous incompleteness theorem. This theorem says that in any formal axiom system AX of elementary arithmetic there are propositions G that are true but unprovable. Gödel's proof assumed that AX is consistent. As Gödel (and John von Neumann) quickly realized, his proof could be formulated in

AX itself. Hence, if the consistency of AX could be proved in AX, G could likewise be proved in AX, which would contradict its unprovability.

Gödel's argument, known as his second incompleteness theorem, has virtually universally been taken to imply a failure of Hilbert's program in its original form. This construal of the consequences of Gödel's results is mistaken. For one thing, it can be shown that the consistency of a suitable first-order system of elementary arithmetic can be proved in the same system (Hintikka and Karakadilar, 2006). As a consequence, we have to reconsider the entire question of the prospects of Hilbert's program.

This claim might easily seem puzzling and even paradoxical. Gödel's second incompleteness theorem is a valid metatheoretical result. In view of its indisputable validity, it seems hopeless to try to get around it.

Several different specific sources of this puzzlement can be diagnosed. The most general perspective can be explained already here. Gödel's argument does not rest on any assumptions that can be challenged in any literal sense. However, the framework Gödel is operating in is not the only possible one, and is turning out not to be the happiest one. Gödel assumes that the logic that is used in his elementary arithmetic is the ordinary first-order logic. Or, rather, we should speak of the received Frege-Russell logic of quantifiers of any order, for the first-order part was separated from it only slowly under Hilbert's influence. It has by now turned out that this logic is too poor, however, and has to be replaced by a logic richer in expressive power. This logic is known as the independence-friendly (IF) first-order logic. (See Hintikka 1996 and 1997(a).) Since it is richer than ordinary first-order logic, it cannot be excluded sight unseen, when it is used as a basis of elementary (first-order) arithmetic, that the consistency of this arithmetic should become provable in the same arithmetic (with the same stronger logic, of course). Such a resurrection of Hilbert's project is thus a part of a general revolution in the foundations of logic and mathematics brought about by IF logic. (Cf. Hintikka 1997(a).) Or perhaps the term "revolution" is inappropriate. Unfortunately it is not easy to find an appropriate term, even a metaphoric one, that would capture the double development of deepening the foundations of logic and *ipso facto* extending the range of its applications.

However, it is far from obvious that the enriching of one's basic logic facilitates a Hilbert-style project. Three prima facie reasons seem to discourage any attempt to salvage Hilbert's project with the help of a richer logic. For one thing, enriching the underlying logic makes one's arithmetic stronger and hence more prone to contradictions than before, and presumably by the same token more difficult to prove consistent.

A closely related indication of the difficulties here is the fact that Gödel's 1931 incompleteness proof apparently turn on reasoning closely related to the liar paradox, even though in Gödel's hands it results in an incompleteness result rather than contradiction. Strengthening the underlying logic thus might sight unseen heighten the danger of liar-type paradoxes.

Most importantly, Hilbert's idea was, as we will see, to prove the semantical (model-theoretical, intuitive) consistency of a formal system of elementary arithmetic by proving its deductive (syntactical, formal) consistency in the sense that no contradictions can be proved by the formal rules of inference from the fully formalized axioms. This apparently presupposes that the formalized logic is semantically complete in the sense that every logical truth in the semantical (model-theoretical) sense can be formally proved in it. The received first-order logic was proved complete in this sense by Gödel in his 1930 dissertation. (See Gödel 1930.) But IF first-order logic is known to be semantically incomplete. The set of its valid formulas is not recursively enumerable in the sense in which fully formalized rules of inference effect an enumeration of the theorems that can be proved by their means from formalized axioms. Hence at first sight a reliance on IF logic might seem to make consistency proofs in Hilbert's style totally impracticable.

2. Formalist or Axiomatist?

It can be shown that all these obstacles can be overcome. As a preparation for doing so, we will in this paper examine Hilbert's motivation more closely so that it can be better understood precisely what Hilbert's program was and what can count as carrying it out.

Hilbert is routinely called a formalist, and his entire school is labeled in the same way. This is in the most relevant sense of the word a radical misunderstanding. If we want to be charitable we can distinguish what might be called philosophical formalism from the doctrine of the formal character of logic. By philosophical formalism in the philosophy of mathematics I mean the view that mathematical activity consists (or should be understood as consisting) primarily of the manipulation of formal symbols. By the formal character of logic I mean the view that the validity of a logical inference, say an inference from S_1 to S_2 is independent of the nonlogical constants occurring in S_1 and S_2. It depends only on the logical structure of S_1 and S_2.

It can be shown that Hilbert was not a philosophical formalist. What is true is that he believed wholeheartedly in the formal character of purely logical reasoning. This point cannot be expressed more forcefully than Hilbert did himself in his inimitable manner. In speaking of the derivation of theorems from the axioms of geometry, he said that in these derivations one might as well speak of tables, chairs and beermugs instead of points, lines and circles. (See Hilbert 1899 and 1918.)

It was because of this belief in the formal character of logical reasoning that Hilbert was originally labeled "formalist". Someone might even defend this label and suggest that this usage merely involves a secondary sense of the term, different from the philosophical formalism defined earlier. I am afraid that such a usage, even if it might accord with ordinary discourse, would still be seriously misleading.

The use of the misleading label was later encouraged by Hilbert's use of formalization as a part of this project of proving the model-theoretical consistency of mathematical theories by showing their formal (proof-theoretical) consistency. As such, such a project does not commit anyone to philosophical formalism. By itself, it does not justify applying the label 'formalist" to Hilbert. (Cf. here Kreisel 1958 and 1983.)

Yet one can perhaps see in Hilbert's project the beginning of a slippery slope that could eventually lead to philosophical formalism. As was mentioned, his project seems to presuppose that we have found a semantically complete axiomatization of the underlying logic. But if so, all derivations of theorems from axioms could be carried out purely formally. And since they are in the last Hilbertian analysis all that is involved in pure mathematics, all mathematics would fit into a philosophical formalist's framework.

This slippery slope is not unavoidable. However, close analysis is needed to establish to what extent Hilbert was or was not aware of it. It is in any case of interest that archetypally non-formalist model-theoretical questions of semantical completeness – and even the very concept of semantical completeness – seem to have come up for the first time within Hilbert's circle. Indeed, the completeness of propositional logic was proved there at an early stage of the game, and Gödel's motivation for his completeness proof (for the received first-order logic) seems to have been done at least in part due to Hilbert's influence. Furthermore, Hilbert's one time right-hand man John von Neumann was certainly paying close attention to questions of completeness so much so that he lost his interest in logic because of Gödel's incompleteness results. Could it be that later thinkers have looked at the fate of Hilbert's program too exclusively through the lenses of someone like von Neumann?

One way of seeing the difference between the two meanings of the term "formalist" is to envisage a situation in which the philosophical formalism fails, that is in which mathematical activity cannot be reduced to the manipulation of formulas, such as formal deductions, in the sense of mechnizable deductions. Indeed, we do not have to imagine such an eventuality, because in a sense we are faced with it in actual mathematics. Much of actual mathematical reasoning can be thought of as being carried out in second-order logic. Such an enterprise cannot be restricted to formal deduction, for there is no complete axiomatization of second-order logic. Hence philosophical formalism cannot yield an adequate account of mathematical practice. Deduction must be complemented by a gradual addition of new principles of proof, presumably on the basis of suitable model-theoretical considerations.

However, this does not make any difference to the formal character of relations of logical consequence in a different sense of "formal". A sentence S_1 logically implies another one, say S_2, if and only if the same relation holds between any two sentences of the same logical form but with different non-logical constants. In this sense, what was called the formal character of logical reasoning is an obvious truth. Philosophers' frequent references to "informal logic" betray a confusion. The charitable way of understanding such phrases is to take them to refer to strategies of reasoning.

Hilbert's own view of the nature of mathematics and of the role of logic in mathematics is not formalistic but strictly axiomatic. Mathematics – perhaps one should say here, pure mathematics – consists of setting up axiom systems and then deriving theorems from the axioms by purely logical means. Hilbert's own mathematical practice should be enough to show the importance of the axiomatic method for him. Not only did he axiomatize geometry, he spent a great deal of time and energy in trying to axiomatize different physical theories. It was in fact believed earlier that Hilbert discovered Einstein's general theory of relativity independently of Einstein. It has been shown that his discovery was not independent of Einstein, but even so it is an impressive proof of the seriousness of Hilbert's work in physics. For several generations mathematical physicists, their most important toolbox was Courant's and Hilbert's *Mathematische Physik* (1931–37). One of Hilbert's famous open problems in mathematics – the sixth problem – was simply the injunction "axiomatize physical theories". In pure mathematics Hilbert inspired or at least encouraged Zermelo's axiomatization of set theory. If I may put it in the form of a slogan, there are no formalists in the trenches of mathematical physics. And this formulation would have been more than a slogan for Hilbert. If even there was a temptation to look at a formal technique formalistically, it was offered by Dirac's use of his delta function (or, rather, non-function) in his version of quantum theory. Yet it was largely Dirac's very use of mathematically uninterpreted symbols that prompted Hilbert and von Neumann to develop the current mathematical techniques for quantum theory which avoids all use of such uninterpretable functions.

Now why is such axiomatization supposed to be important? And why should such an axiomatization be purely logical, as Hilbert insisted? The main part of the answer is obvious. By compressing everything that we know in some field of knowledge into a logical axiom system we can obtain an overview on this entire field.

Ironically, such a view on mathematical theorizing is not only not a formalist one. It is almost diametrically opposite to a formalist philosophy of mathematics in that it is a model-theoretical one. On this view mathematical axiom systems do their job by specifying the class of their models. These models are specific objective structures, not merely symbols on paper (or on a computer screen) or thoughts in one's mind. Of course, model theory as a systematic part of contemporary logic came about only in the late fifties largely through the efforts of Alfred Tarski and his students and associates (Chang 1974). But Hilbert's thinking can be called model-theoretical both in the sense that he conceived of the job description of an axiom system to be the specifications of a class of models and in the historically interesting sense that his thinking was free from the belief in the primacy of the purely formal (syntactical) methods that otherwise dominated logical theory in the first half of the twentieth century.

Hilbert was able to maintain his axiomatic view of mathematics because he made a sharp distinction between pure mathematics and applications of mathematical theories. All that a pure mathematician does in principle is to set up axiom systems and to derive logical consequences from the axioms. But Hilbert was perfectly well aware that when an interpreted axiom system is applied to reality, as in e.g. in the application of geometry to physical reality, the truth of all axioms has to be established empirically. (See Hilbert 1918, pp. 409–410.) Interestingly, in view of the controversies about the worries as to whether purely mathematical techniques might smuggle in substantial assumptions, Hilbert maintains that even the applicability of geometrical continuity assumptions has to be established empirically.

But is Hilbert's axiomatic approach viable in the light of later developments? Some philosophers seem to think that it is refuted by the semantical incompleteness of the logics that are strong enough to codify mathematical reasoning. This is a mistake. The idea of purely logical axiomatization does not presuppose that the underlying logic is semantically complete in the sense that all valid formulas can be recursively enumerated by deriving them from a finite or recursive set of axioms by means of purely formal rules of inference. An axiomatization can be purely logical even when the derivation of theorems from axioms is mediated by semantically valid inferences instead of mechanizable derivations. What is required is that these inferences are purely logical and that they accordingly are analytic, in other words do not introduce new information to the argument.

Indeed, IF first-order logic offers an example of means of making logical inferences which depend only on the logical structure of the propositions involved and hence do not convey any new information, but which are not captured by mechanical rules of inference. The use of IF logic as our generic logic in mathematics does not make impossible the kind of axiomatization that Hilbert intended. What the primacy of IF logic in mathematics would mean is a disproof of philosophical formalism: mathematical reasoning cannot be reduced to a mechanical manipulation of formulas. Thus Hilbert's axiomatic approach, far from being formalist, ultimately leads to a rejection of philosophical formalism.

3. Sentences Providing Their Own Models

But what about other alleged evidence to the effect that Hilbert was a formalist? Suffice it here to discuss the one passage everybody cites who claims that Hilbert was a formalist. It occurs in his 1922 "Neubegründung" paper:

In that I adopt this point of view, the objects of number theory are – in a precise opposition to Frege and Dedekind – symbols themselves.

But a closer examination of this passage quickly shows that to interpret it as a profession of formalism in the normal sense of the word is to read it grossly out of context. For one thing, what is the opposition to Frege and Dedekind that Hilbert flaunts? These logicians and gentlemen were not opposed to the use of symbols in logical proofs. Frege can even be said to have created the paradigm of a purely formal system of logic. Neither in the context in question nor anywhere else does Hilbert emphasize or even mention that Frege and Dedekind should be opponents of formalism. What Hilbert has emphasized and criticized in his lead-in to the quoted passage is that both Frege and Dedekind had been dealing with higher-order entities, such as concepts or their extensions, and perhaps even had been assuming that such entities can be given to us *ohne weiteres*. In their place Hilbert wants to place discrete individual objects that can be given to us intuitively and in immediate experience. Now one of the most classical assumptions in the philosophy of mathematics is that only individuals can be so given. Hence what Hilbert is defending here is not formalism, but first-order logic in contrast to a higher-order one. (I made this point earlier in Hintikka 1997(b).) This is in line with Hilbert's overall view on logic and logical reasoning. It is not an accident that it was in Hilbert's school that first-order logic was separated from the higher-order quantification theories of Frege and Russell & Whitehead. Hence Hilbert's opposition to Frege and Dedekind is not an opposition of a formalist to a non-formalist but an opposition of a nominalist to conceptual realism.

And what is "this point of view" Hilbert refers to? He has been discussing the nature of purely logical reasoning. Now the most firmly entrenched idea of Hilbert's concerning purely logical inference, for instance concerning the derivation of theorems from axioms, is that it is independent of the particular nonlogical concepts involved. As Hilbert puts it elsewhere, in deriving theorems from geometrical axioms, we might as well speak of tables, chairs and beermugs instead of points, lines and circles. This is not a formalist idea. The most rigid model theorist can embrace it. For him or her it is only a reminder that logical reasoning is valid modulo isomorphism.

The same holds with vengeance of the individual objects that there are in the models we are thinking of and perhaps even manipulating. In the sense just explained according to Hilbert – and he is right – we can choose freely the objects we think of as the individuals our theory deals with. What Hilbert is here proposing is to study the intended models of axiomatized number theory by studying those models of number theory whose domain consists of symbols.

Whether or not Hilbert's recommendation should be followed or not, it has nothing to do with formalism. It amounts to proposing a kind of possible research tactic. And this tactic is not only possible. It has been used in a small scale by subsequent logicians. In his completeness proof for first-order logic Leon Henkin (1949) used as models for certain kinds of sets of formulas (symbol combinations) those very same sets of symbols themselves. Later, Hintikka (1955) and Smullyan (1968) generalized Henkin's argument. Henkin's idea obviously belongs to the same ballpark as the idea of Gödel numbering and indeed Hilbert's formalization of metamathematics.

There is no reason to project into Hilbert's statements anything more than such a tactical recommendation. Indeed, the reasons Hilbert himself adduces for his proposal are not philosophical or otherwise theoretical, but purely practical

> As a precondition of the application of logical inference .... something must already be represented, viz. certain nonlogical discrete objects which are present in all thinking intuitively and in immediate experience.

In sum, what Hilbert is saying is that since the particular selection of the domain of a model of number theory is arbitrary, we might as well think of this domain as consisting of symbols, presumably number-theoretical symbols. Such a view can be called formalism only by discourtesy to Hilbert's real intentions. If you take Hilbert's words in the quoted passage to mean literally that according to him number theory is about symbols, then you must be prepared to interpret his other pronouncements equally literally and believe that for him geometry is about tables, chairs and beermugs.

Thus, even though the quoted passage does not show that Hilbert was a formalist, it shows something else, to wit, that he was committed to first-order logic in his thinking about the foundations of mathematics.

One can push this line of thought even further. Hilbert recommends symbols as objects of number theory because they are present in immediate experience, because their individuality and all their parts are given to us, as is their order. What does this mean in the philosophical terminology of Hilbert's time? What it amounts to is a plea for symbols as objects of number theory because of their intuitivity. Hence Hilbert is not trying to banish intuitions from number-theoretical reasoning in favor of a purely formal or symbolic approach. On the contrary, he is trying to make logical inferences and logical operations more intuitive. But what precisely is it that is being made intuitive in this way? Not the symbols per se, for their intuitiveness independently of their representative role is irrelevant. Not the ostensive objects of number theory, numbers as numbers, for they are not made any less abstract by having these or those symbols to represent them. What is intuitive is what the symbols tell us about their objects. Hence what are hopefully made more intuitive are the model-theoretical aspects of number theory. Hence Hilbert's statements referred to here are part and parcel of his axiomatic and model-theoretical approach.

It might seem at first glance that the Henkin-type idea of using configurations of sentences as their own models is an artificial gimmick. This impression is misleading, however. Henkin's technique, and the similar idea I am attributing to Hilbert, are rooted in certain important and influential ways of thinking about the relation of correct symbolism to what it represents. These ways of thinking had been in circulation long before Hilbert and they could scarcely have escaped his attention. The leading idea is that a correct symbolism constitutes an isomorphic replica of what it represents. For instance, Leibniz (1677) had written that

> ... characters must show, when they are used in demonstrations, some kind of connection, grouping and order which are also found in objects, and this is required, if not in simple words ... then at least in their union and connection.

Hilbert was maintaining that the only thing that matters in an axiom system is the structure it imposes on its models. When this idea is combined with the assumption of an isomorphism between language (symbolism) and what it represents, there cannot be any objection to using the expressions of a language as their own models, any more than there according to Hilbert can be any objection to imagining that the models of the axioms of geometry consist of tables, chairs and beermugs. Indeed, symbols are easier to manipulate and to communicate than furniture or tableware. Far from being a formalist gimmick, Hilbert's way of thinking is thus rooted in an old tradition.

Nor is the way of looking at Hilbert's idea which is being advocated here even a novelty. Among others, it has been forward by Diedrich Mahnke (1925 and especially 1927). He views the passages under discussion in the light of Husserlian phenomenology, and relates them also to Leibniz' ideas. Hilbert is "phenomenologically correct" when he evokes "artificially created numerals".

> For these, like all good mathematical "symbols" or "characters" are, as Leibniz often emphasized, formed in such a way that they "express" formal-logically "isomorphically" and yet at the same intuitively-clearly all the purely arithmetical relations, such as ordering and connection.

The telltale words here are "relations" and "isomorphically". They are unmistakably model-theoretical terms. Hence Mahnke is in effect making the same point as I am making.

4. The Axiomatist Motivation of Hilbert's Project

Now we are approaching the motivation of Hilbert's attempted consistency proofs. A pure mathematician is in principle free to set up any axiom system whatsoever. Of course the importance of different axiom systems is different. The axiomatization of quantum theory is more important than that of trigonometry. But this importance or lack thereof is determined largely by outside factors. As soon as a mathematician is somehow given an axiom system, he knows what his job is, assuming that the axiom system has models. The axiom system offers the mathematician an overview on the class of its models and thereby on the information codified by the axiom system. By deducing consequences from the axioms the mathematician puts that information to use by exploring what there is to be known about those models.

This view of pure mathematics as a study of axiom systems is tempting in its simplicity. It is essentially what Russell called the postulational view of mathematics and what he espoused himself for a while. Its apparent simplicity nevertheless hides a number of further issues.

First, if we hope to gain a genuine overview of the field of knowledge summed up in an axiom system, the derivation of theorems from the axioms cannot introduce any substantial new information. And what this requirement amounts to is that the derivations in question are tautological, nonampliative. Otherwise the conceptual tools of the derivation might smuggle in tacit assumptions, for instance through the mathematical symbolism that is being used.

Such a smuggling problem was in fact very much in evidence in Hilbert's background. The most important case in point concerned statistical thermodynamics, where certain mathematical techniques were in fact suspected to introduce nontrivial assumptions, not to say facts, into the evolving theory. The question whether mathematical symbolism can contribute to the knowledge codified in a scientific theory was one of the issues in the famous controversy between Mach and Boltzmann. Boltzmann in effect argued that the use of differential equations tacitly introduces nontrivial assumptions concerning the kinds of phenomena they were supposed merely to describe according to the likes of Mach. This controversy is the unmistakable background of the sixth problem in Hilbert (1900).

It seems most likely that one important motive of Hilbert's use of the axiomatic method was to have a technique that eliminated all such smuggling of substantial assumptions through what looks like purely formal reasoning. This seems to be the reason for his sixth mathematical problem, where he mentions Boltzmann's thermodynamics in so many words. It is likely to have motivated (at least in part) his own work on physical theories, including thermodynamics. In a different direction, the same motivation was unmistakably at work in his struggles to rid geometry of all appeals to intuition, except of course on the pre-axiomatic level of guiding the selection of axioms.

Hilbert's method of banishing such covert introduction of information is to make axiomatization purely logical. In other words, the solution is to require that the derivation of theorems from axioms be purely logical. Naturally this helps only if logical deduction is itself uninformative (philosophers might say, analytic), for instance independent of the subject matter of the axiomatic system in question. This uninformativeness assumption was in fact commonplace in Hilbert's background. It may go back in the German language area to Kant. Thus Hilbert's emphasis on the purely logical character of the reasoning involved in a well-formed axiom system is not a formalist dogma or an expression of his search for rigor, but at least partly motivated by his ideas of how actual axiomatization of theories can enhance our knowledge also in the applications outside pure mathematics.

What is needed for this kind of logical axiomatization to be worthwhile? Very little, apparently. If we are dealing with an obviously important mathematical theory, such as elementary arithmetic, analysis or set theory, nothing seems to prevent a mathematician from setting up an axiom system for it and to derive logical conclusions from it. But then how could the paradoxes and other foundational problems have upset the foundations of mathematics of Hilbert's time? Obviously there is only one way in which they could upset a Hilbertian mathematician, viz. by showing that his theory is inconsistent in the model-theoretical sense. For there is nothing that one can accomplish by means of an inconsistent axiom system. Such a system is literally much ado about nothing, about models that do not exist. Hence on Hilbert's view there is one thing, and one thing only, that is needed to ensure *der sichere Gang einer Wissenschaft* for one's axiomatic theory: its model-theoretical consistency.

It follows that for Hilbert consistency proofs have nothing epistemological about them. They are not what is needed to enforce mathematical rigor. That rigor is enforced through the purely logical character of one's axiomatization. Nor are consistency proofs calculated to enhance the credibility of a mathematician's conclusions. From the vantage point of pure mathematics one can choose one's axiomatic theory freely, of course guided by its potential interest and applicability. There is no room for certainty or uncertainty there. All epistemological interpretations of Hilbert's program are off the track.

There is one apparent exception to this nonepistemic character of Hilbert's project. Of course the means of proving consistency must be beyond doubt. But the doubts that Hilbert seems to have contemplated are not about the epistemic status of different methods of proof, but about their admissibility in the first place.

For Hilbert, the typical question is not so much whether ideal elements in one's models exist or whether their existence is credible. The question is whether we should postulate such elements at all. Can we manipulate them and experiment with them even in thought? Problems on this score are more likely to be caused by the infinity of such elements or the infinity of the processes we would have to carry out to comprehend them and to manipulate them, than by reasons to doubt their existence. In any case, they are model-theoretical problems, not proof-theoretical ones.

Boston University

References

Brouwer, Luitzen E. J. 1921: "Intuitionist Set Theory", in Paolo Mancosu (ed.), *From Brouwer to Hilbert: The Debate on the Foundations of Mathematics in the 1920s*, Oxford University Press, New York, 1998, pp. 23-27.

Chang, C. C. 1974: "Model Theory 1945-1971", in Leon Henkin et al. (eds.), *Proceedings of the Tarski Symposium* (*Proceedings of Symposia in Pure Mathematics* vol. 25), American Mathematical Society, Providence, pp. 173-186.

Courant, Richard and David Hilbert 1931-37, *Mathematische Physik* I-II, Springer, Berlin.

Gödel, Kurt 1930: "Die Vollständigkeit der Axiome des logischen Funktionenkalküls", *Monatshefte für Mathematik und Physik*, vol. 37, 349-360; reprinted and translated in Kurt Gödel, *Collected Works*, vol. 1, Oxford University Press, New York, pp. 44-125.

Gödel, Kurt 1931: "Über formal unentscheidbare Sätze der Principia Mathematica und verwandter Systeme I", *Monatshefte für Mathematik und Physik*, vol. 38, 173-198; reprinted and translated in Kurt Gödel, *Collected Works*, vol. 1, Oxford University Press, New York, pp. 144-195.

Henkin, Leon 1949: "The Completeness in the First-order Functional Calculus", *Journal of Symbolic Logic*, vol. 14, 150-166.

Hilbert, David 1899: "Grundlagen der Geometrie", in *Festschrift zur Feier der Enthüllung des Gauss-Weber-Denkmals in Göttingen*, Teubner, Leipzig, pp. 1-92.

Hilbert, David 1900: "Mathematische Probleme", *Nachrichten von der Königlichen Gesellschaft der Wissenschaften zu Göttingen, Math.-Phys. Klasse*, pp. 253-297. (Lecture given at the International Congress of Mathematicians, Paris, 1900.)

Hilbert, David 1918: "Axiomatisches Denken", *Mathematische Annalen*, vol. 78, 405-415 (lecture given at the Swiss Society of Mathematicians, 11 September 1917); English translation in William B. Ewald (ed.), *From Kant to Hilbert: A Source Book in the Foundations of Mathematics*, vol. 2, Oxford University Press, Oxford, 1996, pp. 1105-1115.

Hilbert, David 1922: "Neubegründung der Mathematik: Erste Mitteilung", *Abhandlungen aus dem Seminar der Hamburgischen Universität*, vol. 1, pp. 157-77 (series of talks given at the University of Hamburg, July 25-27, 1921); English translation in Paolo Mancosu (ed.), *From Brouwer to Hilbert: The Debate on the Foundations of Mathematics in the 1920s*, Oxford University Press, New York, 1998, pp. 198-214.

Hilbert, David 1926: "Über das Unendliche", *Mathematische Annalen*, vol. 95, 161-190 (lecture given in Münster, 4 June, 1925); English translation in Jean van Heijenoort (ed.), *From Frege to Gödel: A Source Book in Mathematical Logic*, 1897-1931, Harvard University Press, Cambridge, MA, 1967, pp. 367-392.

Hilbert, David 1928: "Die Grundlagen der Mathematik", *Abhandlungen aus dem Seminar der Hamburgischen Universität*, vol. 6, pp. 65-85 (addresses given at the University of Hamburg in 1927); English translation in Jean van Heijenoort (ed.), *From Frege to Gödel: A Source Book in Mathematical Logic*, 1897-1931, Harvard University Press, Cambridge, MA, 1967, pp. 464-479.

Hilbert, David 1929: "Probleme der Grundlegung der Mathematik", *Mathematische Annalen*, vol. 102, 1-9 (lecture given at the International Congress of Mathematicians, 3 September, 1928); English translation in Paolo Mancosu (ed.), *From Brouwer to Hilbert: The Debate on the Foundations of Mathematics in the 1920s*, Oxford University Press, New York, 1998, pp. 266-273.

Hintikka, Jaakko 1996: *The Principles of Mathematics Revisited*, Cambridge University Press.

Hintikka, Jaakko 1997(a): "A Revolution in the Foundations of Mathematics?", *Synthese*, vol. 111, 155-170.

Hintikka, Jaakko 1997(b): "Hilbert Vindicated?", *Synthese*, vol. 110, 15-36.

Hintikka, Jaakko and Besim Karakadilar 2006: "How to Prove the Consistency of Arithmetic", in T. Aho and A.-V. Pietarinen (eds.), *Truth and Games: Essays in Honour of Gabriel Sandu*, Acta Philosophica Fennica 78, pp. 1-15.

Kreisel, Georg 1958: "Mathematical Significance of Consistency Proofs", *Journal of Symbolic Logic*, vol. 23, 155-182.

Kreisel, Georg 1983: "Hilbert's Programme", in Paul Benacerraf and Hilary Putnam (eds.), *Philosophy of Mathematics*, second edition, Cambridge University Press, Cambridge, pp. 207-238.

Leibniz, Gottfried Wilhelm 1951 (original 1677): "Dialogue on the Connection between Things and Words", in Philip P. Wiener (ed.), *Leibniz Selections*, Charles Scribner's Sons, New York, pp. 6-11. Original in *Die Philosophische Schriften vom Gottfried Wilhelm Leibniz*, vols 1-7, Berlin 1875-1890, vol. 2, pp. 190-194.

Mahnke, Dietrich 1925: *Leibnizens Synthese vom Universalmathematick und and Individualmetaphysik*, Halle.

Mahnke, Dietrich 1927: "Leibniz als Begründer der symbolischen Mathematik", *Isis*, vol. 9, 279-293.

Reid, Constance 1970: *Hilbert*, Springer, Berlin.

Smullyan, Raymond 1968: *First-Order Logic*, Springer, Heidelberg and New York.

Weyl, Hermann 1918: *Das Kontinuum*, Vert, Leipzig.

Weyl, Hermann 1921: "Über die neue Grundlagenkrise der Mathematik", *Mathematische Zeitschrift*, vol. 10, 37-79; English translation in Paolo Mancosu (ed.), *From Brouwer to Hilbert: The Debate on the Foundations of Mathematics in the 1920s*, Oxford University Press, Oxford, 1998, pp. 86-118.

Marja-Liisa Kakkuri-Knuuttila and Simo Knuuttila

Meaning Finitism – An Antirealist Position?

The aim of the present paper is to analyze a theory of meaning which is one of the acclaimed heirs of Ludwig Wittgenstein's conception of rules and language games presented in his *Philosophical Investigations* (1953). This is the so-called finitist theory of meaning developed since the 1970s by the Edinburgh sociologists of science, Barry Barnes and David Bloor.

Since finitism has been developed in close contact with empirical sociological studies of research practice in several of the natural sciences, one may expect it to shed light on changing social practices of concept formation. As is well-known, the Wittgensteinian conception of language as a constitutive factor of social reality involves an emphasis on the public, inter-subjectively observable aspects of meaning. This is in stark contrast both to the conception of meaning in the conceptual analysis in traditional analytic philosophy and to the phenomenological understanding of necessary meaning relations. In spite of the merits of monitoring and codifying contingent changes of linguistic meaning in social practices of science, there are also problems in the finitist approach. We aim to show that, contrary to what has been sometimes argued, these problems are not associated with an alleged radically antirealist position regarding concepts and linguistic meaning.

In their criticism of what is called 'meaning determinism', the finitists argue against positions to be identified as meaning realism, whether in its classical ontologically founded form (Plato, Aristotle, Thomas Aquinas) or as non-relativist conceptual realism without abstract entities (Ockham, Locke, Leibniz). Their own position which is assumed in describing language learning and related matters is close to nominalism with modifications to allow room for historically changing linguistic practices.[1] This is expressed by an emphasis on

---

[1] Se also Niiniluoto (2002), p. 265.

the openness of the applications of classificatory terms in contrast to what they call 'meaning determinism'. In the first part of this paper we describe the Edinburgh meaning finitism as showing similarities to nominalist philosophy. In the second part we explain why some critics mistakenly read it as an arbitrary position with respect to concept formation and indicate some more genuine problems.

1. Finitism and Constructivist Nominalism

In presenting the main ideas of finitism in *Scientific knowledge: A Sociological Analysis*, Barnes et al. (1996) begin with the empirical observation that alternative classifications of natural phenomena exist. They claim that the lack of a standard conceptual framework reveals that experience alone is insufficient to sustain a given classification adopted within a particular culture. Along with the features of natural objects, classifications are generated by social conventions. Classifications are hence based on two distinct factors: 'the world' or 'intuition of the resemblances and differences presented by things' and 'the authority of tradition' or 'the way of life of the collective wherein the activity of classification is occurring'.[2]

The account of alternative classifications is meant to undermine the credibility of conceptual realism. The rejection of this is tied to the fundamental thesis of the openness of meanings which amounts to the same as a rejection of what is called 'meaning determinism'. Meaning determinism is for them the view that the meanings of terms determine their referents in a unique manner; i.e., the correct applications of a term are given in advance of the application.[3] Relying on Wittgenstein's *Remarks on the Foundations of Mathematics*, Bloor rejects the meaning determinist view as an unnecessary supposition.[4] This rejection yields the conception which gives finitism its title: the application rules of terms are not determined in advance, but continuously generated by the language

---

[2] Barnes, Bloor and Henry (1996), pp. 46–47.
[3] Bloor (1997), p. 3; Kusch (2002), pp. 200–210.
[4] Bloor (1997), pp. 14–15.

users.⁵ Hence the extensions of even mathematical terms are finite; they are no more and no less than the actual applications.⁶

Such openness of linguistic meaning does not sit well with conceptual realism, since the latter position implies that the formal sameness between the essences and concepts or the abilities to grasp the resemblances on which conceptual classifications are based uniquely determine the extensions of terms. In giving up this line of thinking, the finitists characterize the openness of meanings in terms of the open extensions of terms.⁷ They do not, however, like to use the terms 'extension' or 'intension' simply because in traditional use these have been tied to the meaning determinist view. Barnes et al. (1996) even suggests that it would be wise to give up the use of the term 'meaning' altogether.⁸

A further expression of the finitist position finds its expression in their concern with what could be called 'the problem of the next move'.⁹ This is the question of whether the classification of the next instance is uniquely determined by the earlier instances of the term (and the existing conceptual framework) together with the properties of the new instance. While a 'meaning determinist' would, by definition, reply 'yes', the finitist is bound to reply 'no'. Barnes et al. (1996) support the finitist position by referring to the complications met with the hierarchical taxonomies of biological sciences and the existence of terms such as 'fertilizers', 'insulators', 'reagents', which are tied to practical purposes so that their application cannot be fixed merely by nature.¹⁰ The finitists present their more detailed argument against meaning determinism and conceptual realism mainly in their treatment of learning by ostension as an illustration of the problem of the next move.

The choice of ostension as the paradigm of linguistic meaning and language learning is associated with the Wittgensteinian emphasis on the public aspects of linguistic behaviour. An act of ostension is public, as is '[a]ny act whereby a direct association is directly displayed or shown or pointed out between an

---

⁵ Barnes et al. (1996), pp. 54–55; Bloor (1997), p. 6; Kusch (2002), p. 201.
⁶ Bloor (1997), pp. 14–15.
⁷ Bloor (1997), pp. 24–25; Kusch (2002), pp. 207–208.
⁸ Barnes et al. (1996), p. 58.
⁹ Bloor (1997), p. 10. Bloor's book involves a discussion of Kripke's meaning scepticism from a finitist perspective. See also Kusch (2004), pp. 571–591.
¹⁰ Barnes et al. (1996), p. 48.

empirical event or state of affairs and a word or term of a language',[11] since it can be recognized as such by an observer just as by the participants themselves to the extent that they are familiar with this particular social practice. This way of putting the matter shows that language learning by ostension already presupposes fairly developed discriminative capacities which are called inherent or innate dispositions.[12]

The favourite language learning example of the finitists is teaching to use the term 'duck' by pointing to instances of ducks. Ostension is bound to remain indefinite, since no single act of ostension suffices to distinguish the thing (duck) from the background (lake), or to show what is exactly being taught ('duck' or 'bird'). Improvement will be brought by repetition, confusion with the background can be lessened by changing it, and the term being taught can be specified by showing instances of non-ducks, and so on. The learner may also improve her skills by trial and error with appropriate feedback. The success of the whole process depends on the learner having the basic discriminative capacities to sense the sameness in several instances, and the inductive 'disposition' to infer from given cases to new ones and from past associations of words and things as persisting associations.[13] Another precondition of learning by ostension is that the learner accepts the teacher as an authority who represents the tradition acknowledged in the culture in question. Accordingly, it is the communal consensus through which norms are built.[14]

In the treatment of ostension, the finitists assume that there are particular things which are relevant to ostensive meaning and exist independently of our cognition. Furthermore, the finitists seem to take as an empirical fact that these particulars can show similarities to each other, though they are not identical to each other in any respect. Hence they emphasize that '[a] resemblance is not the same as an identity'.[15] The target of the criticism against the realist version of meaning determinism appears to be the Aristotelian rather than Platonic

---

[11] Barnes et al. (1996), p. 49.

[12] Barnes et al. (1996), p. 53; Bloor (1997), pp. 19–20.

[13] Barnes et al. (1996), pp. 49–53; Bloor (1997), pp. 13–15, 19–21; Kusch (2002), p. 203.

[14] Barnes et al. (1996), p. 54; Bloor (1997), pp. 15–17; Kusch (2002), pp. 204–206.

[15] Barnes at al. (1996), pp. 54, 56; for the criticism of Kripke's essentialism, see p. 69.

version in which the empirical things are similar to ideas, but certainly the authors are no more friends of ideas than Aristotle was. By referring to particulars and their similarities and dissimilarities, the finitists use the language of resemblance nominalism, in which the conception of resemblance explains the properties of the particulars.[16]

A few pages later in Barnes et al. (1996) we detect explicit attempts to reject non-relativist conceptualism. The authors reject the possibility of pure, abstract ideas for the reason that they would involve an ahistorical view of concepts; i.e., concepts in 'their final "pure" form'. They state, in contrast, that 'finitism describes science wherever it is "not quite finished"'.[17] Even though one might sympathize with the historical perspective on concept formation, the evidence from ambiguities in mathematical terminology is not very convincing at this point. Why would the double use of the 'minus sign' to serve as a property of a number and an operator, or the term 'circle' as referring to a line and an area show contradictions in mathematical terminology, or the insufficiency of abstract ideas in general?[18] Rather than contradictions, these can be regarded as two different uses of the minus sign and 'circle'. Instead of refuting the possibility of abstract ideas, the argument only shows to an adherent of general concepts that one and the same term may denote two or more abstract ideas in different contexts. So far the main tenets of the finitist position are not in conflict with a conceptualist understanding of concepts, if this allows room for historically changing and hence 'not quite finished' concepts.

The openness of conceptual classifications is expressed in the following five principles, formulated partly for what are typically called 'kind terms', but they are meant to apply to all classificatory terms:[19]

1. The application of terms is open-ended.

   ... On a finitist account there is nothing identifiable as 'the meaning' of a kind term, no specification or template or algorithm fully formed in the present, capable of fixing the future correct use of the term, of distinguishing in advance all the things to which it will eventually be correctly applicable. ...

---

[16] For resemblance nominalism, see Rodriguez-Pereyra (2002).
[17] Barnes et al. (1996), p. 63.
[18] Barnes et al. (1996), pp. 63–64.
[19] Barnes et al. (1996), 55–59; cf. Kusch (2002), pp. 206–207.

2. No act of classification is ever indefeasibly correct.

   ... classification can never proceed on the basis of *identity* of appearance, and has to proceed on the basis of *analogy* of appearance. There is no correct or incorrect way of extending an analogy which can be read off, as it were, from the appearances themselves. People must *decide* what is correct and what is not. From a sociological perspective this usefully highlights the role of collective judgments, negotiated on the basis of a range of more or less compatible individual perceptual intuitions, in defining what will count as correct classification....

3. All acts of classification are revisable.

   ...The finitist view of an act of classification as a contingent judgment implies not just the open-ended and indeterminate character of future acts, but the provisional, revisable character of previous ones. Any act of classification, and hence any pattern or system of classification, may be held to have been mistaken....

4. Successive applications of a kind term are not independent.

   ... As a term is used, so its subsequent use is conditioned. One act affects another. In a collective, terms are applied by different individuals at different times in different contexts: the exemplary instances of proper applications of a term are collectively established. In a collective, therefore, where terms are shared as part of the language of the collective, what one individual does will affect what another subsequently does with a term. ...

5. The applications of different kind terms are not independent of each other.

   ...If we seek to classify on the basis of resemblance, we must operate in terms of maximum resemblance: the next instance belongs amongst those existing instances which it is 'most like'. This means that the next instance must be compared with instances of *all* the relevant kind terms before it is classified into that kind where the greatest analogy is perceived. To be fully competent in the use of any term, therefore, a member must be competent in the use of the whole system of classification: he or she must be aware of relevant instances of all existing kinds. ...

The finitists agree that items to be classified are particular things, not identical to, but resembling and differing from each other. This is where the social practices as determining factors step into the picture.[20] For instance, there are no absolute criteria for how 'strategy' or 'a middle-sized firm' should be defined in economic research. The criteria are determined by researchers in dealing with research problems which are evidently tied to particular practical interests. The merits of meaning nominalism lie in this conception of the openness of classifications, which allows room for social interests of various kinds, including different forms of power relation. This has provided theoretical tools for methodological discussions in various areas including philosophy, as is shown by recent works on the sociology of philosophical knowledge.[21]

In his work *The Social Construction of What?* (1999) Ian Hacking characterizes nominalism as part of the social constructionist position. To quote:

> Constructionists tend to maintain that classifications are not determined by how the world is, but are convenient ways in which to represent it. They maintain that the world does not come quietly wrapped up in facts. Facts are the consequences of ways in which we represent the world. The constructionist vision here is splendidly old-fashioned. It is a species of nominalism.[22]

We have paid attention to similarities between the terms used in Edinburgh finitism and resemblance nominalism. This species of nominalism is not antirealist with respect to particulars and their resemblances, neither does the finitist maintain that judgments of similarity completely subjective, as can be seen from principles (2) and (5).[23] Because of its view of the concepts, finitism

---

[20] Kusch (2002), pp. 204–206.

[21] For the sociology of philosophical knowledge, see Kusch (1995) and Kusch (ed.) (2000). There are some interesting similarities and differences between this approach and the criticism of Lovejoy's 'unit ideas' and other unhistorical ideas in analytic and hermeneutic discussions of philosophical historical semantics. See, for example, Hintikka (1981) and Gadamer (1965).

[22] Hacking (1999), p. 33.

[23] Se also Kusch (2002), p. 203. According to Bloor, the constraints on concept forming involve our biological nature, instincts, sense experience, interactions with other people and so on. See Bloor (1997), pp. 19–20. T. Lewens argues that while many

cannot put forward any ontological theory, such as resemblance nominalism, but this can be characterized as its background orientation.

2. Problems with the Formulations of the Finitist Position

Statements like that of Hacking about social constructivism are applied every now and then to the sociologists of science who allegedly represent an antirealist view of language and science by claiming that conceptual classifications or empirical beliefs are purely social constructions and fully arbitrary because of being in no way restricted by how the world is.[24] There are formulations in the works of the Edinburgh school which are read in this way. Some critics have found this approach not merely antirealist; they have associated it with postmodern irrationalism.[25] Since we take radical antirealism as not representing the official view of the Edinburgh school, an explanation will be provided of how possibly confusing readings arise. In our view these result from merging two different arguments for openness, namely, formal and empirical ones which, to be sure, are not clearly distinguished in Barnes et al. (1996). They seem to transfer conclusions drawn in the formal context to the empirical context where, in fact, they have no business.

The formal arguments are presented in the treatment of ostension illustrated by the following figure portraying the problem of the next move:[26]

---

tenets of the Edinburgh School are compatible with a scientific realism founded on an externalist epistemology, the realist should reject the claim that there is no distinction between being rational and being locally accepted as rational; see Lewens (2005).

[24] Barnes et al. (1996, pp. 73–76) mention H. Collins (1992) as representing an antirealist position. Hacking (1999, pp. 24–25) calls such a view 'universal constructionism'.

[25] See Franklin (2002).

[26] Barnes et al. (1996), p. 51. In the original figure, the items under C_2 are denoted by the same set of symbols $I_1 \ldots I_n$ as those under C_1. Since this is an uninteresting exceptional case, we have corrected the figure so that the items under C_1 are different from those under C_2, presuming this to be the intended situation.

Meaning Finitism – An Antirealist Position?

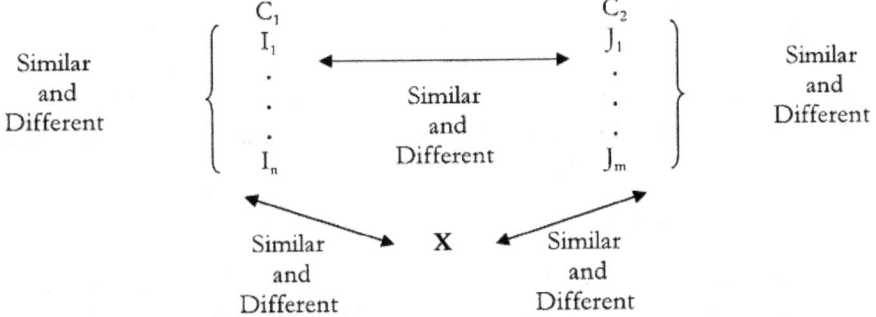

The formal argument claims that the next instance, X, which can be seen to have similarities with both the instances $I_1...I_n$, previously taken as referents of term C_1 (duck), and instances $J_1...J_m$, previously counted as referents of term C_2 (goose), can equally well be classified as a further instance of C_1 or a new instance of C_2. Now one could ask how this is possible considering principle (4) above, according to which successive applications of terms are not independent, or considering principle (5), according to which we should operate on the basis of maximum resemblance, and let X belong to that group which it is 'most like'? These questions do not, in fact, form part of the formal considerations, but are empirical. Here the question is simply whether X can be counted as belonging to C_1 as well as to C_2 without contradiction. The answer of the authors is 'yes': X can be classified as a member of C_1 and C_2 without inconsistency.

> Considered formally, ostensive learning offers us *nothing* by way of guidance in the classification of the next instance. Any classification of the next case can be reconciled with what has already been learned ostensively; however other things have been classified so far, the next thing can be classified in any way without any formal inconsistency with earlier practice.[27]

It seems that the conclusion that the existing instances of application of a term offer no guidance to the next move is fallaciously assumed to deal with ostensive learning and social practices, and that this fallacy results from a shift from purely extensional considerations to conclusions which are of a different

---

[27] *Ibid.*, also p. 73.

type. By extensional considerations we mean the standard (mathematical) set-theoretical treatment of sets, according to which a set is fully determined by its members. Two sets are, by definition, the same if they include the same members, and two sets are different if their members are different.[28] Hence the following two sets are the same, since they have the same members, the order being irrelevant in the set-theoretical context:

{3, 2004, μ, €} {2004, €, 3, μ}.

By removing or adding one member, the result will be a different set. Barnes et al. (1996) describe the class of items exemplifying the application of a term in the extensional idiom. This is shown by their characterization of what they call the 'similarity relation' to refer to the instances of the term. Hence $I_1...I_n$ form the similarity relation of the term C_1 and $J_1...J_m$ form the similarity relation of the term C_2.[29] The fact that this similarity relation is defined in extensional, set-theoretical terms emerges from the following statement:

> To use a kind term, to apply it to a thing, is to add the thing to the existing similarity relation and hence to change that relation.[30]

As far as the similarity relation is meant to capture the meaning of a term, as it obviously is, the meaning is determined fully by the members of its application set. If one thing is taken away or added, the meaning of the term changes, which implies that meanings are identified with the application sets, which are taken in the extensional, standard set-theoretical sense.

The extensionalist reading of the 'similarity relation' may appear to fit neatly with the claim that finitism advocates nominalist assumptions, but more care is needed here, since a nominalist can assume that the similarities and differences between the actual objects are relevant to the classifications. On the purely set-theoretical interpretation of 'similarity relation' in Barnes et al., one's laptop may count equally well as a duck as spectacles. The world poses no restrictions at all on such classifications, and the matter is left solely to social decisions, whatever these may be. This strong antirealist position conflicts with

---

[28] Russell (1993 [1919]), ch. 17.
[29] Barnes et al. (1996), pp. 50, 59.
[30] Barnes et al. (1996), p. 57.

the formulations based on the assumption that both the 'world' and the 'authority of tradition' are relevant to classification.[31]

It is correct to say that no formal inconsistency follows if we classify things one day in one way and in another in another way, but such purely logical possibility is not of much interest to those talking about meanings in the empirical world. While the claim is trivially true by the definition of the set-theoretical notion of 'set', it says nothing about empirical classifications.

Our empirical classifications are not based on such formal considerations, but on actual considerations about the similarities and differences between the objects. In our empirical world, the meaning of the term 'duck' does not change every summer when it is realized that new ducks are born or when others die.

The minimal role of the formal arguments in linguistic practices has not been stressed by the finitists clearly enough to avoid misunderstandings. Their merging conclusions derived in the formal considerations on the empirical context leads to obvious inconsistency, as can be expected. In spite of their often pointing out the ease and facility with which we attach words 'to things not just appropriately, but unproblematically, immediately, without thought, as a matter of course, in the vast majority of cases'[32] we also find statements to the contrary in the finitists' work. Here are some examples:

> ... it reminds us of the status of every act of classifications as a separate and problematic empirical phenomenon.[33]

> We have put our knife into the cake and cut a certain way. But nothing determines how we should continue to cut ...[34]

> Formally speaking, anything goes. Far from telling us anything in particular, current experience can be described as consistent with any extant body of ancestral knowledge ...[35]

> ... every act of categorization is a stab into the semantic dark ...[36]

---

[31] Barnes et al. (1996), pp. 46–47.
[32] Barnes et al. (1996), p. 53; see also pp. 52, 55, 56; Bloor (1997), p. 16.
[33] Barnes et al. (1996), pp. 55–56.
[34] Barnes et al. (1996), p. 70.
[35] Barnes et al. (1996), p. 73.
[36] Kusch (2002), p. 209.

The over-emphasis on the formal argument is the cause of these conflicting statements of the finitists concerning the openness of the next move. But, if we adopt web-footedness, for instance, as one relevant similarity criterion of ducks at the moment, we can no longer say that 'anything goes'. In contrast to the formal argument which leaves the empirical situation intact, we will face a genuine contradiction if we decide to extend 'duck' to include your laptop, for instance:

Ducks are web-footed birds.	Ducks swim in water.
Your laptop is not a web-footed bird.	My spectacles do not swim in water.
Your laptop is a duck.	My spectacles are a duck.
Your laptop is a web-footed bird.	My spectacles swim in water.

As argued by the finitists, we may change the meaning of all these terms to avoid the contradiction involved in both these cases, but this is a trivial kind of claim, because it only indicates the logical possibility of such a change. In practical life things are different. We can no longer accept the formal conclusion that 'anything goes' or that 'ostensive learning offers us *nothing* by way of guidance in the classification of the next instance'.[37] Ostensive learning consists for the most part of learning the relevant criteria involved in the classifications accepted by the community. If one learns by ostension that anything goes, one learns something, but this is not what one is normally supposed to learn.

This seems to be one of the points which the finitists want to suggest by their considerations about formal argument. If meaning is treated extensionally and equated with an existing set, it does not determine anything. Something else is needed to determine the answer to the next move in the process of learning, innate dispositions and social construction. Principles (4) and (5) belong to this context. If the above figure is seen from this empirical point of view, one should interpret the similarities and differences between the objects drawn in the figure in the spirit of resemblance nominalism or some related theory, even though 'nature is indifferent to how it is classified.'[38] When dealing with such more complex empirical terms as 'distance', 'time', 'speed', or

---

[37] Barnes et al. (1996), pp. 73 and 51.
[38] Barnes et al. (1996), p. 73.

'velocity', Barnes et al. (1996) conclude that 'they must always remain cluster-concepts nonetheless, concepts exemplified by non-identical paradigmatic cases of use.'[39] To be a paradigm is something which does not belong to formal argument.

We could conclude that even though the two levels of argument are not always clearly separated in Barnes et al., treating them as suggested here avoids inconsistencies and conclusions which look irrational.

The maximum resemblance principle (5) in concept learning is very demanding. In many cases, we could rather be said to be operating on the basis of relevant similarity criteria.[40] This means that in order for the child to learn what the term 'red' means (in its standard use for colors), for instance, she needs to understand (in some way) that it denotes a specific aspect of the thing, namely, its color. Hence there is no need to trouble oneself with maximum resemblance in general without any pre-specification of what feature of the object counts as relevant. Likewise with the ducks and geese. We learn that the shape of the bird and its color patterns, form of the feet, the nature of the living surrounding, and other matters presented in books for bird watchers are the criteria relevant for identifying them. The conventions forming the culturally specific conceptual frameworks involve such clusters of relevant similarity criteria with clusters of instances of applications. Some of the formulations in Barnes et al. (1996), like the following, come close to the notion of 'relevant similarity criteria':

> Which right way is taught will depend upon the tradition in which the learner is embedded: we considered learning about birds; in another context there might be no class of feathered animals to learn, but instead a class of flying animals, including, as we would say, birds and bats; in scientific subcultures learning might be based upon sameness relations different from either of these.[41]

The idea is also expressed in Kusch's description of Bloor's meaning-finitism:

---

[39] Barnes et al. (1996), p. 62; also pp. 51, 58; Bloor (1997), p. 11.

[40] The acceptance of relevant similarity criteria does not conflict with T. S. Kuhn's claim that 'there is always a point where we have to recognize that some things are the same as others, without being able to say in what respect they are the same.' (Barnes et al. 1996, p. 52.)

[41] Barnes et al. (1996), p. 54.

Obviously, such judgements [of similarity and difference] are underdetermined by any possible learning set. This is where the community comes in: community consensus determines which similarities count as relevant or irrelevant and which concept applications are correct or incorrect.[42]

All the finitists can say with their formal argument already follows from their empirical considerations.[43] The core of the issue is that sometimes the next move causes no problems and the classification takes place smoothly without any effort of thinking, while in some situations it is not possible to reach agreement. The problems in the latter cases may be caused by three different factors: we face with a genuinely new instance which cannot be classified on the basis of the accepted relevant similarity criteria, there is no consensus over the relevant similarity criteria, or the social interests change to the effect of a change in the similarity criteria.[44]

Rejecting the fixed extensions of concepts and referring to dispositions or social conventions where many philosophers speak about intension as another property of concepts is part of the finitist program. If concepts are understood as the ability to classify things in accordance with rules, there is no need to take intensions (application rules) in the form of 'pure' or 'finished ideas' which Barnes et al. (1996) correctly oppose.[45] Intensions as provisional rules are in principle open and may also lose their applicability. In dealing with language learning, the finitists refer to innate and inherent dispositions which explain the reception of language. While argumentative negotiations are not relevant in this context, they are mentioned in the description of the social construction of classificatory terms. How are learning language and learning things through language related to each other? It seems that the normative nature of meaning cannot be constructed in the same way in preconscious contexts and, say, in philosophical debates on what justice is. Bloor argues that meanings as intentional social phenomena are reducible to a non-intentional level. Then we

---

[42] Kusch (2004), pp. 577–578.

[43] Barnes et al. (1996) also mention that the truth of finitism is an empirical matter (p. 63).

[44] Kusch (2002) mentions the first and the third case (p. 207); cf. also Barnes et al. (1996), p. 57.

[45] Barnes et al. (1996), pp. 62–63.

need to be interested 'in our nature as biological organisms, our natural psychological proclivities to react and our instinctive susceptibilities to one another'.[46] Original language learning belongs to this level, but applying reductive ideas to philosophical arguments is only one possible approach to them.

Helsinki School of Economics
University of Helsinki

References

Barnes, B., D. Bloor and J. Henry 1996: *Scientific Knowledge: A Sociological Analysis*, Athlone Press, London.
Bloor, D. 2004: "Institutions and Rule-Scepticism: A Reply to Martin Kusch", *Social Studies of Science* 34, 593–601.
Bloor, D. 1997: *Wittgenstein, Rules and Institutions*, Routledge, London and New York.
Collins, H. 1992: *Changing Order: Replication and Induction in Scientific Practice*, Sage Publications, London.
Franklin, J. 2002: "Stove's Discovery of the Worst Argument in the World", *Philosophy* 77, 615–624.
Gadamer, H.-G. 1965: *Wahrheit und Methode*, Mohr (Siebeck), Tübingen.
Hacking, I. 1999: *The Social Construction of What?*, Harvard University Press, Cambridge, Massachusetts and London.
Hintikka, J. 1981: "Gaps in the Great Chain of Being: An Exercise in the Methodology of the History of Ideas", in S. Knuuttila (ed.), *Reforging the Great Chain of Being: Studies of the History of Modal Theories*, Synthese Historical Library 20, Reidel, Dordrecht, pp. 1–17.
Kusch, M. 2004: "Rule-Scepticism and the Sociology of Scientific Knowledge: The Bloor-Lynch Debate Revisited", *Social Studies of Science* 34, 571–591.
Kusch, M. 2002: *Knowledge by Agreement: The Programme of Communitarian Epistemology*, Clarendon Press, Oxford.
Kusch, M. 1995: *Psychologism: A Case Study in the Sociology of Philosophical Knowledge*, Routledge, London.
Lewens, T. 2005: "Realism and the Strong Program", *The British Journal for the Philosophy of Science* 56, 559–577.
Niiniluoto, I. 2002: *Critical Scientific Realism*, The Clarendon Library of Logic and Philosophy, Oxford University Press, Oxford.
Rodriguez-Pereyra, G. 2002: *Resemblance Nominalism. A Solution to the Problem of Universals*, Clarendon Press, Oxford.
Russell, B. 1993: *Introduction to Mathematical Philosophy*, Routledge, London.

---

[46] Bloor (2004), p. 599.

Martin Kusch

Niiniluoto on Meaning Finitism

Abstract. This paper defends meaning finitism against Ilkka Niiniluoto's criticism. Niiniluoto has four main objections to meaning finitism: that it is self-refuting; that it ignores the role of definitions in mathematics; that it is unable to reckon with future and potential uses of language; and that it fails to explain the possibility of discoveries within social institutions. I argue that Niiniluoto's interpretation and criticism of meaning finitism is defective in a number of respects. First, it is wrong to describe meaning finitism as the view that all concepts are vague. Second, ostensive learning is not essential to meaning finitism. Third, Niiniluoto fails to recognize that Bloor advances a reductive account of meaning. Fourth, Niiniluoto's answer to meaning skepticism has been anticipated and refuted by Saul Kripke. And fifth, it is important to distinguish between Kripke's meaning skepticism and Bloor's meaning finitism.

1. Introduction

In a number of places Ilkka Niiniluoto has criticized Barry Barnes' and David Bloor's "Strong Programme" in the "Sociology of Scientific Knowledge" (Niiniluoto 1991; 1999a; 1999b; 2003). One important part of Niiniluoto's attack is directed against Barnes' and Bloor's theory of meaning and mental content, that is, against "meaning-" or "rule-finitism". Niiniluoto seeks to show that finitism is fundamentally mistaken, and that it poses no threat to his scientific realism. In this paper, I shall defend meaning finitism against Niiniluoto's criticism. In so doing I continue the fifteen-year-old cordial debate that Niiniluoto and I have conducted over such issues as power, realism, relativism, and social studies of science (e.g. Kusch 1991; 1996; Niiniluoto 2003).

2. Niiniluoto's Interpretation of Finitism

Niiniluoto acknowledges the importance of meaning finitism for the Strong Programme by emphasizing that for Barnes and Bloor "social factors influence science primarily through the 'conventional character' of language and conceptual classifications" (1999a, p. 262). Niiniluoto notes that the plausibility of this view depends crucially on how we understand this "conventional character". If we take it in an abstract way – say, as the general claim that language and classifications involve conventions – then there is no direct path from the "conventional character" to the claim that truth is "somehow relative to social interests" or that "particular beliefs formulated in scientific languages have to be explained by social factors" (*ibid.*). After all, scientific realists do not deny that languages and classifications are conventional. And accepting this conventionality does not bar them from holding that at least some scientific propositions can be true in virtue of their correspondence with facts in the world.

Of course there is more to meaning finitism than a general and vague insistence upon the conventional character of language and classification. Niiniluoto mentions a number of overlapping further ingredients.

A first such ingredient is a "narrow" or "radical nominalism" (1999a, pp. 263, 265). This is Niiniluoto's title for Bloor's idea that predicates do not have fixed extensions. Bloor writes that "particular things, or individual objects, exist in advance [of our doing], but not classes of things" (1997, p. 24). Niiniluoto agrees that classes are human constructions but he disagrees with the assumption that classes are constructed by "adding one member after another" (1999a, p. 265).

A second element of meaning finitism is the rejection of all forms of Platonism, including Karl Popper's "World 3". Since Niiniluoto is an advocate of Popper's three-world ontology, he has little sympathy for this tenet of finitism. For Niiniluoto World 3 allows us to be anti-psychologistic, constructivist and realist about mathematical entities – all at the same time (1999a, p. 263).

Third, meaning finitism is built on the model of ostensive learning. We acquire basic classificatory terms by being shown paradigmatic cases. Subsequently we apply these terms in new circumstances on the basis of similarity or resemblance ("Is this creature like the creature for which my teacher used the term 'cat'?"). Our applications are correct if other members of our linguistic community agree with us. Sometimes we find agreement only after extensive negotiation.

Fourth, meaning finitism builds on Ludwig Wittgenstein's definition of "meaning as use". Like many other interpreters of the *Philosophical Investigations* (Wittgenstein 1967), Bloor takes this definition to imply that language is a social institution and thus rule-governed. However, Bloor deviates from other Wittgensteinians by insisting that "meaning is constructed as we go along", and that the future applications of a concept are not "fully determined by what has gone before" (Bloor 1991, pp. 164, 167; Niiniluoto 1999a, p. 263). Put differently, for Bloor use is primary with respect to meaning: the meaning of an expression is determined by how we use it, not *vice versa*. The already-mentioned rejection of fixed extensions (or classes) follows immediately from this position. Meaning finitism is thus committed to the view that the meaning of an expression can never determine any future applications. Bloor seems to say that "we do not have any capacity to construe rules [or meanings] which have implications in advance for the next case" (1999a, p. 265).

Fifth, Bloor's position is really a form of "meaning skepticism". Niiniluoto defines meaning- or rule-skepticism as the view that "there is no coherent notion of a rule which both explains how rules can normatively guide our actions and be applied to an indefinite number of cases". Bloor's version of meaning skepticism is particularly "interesting but problematic" (1999b, p. 83). Niiniluoto does not say much about parallels and differences between Bloor's and, say, Saul Kripke's meaning skepticism (Kripke 1982). Niiniluoto's classification of Bloor's position as meaning-skeptical seems at least in part influenced by the fact that both Bloor and Kripke's Wittgenstein belabor the idea of some kind of underdetermination in mathematics. Kripke points out that any given finite number of calculations involving the sign "+" leaves underdetermined which function the "+" denotes. And Bloor reminds us that no finite sequence of numbers can specify a unique continuation.

Sixth, meaning finitism restricts talk of meaning and use to the past and the presence of the actual word. Since meaning does not determine future use, the concepts of "future or potential uses" (1999a, p. 267) have no application. In other words, meaning finitism denies that "social rules resemble dispositions, causal powers, propensities, and laws of nature (in their non-Humean or anti-nominalistic reading)" (1999b, p. 85).

Seventh, and finally, in proposing that meanings are social institutions, Bloor relies on the following conception of institutions: "Roughly speaking, an institution *I* exists for a group *G* if and only if the members of *G* mutually believe (accept) that *I* exists." (Niiniluoto 1999a, p. 265)

3. Niiniluoto's Objections

Niiniluoto raises a number of objections to meaning finitism. For ease of reference I shall call them the "Self-Refutation Objection", the "Algorithm Objection", the "Potentiality Objection", and the "Discovery Objection".

According to the *Self-Refutation Objection*, if finitism were true, then finitism either would be false, or it could not be stated. Finitism seems committed to the view that "all of our concepts are completely vague", and that future applications of our concepts are always and everywhere "fully undetermined by rules". This is tantamount to saying that all concepts are "semantically completely indeterminate". But if that were the case then one of two things would have to follow: Either "human communication would be impossible" (1999b, p. 86), or "human communication (if possible at all) would have to be explained by merely behavioral regularities in our use of language, which conflicts with the institutional viewpoint where we [i.e. finitism] started" (1999a, p. 265). Niiniluoto draws the conclusion that finitism is false and that the phenomena of semantic determination are real. While he agrees that – at least outside of mathematics – full determination of future applications (or series) is rare, he believes that, as far as natural and scientific languages are concerned, many future applications are at least partially determined. Vagueness is a real phenomenon, but it affects only some uses and some concepts.

Niiniluoto's *Algorithm Objection* is meant to answer rule-skepticism in general. Bloor makes much of the fact that a finite sequence of numbers can be continued in many different ways; Kripke stresses that – due to their finite number – my past calculations with the sign "+" do not determine one unique function as the denotation of "+" (in my idiolect). Niiniluoto is not impressed. He insists that mathematicians often specify algorithms that determine how a given sequence is to be continued and that thus "no reference to social 'negotiations' is needed" (1999a, p. 266). One of Niiniluoto's examples is the recursive definition of the sum of two natural numbers:

... when the number 0 and the successor operation
$S(x)=x+1$

are given in arithmetic, the sum of two natural numbers is defined recursively:

x+0=x

x+S(y)=S(x+y)

for all x and y in N. [This] definition specifies a rule that can be applied to an infinite number of special cases. It can be identified by a finite description in World 1, and by a finite mind in World 2, but it defines the sum function as a public entity in World 3. (1999a, pp. 266–267)

At least in mathematics an infinite number of applications can be fully determined by a rule and this rule can be grasped and used "by a finite human mind or even a computer" (1999b, p. 86).

The *Potentiality Objection* attacks Bloor's neglect of future possibilities. Linguistic meaning must be more than just actual past and present uses; "at least some future and potential uses should be included as well" (1999a, p. 267). Contrary to what Bloor alleges, social rules "resemble" dispositions and propensities.

Finally, the *Discovery Objection* is closely related to *Potentiality Objection*. On Bloor's conventionalist theory of social institutions, something exists as an element of an institution only if the members of the institution believe it to exist. Niiniluoto objects that this view makes genuine discoveries about institutions impossible. He stresses that our social constructions often have "unknown logical consequences" to which "we are normatively committed" (1999b, p. 85). Popper's World 3 explicitly allows for this possibility, Bloor's theory does not. This constitutes an argument in favor of the former, and against the latter.

In light of these objections and corrections, Niiniluoto concludes that finitism is no threat to scientific realism. In particular, nothing Bloor says threatens the "realist principle": "we choose the language L, and the world decides which sentences of L are true" (1999a, p. 267). Nevertheless, Niiniluoto does not deem finitism altogether worthless; it applies in at least one area:

> In my view, the most interesting example of finitism comes from the theory of market prices (Bloor 1997, pp. 75–78): past prices do not determine future prices, but rather prices are at each moment fixed by a complex social system of individual decisions about economic transactions. Here finitism

is a plausible position, as there is no reason to believe in the existence of a lawlike regularity or a rule that makes the present price a function of earlier ones (1999b, pp. 86–87).

4. Vagueness

The remainder of this paper consists of some comments on Niiniluoto's interpretation and critique of finitism. I begin with the issue of vagueness. It is tempting but incorrect to summarize meaning finitism as "all concepts are vague". The distinction between vague and non-vague stands orthogonal to the distinction between meaning as analyzed by finitism and meaning analyzed by its opposite, "meaning determinism". Take a concept like "bald", the prototype of a vague predicate. What makes "bald" vague is not that its application is based on exemplars and judgments of similarity (sometimes socially negotiated judgments). Instead, "bald" is vague because we are collectively willing to accept both "X is bald" and "X is not bald" as assertible of the same X and at the same time. Regarding non-vague terms (like "one-meter-long") no such tolerance is generally displayed. More formally, a concept C is vague if the communal practices of evaluation allow for different speakers to apply both C and non-C to a given entity. A concept C is non-vague if the communal practices of evaluation do not allow for different speakers to apply both C and non-C to a given entity. Put differently, all acts of concept-application are underdetermined by past practice. But this does not make the concepts vague. "Vague" and "precise" are two social statuses that communities attach to concepts (cf. Lynch 1998, p. 61).

5. Ostensive Learning

According to Niiniluoto finitism is committed to the view that we acquire the names for basic categories in situations of ostensive learning. While it is true that Barnes and Bloor often focus their attention on ostensive learning, little reflection is needed to see that the plausibility of finitism does not depend on any hypothesis about the central position of this form of learning. Meaning finitism only needs the weak assumption that learning sets are invariably finite – never mind whether this set is gotten by ostensive learning or by some other mechanism, such as overhearing the conversations of others. As long as the finite character of learning sets is granted, the finitist has all he needs. If learning sets are finite then applying a term in new contexts must involve similarity judgments – must involve judgments concerning the similarity between the exemplars of his learning set and any newly encountered entities.

In conversation Niiniluoto has once suggested to me that if we were to learn expressions by means of verbal (rather then ostensive) definitions then the problems of underdetermination of future use by past use, and of (always contestable) similarity judgments, simply would not arise. I beg to differ. On the one hand, we cannot have definitions "all the way down". Verbal definitions capture the meaning of one expression in terms of the meanings of other expressions. But this process has to start from somewhere. That is, at least some expressions have to be learnt on the basis of a finite numbers of examples (be they given through ostension, be they overheard). On the other hand, the suggestion involves the thought that we can somehow create 'semantic immovability' by tying the fate of one expression (the *definiendum*) to that of another expression (the *definiens*). An obvious analogy is with a ship that is tied up safely at the dock. Alas, this is the wrong picture. Tying the fate of one expression to that of others by means of definitions is more like tying one ship to another. The two ships might float and drift together, or the rope between them might snap. Assume for instance that we have learnt the term "bachelor" on the basis of the definition "a bachelor is an unmarried man". A number of things might happen subsequently. Here are just two. The use of "marriage" might change – perhaps under pressure to legalize same-sex marriages. And once "marriage" changes, so does the use of "unmarried", and so might the use of "bachelor". Or consider the stress on the definitional "rope" that is created by our encountering religious figures like the Pope or Jesus: Is the Pope a bachelor? And is Jesus an unmarried *man*?

6. "Not Fully Determined"

How much or how little determination of future applications does finitism allow for? Niiniluoto correctly quotes Bloor as saying that "future applications are *not fully* determined by what has gone before" (Bloor 1991, p. 164, emphasis added; Niiniluoto 1999a, p. 263). Nevertheless, according to Niiniluoto the expression "not fully determined" fails to capture the real message of meaning finitism. The real message is that future applications are "fully undetermined" (*ibid.*, 265). The compromising evidence consists of two passages in Bloor (1997). Niiniluoto quotes these as "we create meaning as we move from case to case" (1997, p. 19) and as "Bloor gives up the idea that rules exist 'in advance of our following them'" (Bloor 1997, p. 21; Niiniluoto 1999a, p. 265).

In my view Bloor does not hold the extreme position that Niiniluoto ascribes to him here. This can best be seen by putting the two short quotations from Bloor back into their original context:

> According to meaning finitism, we create meaning as we move from case to case. We *could* take our concepts or rules anywhere, in any direction, and count anything as a new member of an old class, or of the same kind as some existing finite set of past cases. We are not prevented by "logic" or "meanings" from doing this, if by these words we have in mind something other than the down-to-earth contingencies surrounding each particular act of concept application. (...) The real sources of constraint preventing our going anywhere and everywhere, as we move from case to case, are the local circumstances impinging upon us: our instincts, our biological nature, our sense experience, our interactions with other people, our immediate purposes, our training, our anticipation of and response to sanctions, and so on ... (1997, pp. 19–20)

> We experience the rule itself as having the character that should rightly be attributed to our own collective behavior. None of our talk about rules existing in advance of our following them should be taken literally. (1997, p. 21)

The upshot of these passages is not that the applications of our terms are unconstrained or undetermined; the upshot is that they are not constrained by meanings or rules *if by meanings and rules we mean something more, or different from, our instincts, our biological nature, our sense experience, and so on*. Bloor is proposing a reduction – reduction as incorporation, not reduction as elimination – of meanings and rules to our instincts, biological nature, and the other listed items. He is rejecting the idea that there is something to meanings and rules over and above these biological, psychological and social constraints. To repeat, Bloor does not deny that future use of our terms is constrained (or partly determined) by meaning *properly understood*.

The same misunderstanding surfaces also in Niiniluoto's criticism according to which meaning finitism fails to reckon with the fact that meanings and social rules "resemble dispositions, causal powers, propensities, and laws of nature" (1999b, p. 85). Again at issue is the question whether meanings or rules – on their correct analysis – constrain or determine. And again I feel that closer attention to Bloor's text can help to make meaning finitism a less implausible viewpoint. Bloor insists in no uncertain terms that "the institutional approach [i.e. meaning finitism] can, indeed, be considered as *a form* of community-wide dispositional theory. ... An institution can be looked upon as the collective product of the interactions between the dispositions of many individuals" (1997, p. 68). Of course, Bloor's frequent reference to instincts points in the same direction; after all, instincts are biological dispositions.

With so much determination in meaning finitism, why then does Bloor distinguish his meaning finitism from a position he calls "meaning determinism"? Part of the answer is implicit in the above. Meaning determinism refuses the reduction of meaning talk to talk about communal dispositions and local circumstances. The meaning determinist believes that meanings have the power to determine future applications in ways that transcend our biology, sociology and psychology. A further reason for Bloor's opposition to meaning determinism has to do with the crucial distinction between *causal* and *normative* determination of future applications (Bloor 1997, pp. 15-17). If I have been trained in the use of a term, then it is likely that I will be disposed – that is, at least partially causally determined – to apply the term in specific ways. But that my training disposes me to act in this way does not safeguard that my applications are, or will be, correct. It does not safeguard this since there is no guarantee that my extrapolations from my learning set will be sufficiently similar to the extrapolations of other speakers from their respective learning sets. That is to say, the normative determination of my application as correct is something that – at least in principle – requires others' agreement. Or rather, nothing about the causal determination of my application guarantees that it cannot be challenged and rejected by members of my linguistic community. (Unless of course my community decides to let my dispositions set the standard for correctness.) Since any application can in principle be challenged, since any similarity judgment between learning set and newly encountered entity can be rejected, there is a second and independent sense in which meaning cannot fully determine future use.

It is important to emphasize the "in principle" character of such challenges. Finitists do not deny that much of our language use is routine. But they emphasize the implicit judgment that is at the root of our willingness to let our unreflective linguistic activity to develop proper usage:

> ... we allow, *as we need not*, our unreflective linguistic activity to develop proper usage: we place our confidence, *as we need not*, in our habits and routines. It is true that if we did not trust ourselves, and our own routine linguistic competences, most of the time, the use of language would become impossibly difficult and discourse would cease. But such trust in what comes routinely and unreflectively is essential in no *particular* case, so that its presence in any particular case can be said to reflect a contingent judgment that usage may be best developed in that way in that case. (Barnes 1982, p. 34)

If the partial defense of meaning finitism offered above is right, then Niiniluoto's *Potentiality Objection* is unfounded. And if the *Potentiality Objection* is misdirected then so is the *Discovery Objection*. For there to be room for discoveries regarding social institutions, there must be a way to distinguish discoveries from collective stipulations and inventions. This room is created by finitists' willingness to allow for some degree of determination of future use by current meaning – meaning properly understood. At the same time, however, finitists would want to add two caveats regarding talk about discovery. First, many new insights concerning social institutions can be described both as discoveries and as inventions. To *discover* that a certain opening in chess leads to a winning position is to *invent* a new opening. To *discover* that a certain theorem can be derived from a specific set of axioms is to *invent* a new proof. And to *discover* that a certain right can be defended on the basis of a constitution is to *invent* a new legal argument. Moreover, and this is the second point, discoveries and inventions are themselves subject to finitism. This does not only mean that 'discovery' and 'invention' are themselves social statuses awarded under contingent circumstances. More importantly, members of the institution need to form a view on whether the new opening, the new proof, and the new legal argument are desirable and should be permitted. Members always have the option of changing the institutions so as to block the new discoveries or inventions.

7. Niiniluoto's Algorithm Objection

Niiniluoto insists against both meaning finitism and meaning skepticism that at least in mathematics we can have concepts that fully determine their future applications. One of his examples is the definition of addition in terms the recursion equations *x+0=x* and *x+S(y)=S(x+y)* (1999a, pp. 266-267; 1999b, p. 86). Niiniluoto's case is most naturally challenged by drawing on Saul Kripke's book *Wittgenstein on Rules and Private Language* (1982). As will be recalled, the 'skeptical challenge' posed there is to identify a type of fact that makes utterances like "I mean *addition* by '+'" true and meaningful. Kripke's skeptic argues that there can be no such type of fact. In order to press home his point, the skeptic asks us to refute his alternative skeptical hypothesis according to which by "+" we mean *quaddition*. *Quaddition* is defined as:

x + y, if x, y < 57
5 otherwise. (Kripke 1982, p. 9.)

Kripke discusses a number of proposals. Second in line is the "Algorithm Response" (1982, pp. 16-17). What makes "I mean *addition* by '+'" true is that I am following, and have followed in the past, some algorithm for calculations with the sign "+". One such algorithmic procedure might be the counting of marbles. I determine the result of *x+y* by forming two heaps of marbles, with *x* and *y* marbles respectively, by combining the two heaps into one, and by counting the overall number of marbles. The skeptic is quick to point out that this answer fails. Trying to identify the mental state that constitutes my meaning addition by "+", the Algorithm Response invokes other of my mental states; to wit, the mental state that constitutes my meaning *counting* by the word "count" or the mental state that constitutes my meaning *heap* by the word "heap". And thus the Algorithm Response does not solve the initial problem, it merely passes it on.

In a footnote, Kripke considers a more sophisticated version of the *Algorithm Response*:

One might ... observe that, on the natural numbers, addition is the only function that satisfies certain laws that I accept – the "recursion equations" for +: *(x) (x+0=x)* and *(x) (y)(x+y'=(x+y)')* where the stroke or dash indicates the successor; these equations are sometimes called a 'definition' of addition.

This is of course exactly Niiniluoto's idea. But Kripke has a ready reply:

> The problem is that the other signs used in these laws (the universal quantifiers, the equality sign) have been applied in only a finite number of instances, and they can be given non-standard interpretations that will fit non-standard interpretations of "+". Thus for example "*(x)*" might mean for every $x<h$, where h is some upper bound to the instances where universal instantiation has hitherto been applied, and similarly for equality.
>
> In any event the objection is somewhat overly sophisticated. Many of us who are not mathematicians use the "+" sign perfectly well in ignorance of any explicitly formulated laws of the type cited. Yet surely we use "+" with the usual determinate meaning nonetheless. What justifies us applying the function as we do? (1982, pp. 16–17, fn. 12)

The upshot for Niiniluoto's Algorithm Objection should be clear: Niiniluoto is wrong to think that we can block meaning skepticism by invoking mathematical definitions. To do so is to presuppose what we are meant to show – to wit, that we can fully determine the future use of our terms by means of meanings and rules.

8. Excursus: Tennant on the Algorithm Response

Since Niiniluoto's *Algorithm Objection* is really the centerpiece of his case against meaning finitism and meaning skepticism it might perhaps be helpful to illuminate this objection and its problems from a further direction. That is, in this section I shall, by way of a short *excursus*, consider Neil Tennant's attempt to improve the Algorithm Response (Tennant 1997). Taking his starting point from Kripke's footnote cited above, Tennant insists that the skeptic overlooks (at least) three ways in which our use of "+" is constrained. Tennant formulates these ways as three problems for the meaning skeptic.

Problem One: The skeptical hypothesis conflicts with our intuition that the universal quantifier must have a uniform interpretation. We cannot uniformly interpret every occurrence of "*(x)*" in mathematics as *(x<h)*. For instance, in some mathematical laws universal quantifiers range over *terms* for numbers rather than over the numbers themselves. And whereas it makes sense to say that a number is smaller than 57, it does not make sense to say so of a term for a number.

Problem Two: Addition (with natural numbers) displays the following pattern. Any given number *n* is the sum of *n+1* ordered pairs of numbers: there are two such pairs for the number *1*, three such pairs for the number *2*, four for the number *3*, and so on:

0+1=1	0+2=2	0+3=3	0+4=4	0+5=5
1+0=1	1+1=2	1+2=3	1+3=4	1+4=5
	2+0=2	2+1=3	2+2=4	2+3=5
		3+0=3	3+1=4	3+2=5
			4+0=4	4+1=5
				5+0=5

This theorem can be given a formal rendering:

> For every [number] *n*, there are exactly (*n+1*) distinct addition sums with result $\underline{n}$ [$\underline{n}$ is the numeral for *n*], including, for every $p < n$, the sum "$\underline{p}$ + $\underline{(n-p)}$" and the sum "$\underline{(n-p)}$ + $\underline{p}$" (Tennant 1997, p. 111).

The important point is that we all accept this result. For instance, we all accept that there are exactly six addition sums with 5. If by "+" we meant some other function, we would most certainly not be happy with this claim. The fact that we are shows that the skeptical hypothesis — according to which we do *not* mean addition by "+" — is untenable.

Problem Three: Our use of numerals is constrained by the following scheme: "For every natural number *n*, there are *n* F's if and only if the number of F's = $\underline{n}$" (1997, p. 112). For instance, there are two grown-ups living in my household if and only if the number of grown-ups living in my household equals <u>two</u>. The underlined "*n*" and "two" are not numerals but numbers. The logical meaning of the occurrence of "two" in the second-to-last sentence is captured by the following logical formula (I use "$\exists(x)$" for the existential quantifier):

$(\exists x)\ (\exists y)\ (Fx\ \&\ Fy\ \&\ \sim x=y\ \&\ (w)\ (Fw \rightarrow (w=x\ \vee\ w=y)))$

There is an *x*, and there is a *y* such that: *x* is F and *y* is F and *y* is the negation of *x*; and for all *y*: if a *w* is F, then it is identical either with *x* or with *y*.

In this formula, no numeral appears, even though the twofold occurrence of the existential quantifier "shows something about the two-ness of F" (1997, p. 113). The same sorts of formula can of course also be constructed for 57 and

68. Assume we have 57 things that are F and 68 things that are G. There are then 125 things that are either F or G. This sum is captured by the following formula:

$$(\exists x_1)(\exists x_2)\ldots(\exists x_{125})(Fx_1 \vee Gx_1) \& (Fx_2 \vee Gx_2) \& \ldots \& (Fx_{125} \vee Gx_{125}) \& (y)((Fy \vee Gy) \rightarrow (y = x_1 \vee y = x_1 \vee \ldots \vee y = x_{125})))$$

Again, this formula "says nothing about the number 125 (the sum of 68 and 57); rather it shows the 125-ness of (F or G)". The quadder arrives of course at a different formula. For him the formula for F+G has only five existential quantifiers. "Yet he will have to maintain that it follows logically from the two formulae saying, respectively, 'There are exactly 68 F's' and 'There are exactly 57 G's', along with the premises 'Nothing is both F and G'." And this forces the skeptic into "very difficult logical gerrymander indeed" (1997, p. 114).

In my view, Tennant's problems can easily be dissolved by the meaning skeptic. Concerning *Problem One* it must be acknowledged that – given our normal understanding of "uniformity" – the skeptical re-interpretation of quantifiers is not uniform. But maybe we are wrong about our past intentions regarding the use of "uniformity". Maybe by "uniform" we really mean *quuniform*, where *quuniformity* is compatible with different interpretations for different occurrences of the universal quantifier. Regarding *Problem Two* we only need to note that the meta-mathematical result is no independent touchstone. The skeptical challenge asks us in virtue of what fact we mean *addition* rather than *quaddition* by "+". What is Tennant's response? Is it this: the fact that makes it so that we mean *addition* is that we can be brought to accept the meta-mathematical result? It would be strange if this was Tennant's proposal. To give this answer would be to claim that before the meta-mathematical result was known no-one could really believe himself to be an adder. To this difficulty we can add another. The meta-mathematical result can of course itself be re-interpreted in familiar ways. For instance, the skeptic might ask how we know that by "addition sums" we did not mean "addition sums in which neither summand is bigger than 57". The skeptic might also propose that by the expression "for every n" we really meant "for every n except 5". Similar difficulties bedevil *Problem Three*. The move from arithmetic to logic is not of much help here – after all, in logic too we rely on the meaning of words and thus on the description of procedures. The skeptic can grant that the two logical formulas with 57 and 68 existential quantifiers respectively do not contain numerals but *show something* about the "fifty-seven-ness" and the "sixty-eight-ness" of the two sets. And yet, anyone who seeks to decide whether the union of these two sets has fifty-seven or sixty-eight elements will still have to count – or quount. And thus we are back in familiar skeptical territory.

9. Meaning Finitism and Meaning Skepticism

Niiniluoto discusses meaning finitism and (Kripke's) meaning skepticism as if these were one and the same position. This ignores some very important differences and fails to do justice to Bloor's overall ambitions. Let me explain. (Cf. Kusch 2006.) The most important proposal discussed in Kripke's book is not the Algorithm Response; the most important proposal is "reductive semantic dispositionalism" (Kripke 1982, pp. 22–37). The reductive semantic dispositionalist believes that meaning attributions have factual truth-makers that can be captured in non-intentional terms. That is, "Jones' means addition by '+'" is true, the dispositionalist says, if, and only if, *ceteris paribus*, Jones has the dispositions to produce the sign "4" in response to the sign-sequence "2+2=", the sign "8" in response to the sign-sequence "6+2=", the sign "125" in response to the sign-sequence "57+68=", and so on. Kripke argues that reductive semantic dispositionalism fails. On the one hand, talk of meaning is normative, whereas talk of dispositions is descriptive. To say that Jones means addition by "+" is to say (amongst other things) that Jones *ought to* respond in specifiable ways to given addition tasks. But to say that Jones has certain dispositions regarding the use of "+" contains no such normative element. On the other hand, to say that Jones means addition by "+" is to attribute to Jones a propositional attitude towards an infinite entity. Again, talk of dispositions cannot capture this: as a finite being Jones can only have a finite number of dispositions. And thus dispositionalism fails.

Enter Bloor's book *Wittgenstein, Rules and Institutions* (1997). For Bloor meaning skepticism is both a friend and a victim. Meaning skepticism is a friend in finitism's battle with *individualism* about meaning. But once this battle is won, finitism can quickly and easily dismiss its former ally. Put differently, the issue is whether Kripke has exhausted the possible candidates for facts that make meaning attributions true and meaningful. Bloor thinks that Kripke has not. As Bloor sees it, all of the facts considered and rejected in Kripke's book are facts about the *individual* to whom meaning is being attributed. For instance, according to the dispositionalist solution, I mean addition by "+" because I have the disposition to produce sums in answer to plus-queries. Bloor urges us to see that Kripke's book is not really a refutation of the idea that meaning attributions correspond to *facts simpliciter*: it merely is a refutation of the idea that meaning attributions correspond to *facts about an individual*. Call the refuted position *meaning individualism*. Bloor believes that Kripke has overlooked an alternative response to the meaning-skeptical challenge: *(collectivistic) meaning finitism*. According to meaning finitism meaning attributions are made true (and meaningful) by *social* facts about the group to whom the individual in question belongs. These social facts are social institutions. Looked at from a different angle, these social facts and social institutions are nothing but meanings. And to see meanings as social institutions, Bloor submits, explains their normative properties. The number "125" is the right answer to "57+68" because it is the answer "which is collectively called 'right'". And meaning "'exists' in and through the practice of citing it and invoking it in the course of training, in the course of enjoining others to follow it, and in the course of telling them they have not followed it, or not followed it correctly" (Bloor 1997, p. 34).

Bloor thus reduces meanings to social institutions. But he does not stop here. Bloor also accepts the further requirement according to which meanings must ultimately reduce to non-intentional facts. A reduction of meanings to social institutions does not yet satisfy this further requirement: social institutions are themselves intentional phenomena. Bloor suggests that the needed non-intentional facts are *communal* non-intentional dispositions. That is to say, Bloor accepts that Kripke has shown individualistic forms of reductive dispositionalism to be untenable, but he believes that communal dispositions – the dispositions shared by a group, and the interacting dispositions of different individuals – survive Kripke's critique. Bloor writes that meaning finitism "can, indeed, be considered *a form* of community-wide dispositional theory" (1997, p. 68).

I disagree (Kusch 2004a & b). I am with Kripke (1982, p. 111) in thinking that a communal reductive dispositionalism does not overcome the problems of the individualistic versions. Making dispositionalism communal does not enable us to circumvent the difficulties with normativity and infinity. Bloor tries to disable the normativity problem in the following short passage:

> To be disposed is one thing; to be disposed rightly or wrongly is another. That was the skeptic's point. By contrast an institution, which arises from

the interaction of individual dispositions, does provide a normative basis for the actions of the individuals who are within it. (1997, p. 68)

As I see it, this is just a promissory note: Bloor tells us *that* meanings ultimately are communal or interacting dispositions, but he does not tell us *how* this is to work in a single concrete case. Which communal or interacting dispositions make it so that I mean *addition* rather than *quaddition* by "+"? At least the individualistic dispositionalist came up with a testable proposal. Bloor does not.

10. Conclusion

In this paper I have tried to defend meaning finitism against Niiniluoto's criticism. In particular, I have attempted to show that none of Niiniluoto's four objections – the *Self-Refutation Objection*, the *Algorithm Objection*, the *Potentiality Objection* and the *Discovery Objection* – do any damage to Bloor's position. Niiniluoto acknowledges that *if* finitism were correct then his "realist principle" – "we choose the language *L*, and the world decides which sentences of *L* are true" – would be under threat (Niiniluoto 1999a, p. 267; cf. Barnes 2001; Bloor 1996; Kusch 2002). I believe that the threat is real.

University of Cambridge

References

Barnes, B. 1982: "On the Extensions of Concepts and the Growth of Knowledge", *Sociological Review* 30, 23–44.
Barnes, B. 2001: "The Construction of Social Reality", *Revue Internationale de Philosophie* 55, 263–268.
Bloor, D. 1991: *Knowledge and Social Imagery*, 2nd edn. with afterword, Chicago University Press, Chicago.
Bloor, D. 1996: "Idealism and the Sociology of Knowledge", *Social Studies of Science* 26, 839–856.
Bloor, D. 1997: *Wittgenstein: Rules and Institutions*, Routledge, London.
Kripke, S. 1982: *Wittgenstein on Rules and Private Language*, Harvard University Press, Cambridge, Mass.
Kusch, M. 1991: "Koko totuus totuudesta" (The Whole Truth about the Truth, in Finnish), *Tiede ja Edistys* 16, 284–297.
Kusch, M. 1996: "Kehitysapua kolmannelle maailmalle: kuinka argumentit ovat olemassa?" (Developmental Aid for the Third World: How Do Arguments Exists?,

in Finnish), in I. A. Kieseppä, S. Pihlström, and P. Raatikainen (eds.), *Tieto, totuus ja todellisuus: Kirjoituksia Ilkka Niiniluodon 50-vuotispäivän kunniaksi*, Gaudeamus, Tampere, pp. 30–37.

Kusch, M. 2002: *Knowledge by Agreement: The Programme of Communitarian Epistemology*, Oxford University Press, Oxford.

Kusch, M. 2004a: "Rule-Skepticism and the Sociology of Scientific Knowledge", *Social Studies of Science* 34, 571–591.

Kusch, M. 2004b: "Reply to my Critics", *Social Studies of Science* 34, 615–620.

Kusch, M. 2006: *A Sceptical Guide to Meaning and Rules: Defending Kripke's Wittgenstein*, Acumen, London.

Lynch, M. P. 1998: *Truth in Context: An Essay on Pluralism and Objectivity*, MIT Press, Cambridge, Mass.

Niiniluoto, I. 1991: "Realism, Relativism, and Constructivism", *Synthese* 89, 135–162.

Niiniluoto, I. 1999a: *Critical Scientific Realism*, Oxford University Press, Oxford.

Niiniluoto, I. 1999b: "Rule-Following, Finitism, and the Law", *Associations* 3, 83–90.

Niiniluoto, I. 2003: *Totuuden rakastaminen: Tieteenfilosofisia esseitä* (The Love of Truth: Essays in the Philosophy of Science, in Finnish), Otava, Helsinki.

Tennant, N. 1997: *The Taming of the True*, Clarendon Press, Oxford.

Wittgenstein, L. 1967: *Philosophical Investigations*, Blackwell, Oxford.

Jan von Plato

Dialogue on Foundations of Geometry

[We enter in the middle of a discussion three mathematicians AP, AR and GE are having over foundational questions of their science.]

AP: ... is set theory and first order predicate logic. End of story.

AR: No, no, as I was saying, we are given the natural numbers and [interrupted].

GE: We've heard *that* story. Our colleagues at the computer science department bought it all, they even changed the names into Discrete-this, Discrete-that.

AR: So what's wrong about that?

GE: Well, I can't *prove* it's wrong, but I could tell you a few examples.

AR: Let's hear some!

GE: We have to go to the seminar room, I need a blackboard.

⋮

GE: It's all nice and clean, I'll take a piece of chalk. Here, this [marking on the board] is a point, I want to call it P.

AR: So you can mark a point.

GE: Yes, and if I want to, I can refer to precisely that point again, I just say, point P. Now I mark another point, Q, as you see, it is distinct from P.

AR: All right, let's have them all at once, how many points do you need, seven? And by the way, before you mark the points, I know what you are doing, it's called discrete geometry. Err ... finite geometry.

GE: It's not finite geometry, and for the time being I'll be happy with just these two points.

I should have brought along a big ruler, one of those we had at school. Now you'll have to rest content with these freely drawn not-so-straight lines. See, I say I can draw a straight line through these two points P and Q. [Draws the line.] Well, not quite, but we can all imagine it went exactly through the points. I'll call the line PQ. Now look at this. [Draws another line.] There were no points, I just drew a line and now I name it l. Here is another one, m, drawn so it intersects l. Now I put the chalk in the intersection and this is the point lm.

By the way, is P on m?

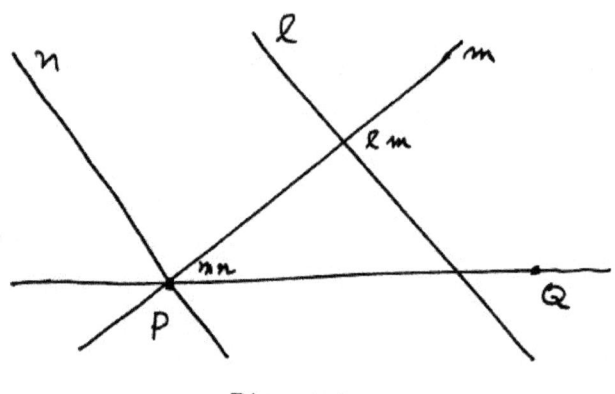

Diagram 1.

AR: I can't see, but it's certainly very close at least.

GE: And here's a third line n, it clearly intersects m, but is P on it?

AR: I can't see that one either! You should choose your lines better.

AP: Yeah, it's called "general position", you should draw your geometrical figures in general position.

GE: Never mind, but I notice that you agree I have an intersection point mn, don't I?

AR, AP: [Nodding.]

GE: And you would say, to use AR's words, that mn is "at least very close" to P?

AR, AP: Yes.

GE: And you agreed that you could not tell if *P* was on *m* and *n*, so I take it that you also agree you can't see if *P* and *mn* are distinct? [Waits a few seconds.]

When I name a point, I later refer to precisely that point: This *P* [points at *P*] can't be distinct from the earlier *P*. I use the forgotten Hilbert notation for distinct points: $P \neq Q$. Now, we have $\sim P \neq P$, *P* is not distinct from itself.

AP: Hilbert notation?

GE: Yes, it's from the first edition of his geometry book, from 1899.

AR: That's boring, what else do you have?

GE: Well, I draw these two distinct points, so we have $P \neq Q$. Now look at *R*. [Draws *R* above *P*.]

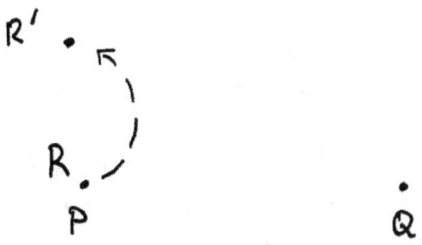

Diagram 2.

AR: What's the matter with you, I can't see if $P \neq R$!

GE: Oh, but you do see instead that $Q \neq R$. So how would you express in numbers what you see?

AR: In numbers? Oh, you mean I see *R* is distinct from one of *P* and *Q*.

GE: Let me replace *R* neatly away from both.

AR: Yes, I correct, *R* is distinct from at least one of *P* and *Q*.

GE: So we all agree that if $P \neq Q$, then $P \neq R$ or $Q \neq R$. Let me write this in symbols, $P \neq Q \supset P \neq R \vee Q \neq R$.

AP: Why do you use that funny implication symbol?

GE: It's from Peano, he inverted the letter "C" that he took from the word "consequentia". Russell also used it, but later others used arrows, simple or double. I use the simple arrow for other things. So this is a law for points, and earlier I already noticed the first law, $\sim P \neq P$. I will say in general, two points P and Q are equal if they can't be distinct.

AR: So for you, $P=Q$ is defined as $\sim P \neq Q$. So $P = P$ is your $\sim P \neq P$, and the other thing gives $P = R \, \& \, Q = R \supset P = Q$ by contraposition. Very good.

GE: Since you want all things to be discrete, how do you define "discrete" itself?

AR: A set is discrete if for any two of its members, say a and b, it can be decided if $a = b$ or... [hesitates]

GE: Yes?

AR: This was a trap! But allright, you have it, speaking of your geometrical points at least, if you place them in that wicked way, indefinitely close to each other, I can't see if $P \neq Q$ or $\sim P \neq Q$, so then I add a double negation in front of the first and it follows that I can't see if $\sim P = Q$ or $P = Q$. To have a discrete geometry I would have to assume all my points distinct.

GE: Yes, and maybe a few more things in addition to that.

AP: So now you have some set with this relation $P \neq Q$ you said comes from Hilbert, with the axioms $\sim P \neq P$ and $P \neq Q \supset P \neq R \vee Q \neq R$. What about symmetry?

GE: Just put $R = P$ in the second axiom.

AP: Oh, it comes out. I wonder why we always had the three axioms for equality.

GE: Actually it comes from Euclid: If two things are equal to a third, they are equal between themselves.

AR: Some more historical remarks? I would instead like to see the standard interpretation. The claim is not that all structures are discrete, of course we get undecidable equalities with the limits of convergent sequences of rationals for example: You can compute arbitrarily long approximations with two such sequences, but this does not decide if the limits are equal. There is a standard interpretation for your plane points, invented by Descartes if you grant me a historical remark. As I said in the beginning, we have the natural numbers, then the rationals, then convergent sequences of rationals, and any two such sequences give you a point in the Cartesian plane!

GE: I don't have anything against that, but I maintain that geometry makes sense to those, including everyone before Descartes and today's children up to some age, who never heard of your Cartesian plane. *My* intended interpretation is this blackboard, this plane, and the points and lines I can draw on it. I of course discard the fact that these are physical things. That step, precisely, is at the heart of the geometrical tradition that goes unbroken from ancient times to our children's school geometry. I mean, the ideality of the objects is what makes it geometry.

AP: You *might* have something, so let's don't leave it to this sermon. Tell us instead the rest, like how you show that your connecting line PQ is unique.

GE: Thank you for this support, AP. So, if $P \neq Q$, we can draw the line PQ, and let's just do projective geometry so things are simple: in projective geometry all parallel lines are equated. Equal lines is just like equal points. I axiomatize it by postulating $\sim l \neq l$ and $l \neq m \supset l \neq n \vee m \neq n$ and defining $l = m$ as $\sim l \neq m$. Then intersection points: I want to call the condition convergence, so l and m are convergent lines is the condition for the construction. Let's write it $l \not\parallel m$, and I can draw, or at least emphasize, the intersection point lm.

AP: And the axioms for convergence?

GE: Just like before, $\sim l \not\parallel l$ and $l \not\parallel m \supset l \not\parallel n \vee m \not\parallel n$.

AR: Not bad, dear GE, you have [goes to the blackboard] $l \parallel l$ and $l \parallel m \,\&\, m \parallel n \supset l \parallel n$.

GE: Yeah, parallels is an equivalence relation. Look why: I draw l and m convergent, and there's no way of placing a third line n on the plane without having $l \not\parallel n \vee m \not\parallel n$.

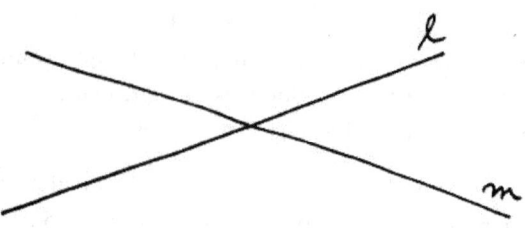

Diagram 3.

And now I say, I get projective geometry by requiring the positive version of $l \parallel m \supset l = m$, so it's $l \neq m \supset l \nparallel m$. Whenever $l \neq m$, I get the intersection point condition from this and then construct lm.

AP: I still miss uniqueness.

GE: Yes, and this is the part of the story I like best. Let's write $P = l$ for: point P is incident with line l.

AR: [interrupts] ... and then $P = l \& Q = l \& P \neq Q \supset PQ = l$!

AP: I start learning the game, let me try the other, it is ... $P = l \& P = m \& l \neq m \supset lm = P$.

GE: These principles are ok, but what if I can't see whether P is on l, and so on?

AR: At least we can combine the two principles, like this: $P = l \& Q = l \& P = m \& Q = m \supset P = Q \vee l = m$; you can't have two different points on two different lines, or to put it in another way, if two points are on two lines, the lines or points are equal.

GE: [writes without a word] $P \neq Q \& l \neq m \supset P \neq l \vee Q \neq l \vee P \neq m \vee Q \neq m$.

AP: So $P \neq l$, a point is *outside* a line is your basic concept, and you can move P or l around a bit in $P \neq l$, and it's still true that $P \neq l$.

GE: What you said can be formalized: So, assume $P \neq l$, and then $P \neq Q$ for any Q, or if you can't see it, you must have $Q \neq l$, too. And the same for l, if $P \neq l$, then $l \neq m \vee P \neq m$. And the contrapositions are ...

AR: The substitution rules for equals, $P \neq l \mathbin{\&} P = Q \supset Q \neq l$, $P \neq l \mathbin{\&} l = m \supset P \neq m$, and also $P = l \mathbin{\&} P = Q \supset Q = l$, and so on.

GE: Yes, I just want to emphasize the justification of the constructive substitution axioms, as so subtly pointed out by AP's observation. I also had a remark about AR's "combined" uniqueness axiom. Let me draw two lines l and m, and I place the points P and Q like this, both points on both lines …

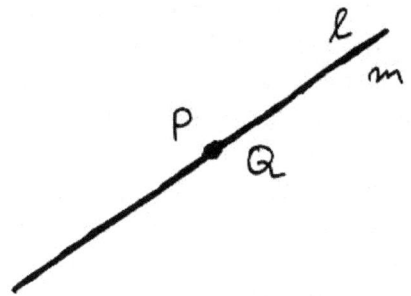

Diagram 4.

AR: The lines are again too close for me, and so are the points! Maybe both pairs are equal?

GE: Just remember what I postulated, or required to be the case: $P = l \mathbin{\&} Q = l \mathbin{\&} P = m \mathbin{\&} Q = m$, the rest must be concluded.

AP: I see the point [laughter]…, we can't decide which is the case with AR's law, $P = Q \vee l = m$. So your axiom with $P \neq l \vee Q \neq l$ and so on is right, but AR's is too strong.

GE: Actually AR's little discovery was made by Skolem in 1920. Now, you can have the disjunction if you use classical logic, but this is just one more example of how the indirect inferences of classical logic lead from computational premises to noncomputational conclusions. By the way, Skolem also had the classical substitution principles written out as explicit axioms.

AR: So where was that?

GE: Aha, your interest in origins seems to have awakened. It was in the famous paper with the Löwenheim–Skolem theorem. He had three different things in there, but people only read the first!

AR: Did *anyone* ever discover anything new? Besides, I am rather skeptical about your remarks on computation. No numbers anywhere and there should be some computational content, do you really mean that?

GE: It is abstract of course. Lots of programming languages have computational things with no numbers, either. I would actually like to show you another example of an abstract axiomatization with computational content. For the time being, let me say that I could formalize computable real numbers, define points to be pairs of such numbers, and lines sets of numbers satisfying a suitable form of equation. Then you would see that the axioms I have proposed can be proved for this particular understanding of the geometrical plane, but your axiom with the disjunction instead, could be refuted. That principle would lead to the decidability of equality of computable reals.

AP: But all of this is just because you have this funny way of understanding disjunctions, it seems to me. What do you need them for, anyway, if you say a disjunction is proved by showing how to decide among the alternatives? And the same for existence: Why do you need existential propositions if you have to prove them by first finding an instance? Couldn't you just quote the instance?

GE: Here is an example that may amuse you. The following is called "the majority vote problem": You have a finite list of votes, where each vote is for one of a finite number of alternatives. Your task is to find out if there is an alternative with a majority of votes and if so, which one it is. Let's agree that $L(s)$ expresses that s is a list of the required kind, and $MV(a,s)$ expresses that alternative a has majority vote in list s. Then I formalize the problem by the following:

$$(\forall x)(L(x) \supset (\exists y)MV(y,x) \vee \sim(\exists y)MV(y,x)).$$

And now I say, the problem is solved by deriving this formula.

AP: So what you have is, if we strip off the logical decoration, more or less like $A \supset B \vee \sim B$.

GE: Precisely, and classically, given a list s, there would be nothing left to prove. But for my understanding of disjunction and existence, a proof of the above formula is something that gives me, for any instance of it, that is, for a given list s of votes of the required kind, a proof of $(\exists y)MV(y,s)$ or a proof of $\sim(\exists y)MV(y,s)$. In the latter case, there is nothing more to say. In the former, the proof furnishes me the instance a that is the majority vote, $MV(a,s)$.

AP: But this is very nice, something that is completely empty merely by its logical form, turns into an effective program for finding out the thing sought when you interpret logic in this new way.

GE: It's not that new, but there was little practical use for it until a couple of decades ago. I take it that your use of the words, "effective program", was not casual. In fact, the above problem has been solved in a completely formal way, and the solution is a computer program that does the job.

AR: How do you mean, a computer program?

GE: Yes, these things are fairly new, and I won't go to them. Let me just say that you can think of programming as consisting of a problem and its solution, the program that executes the thing stated in the problem. Similarly, a mathematical claim is a problem that is solved by finding a proof. Say, if we have a theorem of form $A \supset B$, it is solved by finding a way of transforming any proof of A into some proof of B. It is the same with universal quantification, and I used it already above, in the majority vote problem. The problem was of universal form, so I assumed an arbitrary value s for x, and if the problem is solved in this way, the universal claim can be concluded. Then, proceeding in the other direction, if I am given an instance, I get the proof for that instance.

But we were discussing foundations of geometry, so let me tell you a story: When I was small, I used to sit in my father's feet in his office. He had this big engineer's drawing board, slightly tilted from vertical. It had a moving ruler, with an orthogonal too, counterbalanced carefully so it could be slid around easily, but the ruler always kept parallel. Then you could unlock it and turn it around, and then lock it again. So you could mark two distinct points on the board, unlock the ruler, and align it with the points. Now, lock the ruler and you can draw a line exactly through the two points. Next mark a third point anywhere, and you can slide the ruler until it is aligned with the new point. This way you draw exact parallels and orthogonals to the line you drew previously, through the third point. Now the principles behind all this, these are the interesting part of the drawing board, not the physical setup. It always works on the same principles. So one way of looking at what I have told you so far is that I have taken the first steps in axiomatizing the drawing board!

AP: Sorry, GE, if I interrupt you but I have to say it now or else I forget it. Namely, this "exact" parallels through a point brought to my mind something you must have missed: Let's assume $P \neq Q$, $l \not\parallel m$. Shouldn't we have the incidences $P = PQ$, $Q = PQ$, $lm = l$, $lm = m$?

GE: You are perfectly right, this is what happens when one starts talking out of one's head. So let's resume: We have the relations $P \neq Q$, $l \neq m$ and $l \not\parallel m$ all obeying the same kind of axioms, I call them irreflexivity and splitting. Then we have the constructions PQ and lm where we need $P \neq Q$ and $l \not\parallel m$ as conditions, with the constructed objects required to fulfil the four incidences you mentioned. Then we have the, if I may say, rather elegant single uniqueness axiom for both constructions, and the two substitution principles. That's all.

AP: Let me try a bit with these parallels and orthogonals: We have the parallel to line l through point P, written, say, Pl, where $P \neq l$, and the axioms are $P = Pl$, $l \parallel Pl$. What else?

GE: I would just drop the condition $P \neq l$. People were used to requiring a "point not incident with the line" when putting down the parallel line construction, because they thought Pl sort of was already there if $P = l$; there was simply not anything to be constructed. But we need not know if $P = l$, so we just construct Pl and it might be the case that $Pl = l$, or it might not. And Pl is of course different from l in some way even if $P = l$, in the sense of being given in a different way. Equality, $l = m$, only means that the lines l and m are in the same place, that they are geometrically congruent as one could also express.

AP: I accept that about dropping $P \neq l$, it's rather conventional anyway, and the condition would actually prevent parallelism from being an equivalence relation. So maybe the treatment is more uniform the way you suggest.

AR: I presume we still miss quite a few axioms for parallels! Are we going to see again as many new axioms as before?

GE: Come on, they are not that many, and I will just add one. Look at it this way: People are used to keep in mind, without any notice, that $l = l$, $l = m$ & $m = n \supset l = n$, and the same with the substitution rules, $P = l$ & $P = Q \supset Q = l$, and so on. But all these principles would have to be there if we pretended any exactness about formalization. Just like Skolem did, and I don't have any more axioms than he had. But now, to the last axiom of the geometry of convergent lines. We have the axiom of uniqueness, $l \neq m \supset P \neq l \vee P \neq m \vee l \nparallel m$.

AR: Now it is *my* turn to say your logic is faltering. I accept we get by contraposition $P = l$ & $P = m$ & $l \parallel m \supset l = m$, so by $P = Pl$ & $l \parallel Pl$, we get $P = m$ & $l \parallel m \supset m = Pl$, classical uniqueness. But you said we can't decide $l \parallel m \vee l \nparallel m$, it's all infinitely precise sort of. But I say that if $l \neq m$, then requiring $P \neq l \vee P \neq m$ for any point is just your 'positive' way of stating that l and m have no points in common. So we have $l \parallel m$, and your axiom is $l \neq m \supset l \parallel m \vee l \nparallel m$, parallelism is observable for distinct lines. What do you say about that?

GE: It troubled me, too, until one Swedish logician colleague of mine advised me to be more careful about my logic. Notice that I don't have quantifiers anywhere, just free parameters and constructions. So when I say, $P \neq Q \supset P = PQ$ – and I admit this looks a bit strange so devise your own notation if you wish, like, say, $P \doteq l$ and $P \neq l$ for 'outside' and incidence, then you have the notation $P \neq PQ$, so what was I saying… yes, quantifiers, so Hilbert & co would say, with $X, Y, Z, \ldots$ ranging over points and $x, y, z, \ldots$ over lines, they would say $\forall X \forall Y (X \neq Y \supset \exists z (X \neq z))$, and then the uniqueness and so on …

AR: … aren't you instead just implicitly using the functions of the Norwegian you mentioned? If for all x there is some y such that some property $A(x,y)$ holds, there is a function f such that $A(x, f(x))$. Skolem functions.

AP: I call that the axiom of choice.

GE: I'll talk about existence in a while, but let us first clear the matter of logic AR raised. So we have, with quantifiers as before, and the improved notation,

$$\forall x \forall y \forall Z (x \neq y \supset Z \doteq x \vee Z \doteq y \vee x \nparallel y).$$

This boils down to the failure of the formula, $\forall x (A(x) \vee B) \supset \forall x\, A(x) \vee B$. But usually one reasons quite confidently with parameters and constants.

AR: Defeat accepted, I have seen enough of intuitionistic logic to know that the quantifier formulas don't work in the usual fashion.

AP: The constructivists always say that if you prove a disjunction, you must be able to prove one of the disjuncts, and the same for existence. If you prove there exists and-so-on, you have to be able to find some object that has this and-so-on. Are these still valid points of view?

GE: It used to be required that constructive theories have the disjunction property you mention, but this seems to me a tricky subject so I'll talk in terms of an example that occurred to me the other day. Imagine you add to all lines little arrowheads indicating direction. In my way of speaking, to have two equally directed lines is infinitistic, unobservable, and the same if you have two exactly oppositely directed lines. But to have two unequally directed lines is not, and neither is having two lines that positively avoid being oppositely directed. Then, we have to admit that, given any two lines, they are unequally directed, or if they were given so that you are unable to decide that, they are very very far from being oppositely directed.

AP: This means that you have a disjunction, any two lines are unequally or, how should I say it, yes, inoppositely directed. But you don't know which is the case before the lines are actually given to you.

GE: It seems to me that some disjunctions in a constructive theory work in a logical manner, and then the disjunction property should follow, and other disjunctions can come about through some nonlogical disjunctive axiom, and the disjunction property can fail.

AR: Now the other thing about existence that was taken up, don't you think your geometry is rather uninteresting as you can't even prove the existence of a single point or line?

GE: This is not a real objection, but a question of how one arranges things. I have written down the minimum of things you have if you take the connecting line, intersection point and parallel line constructions. But look, now [wipes the blackboard clean] I cancel all these figures, and there are no points, no lines:

Diagram 5.

Still I want my geometry to be valid. If I just have these two points [draws P, Q], P and Q, with $P \neq Q$, I want my geometry to hold. A second matter is that for me, existence is explained through construction. If I say, for every line x, there are two points Y, Z, with $Y = x$, $Z = x$, $Y \neq Z$, or there is a point Y outside x, where is the construction? I accept that a geometry with just two points and their connecting line is rather uninteresting. If I have three, P, Q, R, with $P \neq Q$ and $R \doteq PQ$, I can construct any number of distinct points.

AP: Can you actually prove that?

GE: I first construct a parallelogram, so that's a fourth point, then just continue producing intersection points by doing more and more parallels.

AR: So that is why you said yours are not finite geometries.

AP: And neither are they discrete. But you said, rather mysteriously to me, that existence is 'a matter of organization'.

GE: Here we can borrow some concepts from computer science, even if I know you think that whole business is nothing but calculations without imagination and actually a threat to pure mathematics. So, each time you do geometry, say with the above axioms, and you are in a concrete situation, what is that situation? And let's not talk about proving theorems, but the more basic situation of a geometric problem.

AP: Well, you have what is called the given, and then you have to find what is traditionally called the thing sought, and of course, when you have found your way from the given to the sought, you have to show that it all went correctly.

GE: Very good, AP, that is the way it is presented in the school tradition. So the given, in any problem we have the given, and it is introduced as a *context*. A context is a list of assumptions, computer scientists call these variable declarations. Say, the connecting line construction can be executed in any context that has two distinct points, lets write it $\Gamma = (X \in \text{Point}, Y \in \text{Point}, X \neq Y)$. To be complete, I should begin the context by assuming that points and lines form sets. On the other end, we may enlarge a given context Γ by adding assumptions, say $Z \in \text{Point}, Z \doteq XY$. Note that here it is legitimate to form the line XY only if the context already has $X \neq Y$. So the assumptions have to be listed in an orderly manner.

Instead of existence axioms, I can consider contexts that declare objects and their properties, and we have a perfectly good understanding of these contexts, they are just geometric situations in which the facts stated in the context hold.

AR: So this is your way of avoiding the explanation of, say, the existential quantifier in a formula like $\forall x \exists Y (Y \doteq x)$. Instead of this general axiom you assume a context, that is, some concrete situation for which $x \in \text{Line}, Y \in \text{Point}, Y \doteq x$, and if you later need more points outside some line, you just add it to the context.

GE: The solution of geometric problems becomes, from the data of the problem, represented as a context, to construct the sought thing and to prove that it has the required relation to the data. The sought can also be given as a context, and the solution of a geometric problem as a mapping from the data context to this latter goal context.

AP: And you use these constructions and logical inferences and the axioms in putting up this mapping. I wonder if it could be done axiomatically, without the constructions. And what about the existence property?

GE: When in the Hilbertian style of existential axioms we put down,

$$\forall X \forall Y (X \neq Y \supset \exists z (X \doteq z\ \&\ Y \doteq z)),$$

that is, postulate merely the existence of connecting lines, we lose track of the information that was used to establish that X and Y are distinct points. In terms of the standard interpretation that AR suggested, we have some computation that established X and Y have different coordinates. Now we infer there exists some line z, but referring later to it, this line z doesn't "know" on what reason it came about. For any two given distinct points, we could introduce a new constant that is the line connecting them, but there is no uniform way of doing this, except to accept that the connecting line construction goes beyond AP's first-order logic.

AP: Don't you think it possible that at some stage, an existential axiom would be justified, even if you don't find any object from which to conclude the existential claim? This would be just like the case with the disjunctive axiom for directed lines. You would only find the object once all the objects in your situation are actually given.

GE: The way I do it is satisfactory to me, but I can't exclude the possibility you mention.

AR: Good, then we have reason to return to these discussions some other time.

GE: Yes, and thank you for your patience.

University of Helsinki

Panu Raatikainen

Truth, Correspondence, Models, and Tarski

In the early 20th century, scepticism was common among philosophers about the very meaningfulness of the notion of truth – and of the related notions of denotation, definition etc. (i.e., what Tarski called semantical concepts). Awareness was growing of the various logical paradoxes and anomalies arising from these concepts. In addition, more philosophical reasons were being given for this aversion.[1] The atmosphere changed dramatically with Alfred Tarski's path-breaking contribution. What Tarski did was to show that, assuming that the syntax of the object language is specified exactly enough, and that the metatheory has a certain amount of set theoretic power,[2] one can explicitly define truth in the object language. And what can be explicitly defined can be eliminated. It follows that the defined concept cannot give rise to any inconsistencies (that is, paradoxes). This gave new respectability to the concept of truth and related notions. Nevertheless, philosophers' judgements on the nature and philosophical relevance of Tarski's work have varied. It is my aim here to review and evaluate some threads in this debate.

1. Early Tarski and Model Theory

It has been common (see e.g. Vaught 1974; 1986) to trace the key notion of model theory, the *satisfiability-in-a-structure*, or *truth-in-a-model*, back to Tarski's seminal paper "The concept of truth in formalized languages" (Tarski 1935, henceforth CTFL), and more generally, to associate Tarski's contribution to the

---

[1] For more of the historical background, see e.g. Niiniluoto (1999b), Sluga (1999), Woleński and Simons (1989).

[2] But one should not exaggerate the amount of set theory needed. Relatively little suffices. For example, if the object language is the language of first-order arithmetic (i.e., that of PA), the relatively weak and predicative subsystem of second-order arithmetic ACA is sufficient. Tarski certainly thought that this much set theory is quite unproblematic, especially when compared to the semantical notions in question.

theory of truth with model theory. Wilfrid Hodges (1986), however, reports "a disconcerting experience" while reading CTFL to see what Tarski says about the notion of truth-in-a-structure: "The notion was simply not there." According to Hodges, it appears in Tarski's writings only in the 1950s. In agreement with Hodges, Peter Milne (1999) adds that even then Tarski was reluctant to use the term "truth" in the model-theoretic context (see also Feferman 2004).

Ilkka Niiniluoto has repeatedly expressed his disagreement, and suggested that Tarski's early account can be seen as a special case of the model-theoretic approach (Niiniluoto 1994; 1999a; 1999b; 2004). Niiniluoto refers to Tarski's remark, in CTFL, which mentions "the Göttingen school grouped around Hilbert" and recognizes the relative notion of a "correct or true sentence in an individual domain a" (Tarski 1935, p. 199). Niiniluoto also proposes that in his paper on logical consequence from the same period (Tarski 1936), Tarski clearly presupposes the general concept of *truth-in-a-model* when he writes: "The sentence X *follows logically* from the sentences of the class K if and only if every model of the class K is also a model of the sentence X" (Tarski 1936, p. 417).

Solomon Feferman (2004) goes even further. He suggests that the notion of *truth-in-a-structure* is present implicitly already in Tarski's 1931 paper on definability, since Tarski's explication of the concept of *definability-in-a-structure* makes use of satisfaction. Feferman points out that in a footnote to the introduction to this paper Tarski says of the metamathematical definition that "an analogous method can be successfully applied to define other concepts in the field of metamathematics, e.g., that of *true sentence* or of a *universally valid sentential function*" (Tarski 1931, p. 111, fn1). Universal validity can, Feferman adds, only mean valid in every interpretation, and for that the notion of *satisfaction-in-a-structure* is necessary. Feferman also presents an impressive body of evidence that Tarski, just like the early model-theorists who preceded him, had been working comfortably with the informal notion of model at least since 1924.

A somewhat related debate has now been going on in the literature on Tarski's account of logical consequence. Namely, John Etchemendy (1988; 1990) has criticized Tarski for advancing a fixed-domain conception of logical consequence (i.e., the idea that all models share a single domain), which creates all sorts of problems. Gila Sher (1991; 1996), Mario Gómez-Torrente (1996; 1998; 1999), and Greg Ray (1996) have all suggested a more charitable interpretation of Tarski and have proposed that Tarski just could not have intended such an implausible conception. Scott Soames (1999) refers to these replies approvingly.

Timothy Bays (2001), in turn, argues that Tarski definitely adhered to a fixed-domain conception, but also that it does not cause any of the problems which both Etchemendy and his critics assume it causes. Bays' arguments are quite persuasive. Recently Paolo Mancosu (2006) has offered new arguments to show that Tarski indeed upheld a fixed-domain conception of model in his 1936 logical consequence paper and that he was still propounding that view in 1940. In particular, he provides new evidence from an unpublished lecture by Tarski from 1940 which shows very clearly that Tarski even then held a fixed-domain conception.

To recap, evidently Tarski was, to some extent, thinking in model-theoretical terms and had some kind of notion of *truth-in-a-model* from early on. However, his early (pre-1950s) view was not quite the full-blooded model-theoretical view with variable domains, for it now seems clear that Tarski held onto the fixed-domain conception for quite some time. At best, Tarski may not have always succeeded, in his mathematical work, to be completely faithful to this official view of his. And contrary to appearances, even in his logical consequence paper (Tarski 1936), Tarski did not yet have a wholly *general* notion of *truth-in-a-model* (as e.g. Niiniluoto (1994; 2004) seems to suggest).

Nevertheless, one may grant that whatever is really original in Tarski's formal definition of satisfaction was already there in the 1930s, even if, at the time, the satisfying sequences were picked up from a single comprehensive universe (although, one may think that even there, Tarski was really "belaboring the obvious" and that the definitions of satisfaction and truth are "practically forced on us"; see Feferman 2004). The relativization of this notion to arbitrary domains was, once one gave up the philosophical obstacles to it, certainly a routine move.

2. The Concept of Truth and Truth-in-a-model

But what is, more exactly, the relation of Tarski's work on truth to model theory? In modern model theory, the standard approach is now the following: Given a language L and a structure W with a domain D, one fixes an interpretation function I which maps the non-logical symbols of L to elements of

D (that is, the function I maps individual constants to elements of D, predicates to subsets of D, etc.). Consequently, an L-structure W is often defined as a pair (D, I), consisting of the domain D and the interpretation function I. In such a model-theoretic setting, a language L is completely a uninterpreted and syntactic formal language. Niiniluoto adds that "an interpreted language could be defined as the pair (L, I)" (Niiniluoto 2004, p. 64).

Tarski's approach to truth differs from such a model-theoretic approach in several important respects. We have already discussed Tarski's early commitment to a single and fixed comprehensive universe. This is, of course, quite different from modern model theory, where one is free to choose any arbitrary set as the domain. But there are also other differences.

In model theory, languages are uninterpreted, and when a model is switched to another, one varies the interpretation, but the language remains the same. In his writings on the concept of truth, Tarski, on the other hand, repeatedly insisted that the 'formalized' languages whose truth is under consideration were, and had to be, always already interpreted languages:

> It remains perhaps to add that we are not interested here in 'formal' languages and sciences in one special sense of the word 'formal', namely sciences to the signs and expressions of which no meaning is attached. For such sciences the problem here discussed has no relevance, it is not even meaningful. We shall always ascribe quite concrete and, for us, intelligible meanings to the signs which occur in the languages we shall consider. (Tarski 1935, pp. 166-167)

> I should like to emphasize that, when using the term 'formalized languages', I do not refer exclusively to linguistic systems that are formulated entirely in symbols, and I do not have in mind anything essentially opposed to natural languages. On the contrary, the only formalized languages that seem to be of real interest are those which are fragments of natural languages (fragments provided with complete vocabularies and precise syntactical rules) or those which can at least be adequately translated into natural languages. (Tarski 1969, p. 68)

Furthermore, this was not just an accidental philosophical opinion from Tarski's side, but it is an essential part of Tarski's whole approach to truth that the meanings of the object language must be fixed. Only that way can a truth definition (applied to sentences) make any sense at all:

[W]e must always relate the notion of truth, like that of a sentence, to a specific language; for it is obvious that the same expression which is a true sentence in one language can be false or meaningless in another. (Tarski 1944, p. 342)

We shall also have to specify the language whose sentences we are concerned with; this is necessary if only for the reason that a string of sounds or signs, which is a true or a false sentence but at any rate meaningful sentence in one language, may be a meaningless expression in another. (Tarski 1969, p. 64)

[T]he concept of truth essentially depends, as regards both extension and content, upon the language to which it is applied. We can only meaningfully say of an expression that it is true or not if we treat this expression as a part of a concrete language. As soon as the discussion concerns more than one language the expression 'true sentence' ceases to be unambiguous. If we are to avoid this ambiguity we must replace it by the relative term 'a true sentence with respect to the given language'. (Tarski 1935, p. 263)

Therefore, it is necessary in Tarski's setting to focus on an interpreted language with constant meanings. If one changes the interpretation of the symbols of the object language, the language changes to a different language, and a former truth definition is not a truth definition for this latter language.

Can this difference be overcome by following Niiniluoto's above-mentioned suggestion that an interpreted language is defined as the pair (L, I)? Now although this idea works perfectly in the ordinary model-theoretic context, I do not think that it is an acceptable line to take in a Tarskian approach to truth. Namely, Tarski expressedly aimed to define truth (or, rather, "true-in-L") without assuming any semantic notions: "In this construction [of the definition of truth] I shall not make use of any semantical concept if I am not able to previously reduce it to other concepts" (Tarski 1935, p. 153). The interpretation function I, however, establishes a link between the language and a domain of extra-linguistic objects, and hence is a semantical concept in Tarski's sense (see also below). Hence, it would be problematic to presuppose it in the definition of truth. Although Tarski assumed that the object language must be an interpreted language, its interpretation cannot be specified by leaning on the model-theoretic interpretation function.

But the question then arises how Tarski can, and indeed can he, specify the object language as an interpreted language with meanings, without begging the question. Rudolf Carnap, in his logical semantics, assumed that the interpretation of the object language is fixed with the help of truth conditions, which in turn appeal to the definition of truth. Whether this makes Carnap's approach viciously circular or not, it is important to note that this is not Tarski's approach. Tarski here explicitly points out the difference between his own approach and that of Carnap (see Tarski 1944, p. 373, note 24). For Tarski, the interpreted object language is instead specified simply through its metalinguistic translation (see e.g. Tarski 1935, pp. 170-171; cf. Fernández Moreno 1992; 1997; Milne 1997; Raatikainen 2003; Feferman 2004).

However, Tarski's approach still assumes the notion of *meaning*, in the disguise of translation or the sameness of meaning. Does this mean that, at the end of the day, Tarski fails to achieve his aim, that is, to define truth without assuming any *semantical* concepts? It has been repeatedly suggested that this is indeed the case (see Davidson 1990; 1996; Field 1972; Soames 1984). But it is not necessarily so. In order to find out, we need to take a closer look on what Tarski meant by 'semantical'. Tarski's paradigm examples of semantical concepts were satisfaction, denotation, truth and definability (see Tarski 1935, pp. 164, 193-194; 1936, p. 401). He explained his understanding of 'semantical concept' as follows:

> A characteristic feature of the semantical concepts is that they give expression to certain relations between the expressions of language and the objects about which these expressions speak, or that by means of such relations they characterize certain classes of expressions or other objects. (1935, p. 252)

Now the model-theoretic interpretation function I discussed above is definitely a semantic notion in this sense. Hence, it would be against Tarski's explicit commitments to assume it in defining truth. This cannot be the way in which the object language is interpreted. But how about translation? I submit that it is possible to view translation, in this context, as a purely syntactic mapping between two languages, without assuming any relations between either language and the external objects. Translation, so viewed, is not a semantical concept in Tarski's sense. Hence, it is admissible for Tarski to presuppose it in this approach (see also Milne 1997).

But let us take a closer look at the details. The interpretation, or translation, of the object language in the metalanguage is specified, in Tarski, through primitive denotation. Let us then recall how exactly Tarski specifies primitive denotation in the object language. For names, this is done by a simple list-like explicit definition such as:

Denotes$_{OL}$(x, y) ↔
   [(x = 'Frankreich' & y = France) ∨
   (x = 'Deutchland' & y = Germany) ∨
   ⋮
   ⋮
   (x = 'Köln' & y = Cologne)].

An analogous definition can be given for the denotation (or, application) of predicates. Such an enumerative characterization of primitive denotation may be philosophically disappointing (cf. Field 1972), but at least it frees Tarski from any charge of begging the question. The interpretation of the object language is fixed through fixing primitive denotation, which in turn can be done by explicit definitions. And what can be explicitly defined can be eliminated. Certainly Tarski's realiance on such notions is unlikely to be problematic.

3. Is Tarski's Truth Definition a Correspondence Theory?

The question of whether Tarski's account is a version of the correspondence theory of truth or not has resulted much debate among philosophers. Karl Popper famously declared that Tarski had "rehabilitated the correspondence theory of absolute or objective truth" and "vindicated the free use of the intuitive idea of truth as correspondence to the facts" (Popper 1963, p. 223). Also, Ilkka Niiniluoto (1994; 1999a; 1999b; 2004), Gila Sher (1998) and Luis Fernández Moreno (2001) have argued that Tarski's definition of truth is a correspondence theory.

Susan Haack, among others, disagrees: "Tarski did not regard himself as giving a version of the correspondence theory" (Haack 1978, p. 114). According to Haack, Tarski's notion of satisfaction at best "bears some analogy to correspondence theories" (*ibid.*). However, she adds, "Tarski's definition of truth makes no appeal to specific sequences of objects, for true sentences are satisfied by all sequences, and false sentences by none" (Haack 1978, p. 113). A. C. Grayling (1998, p. 156) largely repeats Haack's criticism. Earlier, Donald Davidson (1969; 1983) took Tarski's account as a variant of correspondence theory, but later changed his mind for reasons similar to those of Haack: "[T]here is nothing interesting or instructive to which true sentences might correspond" (Davidson 1990; see also Davidson 1996).

I think that the objection of Haack and others, which leans on the fact that truth amounts to satisfiability by *all* sequences, is less conclusive than it may appear to be. To begin with, one can define truth for atomic sentences without the notion of satisfaction. In their case, it is particular individuals and their properties and relations which make a sentence true (cf. Niiniluoto 1999a; 2004). Further, I think that even in the case of quantified sentences the situation is not as desperate as Haack and others suggest. To be sure, a sentence is true if and only if it is satisfied by *every* sequence of objects. However, this is more a consequence of a technical trick Tarski used in his definition of satisfaction in order to handle quantification.

Namely, let us consider, for example, the existentially quantified sentence $(\exists x_1) P(x_1)$, and let us assume that the intended interpretation of $P(x)$ is, say: "x is a Ph.D. student of Leśniewski". According to Tarski's definition of satisfaction, a sequence σ satisfies $(\exists x_1) P(x_1)$ if and only if *some* sequence σ', which agrees with σ except possibly at the variable x_1, satisfies $P(x_1)$. The only sequences σ' that will do (i.e., satisfy $P(x_1)$) are ones which have Tarski as their first member.[3] Moreover, the rest of the sequence is irrelevant and could be omitted (Tarski assumed, for simplicity, that all such satisfying sequences are infinite; but it is well known that one could manage with just finite sequences; in such a case, the finite sequence with Tarski as its first and only member would be the only relevant sequence σ'). Hence, it is quite plausible to consider Tarski, and nothing else, as the "truth-maker" of the existentially quantified sentence $(\exists x_1) P(x_1)$, even if the sentence is, according to the technical definition of satisfaction, satisfied by every sequence of objects. Given an arbitrary sequence σ, we are, so to say, allowed to switch its relevant member (here, the first member) to a relevant object (here, only Tarski is suitable) and produce a sequence σ' which does the real work.

---

[3] Tarski was Leśniewski's one and only Ph.D. student.

But the question of whether Tarski's account is a full-blown substantial correspondence theory of truth, in contradistinction to deflationist views on truth, is different and more complicated. Part of the difficulty is, of course, that it is not altogether clear what exactly is the essence of deflationism. For example, Stephen Leeds (1978), Paul Horwich (1982) and Scott Soames (1984) have all suggested that Tarski's truth definition amounts in fact to a deflationary theory of truth (cf. Davidson 1990). As we have already seen, Fernández Moreno, Niiniluoto and Sher, for example, disagree and argue that it can be instead seen as a correspondence theory.

It is useful to distinguish, in this context, between weak and strong correspondence theories (Woleński & Simons 1989), or, correspondence-as-congruence and correspondence-as-correlation (see Kirkham 1992, p. 119). According to the *weak correspondence theories* or *correspondence-as-correlation* views, every truth-bearer is correlated to a state of affairs, and if that state of affairs to which a given truth bearer is correlated actually obtains, the truth bearer is true; otherwise it is false. The *strong correspondence theories*, or *correspondence-as-congruence* views, require further that there is a structural isomorphism between truth bearers and the facts to which they correspond, if true; a truth bearer mirrors or pictures the state of affairs to which it is correlated. Nothing of the sort is assumed by the former, weaker idea of correspondence. According to it, a truth bearer as a whole is correlated to a state of affairs as a whole. Weak correspondence involves only the idea that truth depends on how things are in the world.

Jan Woleński and Peter Simons (1989) submit that Tarski's theory is a correspondence theory only in the weak (or correlation) sense. Sher (1998), on the other hand, argues that it is a correspondence theory even in the strong sense (or this is at least how Patterson (2003) interprets her). Niiniluoto (1999a; 2004) in turn argues that in the case of atomic sentences, Tarski's theory is a strong correspondence theory, but with compound and quantified sentences, only a weak correspondence theory.

But what are the grounds for thinking that Tarski's truth definition really is a version of correspondence theory? Popper (1960, p. 224) seems to think that T-sentences state correspondences between sentences and worldly facts. Similarly, Niiniluoto writes that a T-sentence "states something about the relation between the language L and the world", and hence, "Tarski's semantic definition of truth is not merely disquotational" (Niiniluoto 1994, p. 63). Also

Sher (1998) makes analogous claims. So did Davidson at one point (Davidson 1983; but see Davidson 1969; 1990).

But, as Douglas Patterson (2003) points out, if it is assumed that T-sentences as such state correspondences between the sentences they mention and something extra-linguistic, then even deflationary and disquotational theories are correspondence theories, at least in the weak sense. This, however, is far too weak a notion of correspondence to be of any interest if we wish to understand what is at issue between contemporary deflationists and their correspondence theoretic opponents. Only strong correspondence theories will be interesting from this perspective, Patterson concludes. One must agree with Patterson's main point. However, contrary to what he may seem to suggest, having a genuine and substantive correspondence theory does not necessarily require a general strong or congruence view of correspondence.[4] The two distinctions substantial/deflationary and weak/strong (correspondence) do not coincide. Patterson also points out that T-sentences simply are not of the right form to state a relation at all, and so cannot state a correspondence relation. A T-sentence is a biconditional and does not predicate a relation between the sentence it mentions and some other objects.

Michael Devitt (2001) in turn argues that although Tarski seemed to view himself as a correspondence theorist about truth, the theory he actually presented is deflationary. Namely, he first reminds us that – as especially Hartry Field (1972) has emphasized – Tarski's truth definition rests on a list-like definition of primitive denotation (see also above). "But such list-like definitions are in no way explanatory, but are essentially deflationary and so could not yield anything substantial about reference." Consequently, Devitt maintains, Tarski's truth definition itself does not show us anything substantial about truth: "Tarski's definition tells us a lot about 'true-in-L'. It tells us nothing about truth-in-L because it is implicitly committed to the view that there is nothing to tell." I think we must accept Devitt's conclusion.

---

[4] For example, a broadly Tarskian theory supplemented with a substantial theory of reference along the lines that Field has suggested (mentioned at the end of this paper) is agreed by all parties to be a substantial theory of truth; however, there is no reason to think that it has to be a strong correspondence theory (correspondence-as-congruence).

However, a fix is now available. Devitt too adds that his conclusion, that Tarski's definition tells us nothing about truth, concerns only Tarski's definition as it stands. However, if we revised it by dropping its list-like definitions, then we could see it as yielding an explanation of truth in terms of reference, as Field points out. If this were then supplemented by a substantial theory of reference,[5] we would have a genuine correspondence theory of truth, Devitt concludes. Patterson too agrees that this theory is indeed a real correspondence theory. There seems to be no question that such a modified Tarskian theory of truth is a robust and substantial correspondence theory.

This move has, however, its price. One must then relax Tarski's initial requirement that no semantical concepts are presupposed. But this just is the price one necessarily has to pay if one wants to turn Tarski's definition into a substantial account of truth. However, unlike the more general semantical notions, primitive denotation is a very elementary notion and does not lead to any paradoxes.[6] Hence, it is a rather harmless concession.

University of Helsinki
Academy of Finland

References

Bays, T. 2001: "On Tarski on Models", *The Journal of Symbolic Logic* 66, 1701–1726.
Davidson, D. 1969: "True to the facts", reprinted in D. Davidson, *Inquiries into Truth and Interpretation*, Oxford University Press, Oxford, pp. 37–54.
Davidson, D. 1983: "A Coherence Theory of Truth and Knowledge", *Kant oder Hegel*. Edited by D. Henrich, Klett-Cotta, Stuttgart, 1983, pp. 423–438. Reprinted in E. LePore (ed.), *Truth and Interpretation: Perspectives on the Philosophy of Donald Davidson*, Blackwell, Oxford, 1986.
Davidson, D. 1990: "The Structure and Content of Truth", *Journal of Philosophy* 87, 279–328.
Davidson, D. 1996: "The Folly of Trying to Define Truth", *Journal of Philosophy* 93, 263–278.

---

[5] However, I don't think that it is likely that we will ever have a strictly physicalistic theory of reference, as Field demands – but that is a wholly different and independent issue.

[6] The situation is, of course, very different with the general notion of denotation, which easily leads, e.g., to Berry's paradox.

Devitt, M. 2001: "The Metaphysics of Truth", in Michael Lynch (ed.), *The Nature of Truth*, MIT Press, Cambridge, MA, pp. 579-611.

Etchemendy, J. 1988: "Tarski on Truth and Logical Consequence", *Journal of Symbolic Logic* 53, 51-79.

Etchemendy, J. 1990: *The Concept of Logical Consequence*, Harvard, Cambridge.

Feferman, S. 2004: "Tarski's Conceptual Analysis of Semantical Notions", in A. Benmakhlouf (ed.), *Sémantique et épistémologie*, Editions Le Fennec, Casablanca [distrib. J. Vrin, Paris], pp. 79-108.

Fernandez Moreno, L. 1992: "Putnam, Tarski, Carnap und die Wahrheit", *Grazer philosophische Studien* 43, 33-44.

Fernandez Moreno, L. 1997: "Truth in Pure Semantics: A Reply to Putnam", *Sorites* 8, 15-23.

Fernandez Moreno, L. 2001: "Tarskian Truth and the Correspondence Theory", *Synthese* 126, 123-147.

Field, H. 1972: "Tarski's Theory of Truth", *Journal of Philosophy* 69, 347-75.

Gomez-Torrente, M. 1996: "Tarski on Logical Consequence", *Notre Dame Journal of Formal Logic* 37, 125-151.

Gomez-Torrente, M. 1998: "On a Fallacy attributed to Tarski", *History and Philosophy of Logic* 19, 227-234.

Gomez-Torrente, M. 1999: "Logical Truth and Tarskian Logical Truth", *Synthese* 117, 375-408.

Haack, S. 1978: *Philosophy of Logics*, Cambridge University Press, Cambridge.

Hodges, W. 1985/6: "Truth in a Structure", *Proceedings of the Aristotelian Society* 86, 135-151.

Leeds, S. 1978: "Theories of Reference and Truth", *Erkenntnis* 13, 111-127.

Mancosu, P. 2006: "Tarski on Models and Logical Consequence", in J. Gray and J. Ferreiros (eds.), *The Architecture of Modern Mathematics*, Oxford University Press, forthcoming.

Milne, P. 1997: "Tarski on Truth and Its Definition", in Childers, Kolár and Svoboda (eds.), *Logica '96: Proceedings of the 10th International Symposium*, Filosofia, Prague, pp. 189-210.

Milne, P. 1999: "Tarski, Truth and Model Theory", *Proceedings of the Aristotelian Society* 99 (1998-1999), 141-167.

Niiniluoto, I. 1994; "Defending Tarski against his Critics", in B. Twardowski and J. Woleński (eds.), *Sixty Years of Tarski's Definition of Truth*, Philed, Warsaw, pp. 48-68.

Niiniluoto, I. 1999a: "Tarskian truth as correspondence – replies to some objections", in J. Peregrin (ed.), *Truth and its Nature (if any)*, Kluwer, Dordrecht, pp. 91-104.

Niiniluoto, I. 1999b: "Theories of Truth: Vienna, Berlin, and Warsaw", in J. Woleński and E. Köhler (eds.), *Alfred Tarski and the Vienna Circle*, Kluwer, Dortrecht, pp. 17-26.

Niiniluoto, I. 2004: "Tarski's Definition and Truth-makers", *Annals of Pure and Applied Logic* 126, 57-76.

Patterson, D. 2003: "What is a Correspondence Theory of Truth?", *Synthese* 137, 421–444.
Popper, K. R. 1963: *Conjectures and Refutations*, Routledge, London.
Putnam, H. 1985–86: "A Comparison of Something with Something Else", *New Literary History* 17, 61–79. Reprinted in H. Putnam, *Words and Life*, Harvard University Press, Cambridge, 1994.
Raatikainen, P. 2003: "More on Putnam and Tarski", *Synthese* 135, 37–47.
Ray, G. 1996: "Logical Consequence: A Defence of Tarski", *The Journal of Philosophical Logic* 25, 617–677.
Sher, G. 1991: *The Bounds of Logic*, MIT Press, Cambridge.
Sher, G. 1996: "Did Tarski Commit Tarski's Fallacy?", *Journal of Symbolic Logic* 61, 653–686.
Sher, G. 1998: "On the Possibility of a Substantive Theory of Truth", *Synthese* 117, 133–172.
Sluga, H. 1999: "Truth before Tarski", in J. Woleński and E. Köhler (eds.), *Alfred Tarski and the Vienna Circle*, Kluwer, Dordrecht, pp. 27–41.
Soames, S. 1984: "What is a Theory of Truth?", *Journal of Philosophy* 81, 411–429.
Soames, S. 1999: *Understanding Truth*, Oxford University Press, New York.
Tarski, A. 1931, "Sur les ensembles définissables de nombres réels, I", *Fundamenta Mathematicae* 17, 210–239; English translation in Tarski (1983), pp. 110–142.
Tarski. A. 1933: *Pojecie prawdy w jezykach nauck dedukcyjnych* (The Concept of Truth in the Languages of Deductive Sciences), Prace Towarzystwa Naukowego Warszawskiego, wydzial III, no. 34.
Tarski, A. 1935, "Der Wahreitsbegriff in den formalisierten Sprachen", *Studia Philosophica* (Lemberg), 1, 261–405. English translation in Tarski (1983), pp. 152–278.
Tarski, A. 1936: "Über den Begriff der logischen Folgerung", *Actes du Congrès International de Philosophie Scientifique* 7, Actualités Scientifiques et Industrielles, Herman, Paris, pp. 1–11. English translation in Tarski (1983), pp. 409–420.
Tarski, A. 1940: "On the Completeness and Categoricity of Deductive Systems", unpublished typescript, Alfred Tarski Papers, Carton 15, Bancroft Library, U.C. Berkeley.
Tarski, A. 1944: "The Semantic Conception of Truth and the Foundations of Semantics", *Philosophy and Phenomenological Research* 4, 341–376.
Tarski, A. 1969: "Truth and Proof", *Scientific American* 220, 63–77.
Tarski, A. 1983: *Logic, Semantics, Metamathematics*, Oxford University Press, Oxford. Second edition. (First edition, 1956.)
Vaught, R. L. 1986: "Tarski's Work in Model Theory", *Journal of Symbolic Logic* 51, 869–882.
Vaught, R. L. 1974: "Model Theory before 1945", in L. Henkin, J. Addison, C. C. Chang, W. Craig, D. Scott, R. Vaught (eds.), *Proceedings of the Tarski Symposium*, American Mathematical Society (Providence), pp. 153–172.

Woleński, J. and P. Simons 1989: "*De Veritate*: Austro-Polish Contributions to the Theory of Truth from Brentano to Tarski", in K. Szaniawski (ed.), *The Vienna Circle and the Lvov-Warsaw School*, Dordrecht, Kluwer, pp. 391–442.

Veikko Rantala

Perception in Context and Its Logic

It is rather commonly – but not unanimously – maintained in current analytic philosophy, psychology, and art history that perception is contextual, dependent on the perceiver's experience, beliefs and knowledge, and scientific theories the perceiver is holding. This means that it is a cognitive process in a relative or even subjective sense, so that, for example, observing a cultural object may involve seeing in it something more than its mere spatial shape, and two persons may see different things in it. There is some psychological and neuropsychological evidence in support of this theory of perception, and several recent philosophers draw from Wittgenstein in order to theoretically ground it. Partly at least – in so far as the philosophy of science is concerned – the theory can be regarded as a reaction against radical empiricists' view according to which objective or neutral observation is possible under appropriate circumstances.

In Sections 1–3 below, I consider certain philosophical and neuro-psychological views of the distinction between context dependence and neutrality. Finally, in Section 4, I study how in the logic of perception the distinction could be taken into account. So, I try to solve a problem presented in Niiniluoto (1982).

1. Perception and Relativity

A well-known example of this relativist view of perception is the argument concerning how we see the duck-rabbit picture, purporting to show that what we immediately see is either a duck or a rabbit rather than just a certain geometrical figure. Hanson (1958) argues that two spectators belonging to different scientific traditions may see different things when watching the same object. Thus Tycho Brahe and Johannes Kepler see different things when experiencing the dawn, since Brahe sees that the sun is raising and Kepler that the earth is moving. Seeing is an epistemic process. An important aspect of

Hanson's theory is that seeing is propositional, that is, seeing is to *see that* something is the case. Kuhn (1962), in turn, emphasizes the importance of processes by means of which one learns to see, that is, learns to *know how to see*. This is done within the conceptual framework of the scientific paradigm, or, more generally, the culture, to which one belongs.

Hanson's theory has been criticized and it has been defended on various grounds. For example, it has been argued that Kepler and Brahe see the same thing, but they see it *as* different things. Some say that Hanson has not succeeded in showing that seeing involves seeing that something is the case, but only that visual observation involves seeing that something is the case. According to Suppe (1968), even this is not conclusive, but only plausible. Fodor (1984), in turn, argues that perceptual processes are isolated, and evidently this is a consequence of his notion of the modularity of mind. The following kind of argument is often presented to defend the Hansonian view. It is a linguistic or logical argument to the effect that the way in which we speak about *seeing that* shows that the language we use to talk about propositional seeing is intensional. Thus one may consistently claim that Mary sees that the Evening Star is shining in the sky but does not see that the Morning Star is shining there, thus violating the principle of the Substitutivity of Identicals.

In current aesthetics and philosophy of art the dominant view concerning perception is very much analogous to that in current philosophy of science. It is well known, for example, what Gombrich (1959) says of it. When we receive a visual impression, we response to it by "docketing it, filing it, grouping it in one way or another. ... It is the business of living organism to organize, for where there is life there is not only hope, as the proverb says, but also fears, guesses, expectations which sort and model the incoming messages, testing and transforming and testing again. The innocent eye is a myth" (p. 251). This is, again, contrary to the view that when observing an object we can eliminate the influence of the context and knowledge and see the object as it is, and when depicting an object we can picture it as such. Goodman (1968) argues similarly by saying that "[the eye] selects, rejects, organizes, discriminates, associates, classifies, analyzes, constructs. It does not so much mirror as take and make; and what it takes and makes it sees not bare, as items without attributes, but as things, as food, as people, as enemies, as stars, as weapons" (pp. 7-8). It is difficult to say it better.

It follows, as Gombrich and Goodman remark, that in art an object cannot be depicted without an influence of knowledge, experience, and interests. But an artist may sometimes try to get rid of that influence and strive for the innocent eye, since in this manner it may be possible to reach new ideas of representation – even though the goal were not completely attainable. It seems that such a process is an example of the idea that what is often called disinterestedness and detachment can be regarded as theoretical and idealized outcomes and limiting cases – that can actually be beyond reach – of more realistic aesthetic activities.

2. Subjective vs. Objective Perception

We have to ask now what might be the relation between the two views concerning perception, the one saying that it can be objective and the other arguing for context dependence and subjectivity or relativity. We may outline an answer that is based on some psychological theories according to which visual representations of the external world proceed from subjective representations towards ones that are more objective. Thus, Marr (1982) argues that a representation of a visual image is transformed into a viewer-centered representation, which is in turn transformed into an object-centered one. Marr's view of this process relies on the symbolic model, meaning that all the stages are describable (within a symbol system) by means of symbols and the transformations are symbol manipulations (computations) according to rules. The last stage, which is objective and object-centered, consists of a description of the three-dimensional shape of the object being viewed and its spatial arrangement that is independent of the direction from which the object is viewed.

This theory of the human visual processor is computational, and as such, it yields an abstract picture of its subject matter, as all theories do. One problem, which Marr is of course aware of, is whether every stage and transformation occurring is describable so as to match human performance, or even describable at all. But we are here more interested in the manner in which the final stage is described in the theory. It objectifies vision by suggesting, for example, the direction-independence of the outcome. It is argued, furthermore, that the final, objective stage is typical of advanced visual systems, such as the human one – more advanced than, for example, the fly's system.

Consider next how this process from a subjective representation towards a more objective one would be described in terms of the connectionist model of vision. Roughly, a visual image is represented (in an encoded form) as an input vector of a network. The vector represents visual features, such as color and shape or their microfeatures, sensations that fall on the retina when the object is surveyed. If the network recognizes the pattern, the output vector represents a more or less integrated view of what there is, that is, its activation values are close enough to the correct values.

It is known that the eye usually examines an object by checking its different parts. According to Hochberg (1972), an object is usually examined by a succession of multiple glimpses. Now a connectionist system takes care of the different directions from which the object is viewed by varying activation values of input units, or adding or dropping some input units. In particular, the network may be able to recognize a pattern it has learned, if the stimulus caused by the object is partial, by completing the input vector. If the number of units and connections is large, the network is flexible, being able to generate a sufficiently good output in each of the above cases, and for the same reason generalizing from what it has learned becomes possible.

The above recognition process in connectionist visual systems can also be described by referring to the distinction between nonpropositional and propositional seeing. The latter is a propositional attitude; it is conceptual and can be described, as we already noted, by using the phrase *seeing that*. The former means processes that lead to propositional seeing; it is a skill that makes the latter possible, and it can often but perhaps not always be described by the phrase *knowing how to see*. Nonpropositional seeing involves an interaction with an object, the environment, and so on, whereas propositional seeing results as soon as the viewer recognizes that such and such states of affairs hold.

It is pointed out in Rantala and Vadén (1994) that on certain very general conditions an output of a connectionist system can be construed as approximating the ideal or objective solution to a recognition problem, which is performed by a symbol system. Furthermore, the more complex the network is the closer its solution is to the objective one. If this is applied to modelling vision, we can feasibly argue that complexity implies visual objectification in the sense that greater complexity means that the viewer's propositional but subjective visual information, propositional seeing, can more closely approximate objective information. However, in many cases of "real life" information of the latter kind may be theoretical and unattainable.

When he says that the outcome in the process of advanced seeing is objective, or rather object-centered, Marr (1982) maintains that the respective representation is a representation of the three-dimensional shape and spatial arrangement of the object viewed (together with a description of its surface structure). How does this relate to the view that what we immediately see depends on the context, knowledge, and experience – and more broadly, on the culture – the view advocated by Hanson, Gombrich, and other representatives of current philosophy of science and philosophy of art? If one is looking at the duck-rabbit picture and immediately sees a duck or a rabbit, we are evidently allowed to argue – if we refer to the distinction between propositional and nonpropositional seeing – that this is the stage of subjective propositional seeing. If this is what takes place first, then it is obvious that one has to make an additional effort, an additional computation, in order to see only the objective geometrical shape of the object.

Examples such as seeing a duck or a rabbit do not seem to play any role in Marr's description of the process of vision, and he argues that shapes can be determined by vision alone. Research of various kinds of lesion people may suffer suggests, according to Marr, that the representation of the shape of an object is stored in a place in the brain that is different from the place where the representation of its use and purpose is stored. According to Churchland and Sejnowski (1996), the brain performs a variety of tasks simultaneously, since it is a vast network of networks such that different regions specialize for different tasks. That is, the brain is modularized in a biological and interconnected style of modularity, and this evidently explains why, theoretically speaking, one can recognize the shape of an object but not its use and purpose. This may also explain why Hanson's and others' view of how culture relativity and knowledge influence what one immediately sees is diminished in Marr's theory.

On the other hand, if the so-called sense-data theory – saying that when two persons are watching something they see the same thing (the same sense datum), but their interpretations of what they see may be different (see Suppe 1977, p. 153) – is correct, then Marr's theory can also be explained by arguing that though patients suffering a lesion of a certain kind can recognize the shape of an object, their deficiency prohibits them to interpret what they see. But probably this explanation of the theory also requires the assumption of modularity. The idea that the recognition of the shape of an object comes first and its recognition as a cultural or natural object would only come subsequently might mean, analogously to what we said above, that in the brain there is a distinctive region that performs interpretation – and this region is damaged in the patients under discussion. Both of these possibilities concerning vision are compatible with the connectionist assumption, made, e.g., by Bechtel and Abrahamsen (1991, p. 163), that in a network there is a separate ensemble of processing units together with their connections in which cultural knowledge is stored and which is influential when it learns to recognize an object.

3. Indirect vs. Direct Theories of Perception

Many of the above distinctions concerning perception can be conveniently summarized in terms of two psychological paradigms of visual perception, pertaining to the alleged role of knowledge in perception. The more traditional one, often called constructivist or Helmholtzian, takes perception as being computational and indirect, processing information that is not limited to the information directly obtained in the sensory stimulation. According to Norman (2002), constructivist theories argue that higher mental processes are involved in perception when apprehending the environment, so that memory, knowledge, past experience, and stored schemata play an important role in perception. Thus it seems that what we have here is the idea that perception is contextual, and it even covers what we called propositional perception and theory-laden perception.

Theories of the other paradigm, called ecological or Gibsonian, say that perception is more direct and that its processes are limited to information in the stimulation. Norman describes it by saying that the subject needs no higher mental processes in order to pick up relevant information about the environment. Thus memory and related phenomena play no role in perception, but movements of the body of a perceiver (including movements of the eyes, head, etc.) are an important part of the perceiver's perceptual processes.

Now Norman and some others propose that both kinds of perception exist and the two paradigms, rather than each covering perception as a whole, complement each other and can be incorporated into a common framework. The two approaches are theories about different aspects of perception rather than exemplifying two incompatible paradigms. Norman points out that two visual systems, called ventral and dorsal, which are known to exist and which function simultaneously, parallel the two theories, and naturally this would provide evidence for the aspect approach. The former system utilizes visual information for identifying and recognizing objects and storing new information, whereas the latter utilizes visual information for the guidance of behavior in one's environment. Furthermore, the former is concerned with object-centered information, whereas the latter with body-centered information. This distinction looks similar to Marr's distinction between object-centered and viewer-centered representations, but I am not able to study here how close this analogy really is.

Hatfield (1990), on the other hand, attempts to find a common ground in terms of connectionism. Since indirect theorists, some of them at least, talk about deductive inferences in connection with visual information processing, their paradigm can be discussed in terms of the symbolic paradigm in cognitive science. This seems to be Hatfield's position (though he uses the term 'symbolism'). However, indirect theories are compatible with connectionism since the rules of a symbolic system need not be cognitively represented, says Hatfield. "A system may be rule-instantiating without being rule-following" (p. 249). For the very same reasons, direct theories, too, are compatible with connectionism, for we can think of "...the rules of the detection device as being instantiated in the connection weights without violating Gibsonian strictures against mechanisms of cognitive mediation" (*ibid.*).

I would rather describe the common ground achievable by means of connectionism by referring to the above-mentioned relation between nonsymbolic (i.e. subsymbolic) and symbolic representations, which implies that visual systems described by direct and indirect theories approximate each other, in the sense described above, in so far as the former systems can be modelled in terms of very complex connectionist systems – as we have assumed. On the other hand, this modelling provides a way of pondering the question of the context dependence of perception. In connectionist models of pattern recognition, which is often an important feature in perception, both so-called top-down and bottom-up processes are needed, especially when learning to recognize a pattern is in question. According to Bechtel and Abrahamsen (1991), processes of the former kind take place when, for instance, higher-level units of a network influence the responsiveness of lower-level perceptual units that recognize objects, and this leads to a revision of the weights of connections at the lower levels, that is, higher-level units are positively connected to lower-level ones. Processes of the latter kind are exemplified by positive connections in the opposite direction. If the higher-level units, especially the output ones, encode the information provided by a theory or a piece of propositional knowledge, or propositional context of another kind, then in the former case we may say that perception is theory-laden and in the latter that perception plays a role in theory formation.

There is even empirical evidence, provided by cognitive psychology, for the above kind of relation between the context and perception. Thus, e.g. Martindale (1991) and Estany (2001) argue that there is evidence showing that both top-down and bottom-up processes occur in humans; both processes are needed since both have advantages and disadvantages. However, according to Raftopoulos (2001) there is also evidence, available from cognitive neuroscience, showing that part of perception – more exactly, early vision processing – can be considered theory neutral despite top-down processes in the brain. He grounds his argument on Leopold and Logothesis (1996) in which it is shown that in the primary cortex only 18% of neurons changed their response according to the image perceived by an animal. This suggests, Raftopoulos argues, that most of the neurons in early vision processing are not influenced by higher cognitive functions. At the higher levels of processing, the relative number of neurons that change their response is higher, and at the highest level, almost all neurons did so. Now we can see that this is an argument against the theory-ladenness of lower-level perception, whereas Leopold and Logothesis also give rise to an argument for a strong theory-ladenness of higher levels, that is, of the cognitive parts of perception, such as, for example, a level at which a pattern is recognized or an observation made. It is obvious that this is not an argument against the theory-ladenness of perception *in toto*, even though it may weaken its role a bit. As Raftopoulos himself remarks, this may weaken the threat of relativism in epistemology, since it makes early vision more objective.

4. On the Logic of Vision

It follows from what we have learned above that it is justified to consider higher levels of vision, e.g. propositional seeing and observation, context dependent. We have also learned that sometimes propositional seeing depends on a propositional context, such as a theory, sometimes on a nonpropositional one, such as early levels of learning, skills, etc. I consider here some logical aspects of the distinction between the notions of context dependent and context independent vision. The latter may not exist at all at higher levels, and our investigation below seems to confirm this doubt by showing that logically its existence is not grounded.

The pioneering work in the logic of perception was done by Hintikka (1969), and references to logical investigations that are more thorough than what I provide here can be found, e.g. in Niiniluoto (1982). Niiniluoto is interested, among other things, in the logical aspects of how background beliefs and knowledge influence our perception, i.e. in its theory-ladenness, and he proposes to add the epistemic or doxastic operator to the language of the logic of perception. Since, however, sometimes perception and knowledge are relatively independent of each other – as in cases of conscious illusion – he argues: "It is reasonable to first develop epistemic logic and the logic of perception separately and then try to find principles which express connections between epistemic and perceptual notions" (Niiniluoto 1982, p. 127). Another way of logically studying the role of context in perception is to take formal advice from dynamic logic and the logic of learning so as to model the notion of context by means of a set of symbols embedded in the modal operator representing perception. That is, instead of formally considering a phrase of the form *an agent perceives that* one studies phrases *an agent perceives in the context α that*, where α indicates a propositional or nonpropositional context, as the case may be.

In the sequel, I restrict myself to propositional seeing, that is, to *seeing that*.[1] Formally, one adds to a language of propositional or predicate logic symbols $a_0, a_1, a_2, \ldots$, representing *atomic contexts* (of vision) for an agent. By means of atomic contexts compound ones can be formed by *context operators*, for instance, by the following operator: if α and β are contexts, then $\alpha\beta$ is a context. $\alpha\beta$ can be given different intuitive interpretations, e.g. that (i) first α is a context and then β, or (ii) α and β together form a context. This language is further extended by adding the intensional operator $\mathbf{S}(\alpha)$ for each context symbol α. Assume that we consider a single agent. Then $\mathbf{S}(\alpha)\varphi$ formalizes the idea that the agent sees in the context α that φ.

---

[1] For a nonmodal logic of perception, see, e.g. Bell (2002).

If contexts are considered propositional, then compound contexts can be intuitively interpreted as compound propositions, and if contexts are considered nonpropositional, interpretations are, e.g. relations or sequences. The latter kind of interpretation are given in dynamic logic, where symbols in the context set $\Lambda = \{\alpha, \beta, \ldots\}$ represent programs (of digital computers, for example), and in the logic of learning, where they represent learning processes (for the latter, see Rantala 2006).[2] In the propositional case, we might even suppose on some occasions that the symbols correspond to formulas of some relevant language. Such a language could express, for instance, the agent's epistemic or doxastic background. We are not going to study here a logic of this kind in detail, but only indicate here and at the end of the present section, by means of a couple of examples, how the language can be used to make some distinctions.

We can say, for instance, that contexts α and β are incompatible (with respect to what can be seen) if for some sentence φ of our language the sentence $S(\alpha)\varphi \land S(\beta)\neg\varphi$ is valid (provable) in the respective semantics (system). The above Brahe–Kepler example is a case in point. But if we assume that $\alpha\beta$ is interpreted as the conjunction of the propositions represented by α and β, then the sentences of the forms

($S\alpha\beta$) $S(\alpha)\varphi \to S(\alpha\beta)\varphi$;
$S(\beta)\varphi \to S(\alpha\beta)\varphi$

(for all α, β, φ) express the idea that adding propositions to a context preserves what is already perceived. If contexts mean propositions of some science, then ($S\alpha\beta$) expresses the idea that the progress of this science is cumulative (with respect to observations). Other context operators can be added to the language (see below), analogously to what is done in dynamic logic, in order to increase the expressive power of the language.

If in the axioms and rules of the system T of (propositional) alethic modal logic the necessity operator is replaced by the operators $S(\alpha)$, and ($S\alpha\beta$) is added as an axiom schema, we obtain a system for contextual and cumulative

---

[2] Ari Virtanen and his research group from the University of Tampere have generalized the logic of learning (in the spirit of dynamic logic) by studying rules by means of which new learning processes are formed from given ones in various ways (see Rantala 2006, for a brief description of their research programme).

perception (seeing). Let it be **T**Λ. A complete and sound semantics is obtained in the usual fashion by means of possible worlds models, in each of which, however, there is an alternativeness relation, say $R(\alpha)$, for each context symbol α, satisfying the following conditions:

(Rαβ) $R(\alpha\beta) \subseteq R(\alpha)$;
         $R(\alpha\beta) \subseteq R(\beta)$.

In view of conscious illusions, the axiom corresponding to axiom (T) of system T would be problematic for the logic of perception, as noted by Niiniluoto (1982) and many others investigating the logic of perception, and so would be the respective axioms here:

$\mathbf{S}(\alpha)\varphi \to \varphi$.

Either we have to give up (T), as Niiniluoto does, and the condition that the alternativeness relation be reflexive, or we preserve the axiom and claim that we are studying the logic of successful seeing. Or, perhaps it is enough to argue that Hanson is right. On the other hand, it may be difficult to understand how anyone could see that a stick in water is straight even though the context includes knowing that it is straight, or how Kepler could see that the earth is moving.

I take here no definite standpoint in this question but consider a more difficult logical problem noticed by Niiniluoto. It is the much-discussed problem of logical omniscience, but now related to perception; for example, whether it is intuitively acceptable to say that $\mathbf{S}(\alpha)\varphi$ is provable or valid if φ is, or that $\mathbf{S}(\alpha)\psi$ is provable or valid if $\mathbf{S}(\alpha)\varphi$ and $\varphi \to \psi$ are – both principles in the above system and the respective possible worlds semantics. It is not even clear what it means to say that an agent sees, e.g. that Socrates is mortal or Socrates is not mortal, unless what we mean by seeing is closer to understanding or knowing than sensation, that is, unless what is meant pertains to a cognitive level of vision.

Be this as it may, logical omniscience in the logic of vision can be avoided in many ways, as in epistemic logic, as, for instance, by means of employing nonnormal possible worlds in semantics. Then we can imitate the semantics of the logic of learning, discussed in my (2006),[3] by first separating a subset (satisfying certain conditions), say Ω, from the sentences of the above formal language and then introducing a partial valuation, that is, defining ordinary truth conditions at nonnormal worlds only for the sentences belonging to Ω. The rest of the sentences of the language can either be regarded as having no truth-value at all at nonnormal worlds or having an intermediate or nonstandard truth-value. If now in the system $T\Lambda$ the rule corresponding to the rule of necessitation is replaced by the following weaker rule:

For all $\varphi \in \Omega$, if φ is provable, then $S(\alpha)\varphi$ is provable

and the following axiom is added:

For all $\varphi \notin \Omega$, $\neg S(\alpha)\varphi$,

we have a system, let it be $T\Omega\Lambda$, for which the nonnormal worlds semantics is sound and complete.[4] The set Ω represents those propositions towards which the agent can already have the attitude of seeing, for whatever reason, and therefore its sentences have a definite truth-value at all visually possible worlds. Instead, the rest of the sentences of the language represent propositions with respect to which this is not the case, hence the respective sentences possess no truth-value or have an intermediate truth-value at some visually possible worlds, those which we have called nonnormal.

One may ask, how this system and its (formal) semantics can be similar to those of the logic of learning and epistemic logic, even though intuitive interpretations are different. One answer, already hinted at above, is, of course, that higher-level vision is cognitive very much in the same way as other epistemic processes and attitudes, such as learning and propositional knowledge, hence their logics can also be similar, at least in some surface sense of similarity. Since seeing and learning are processes, then it seems to be appropriate to model them by using some tools that are similar to those in dynamic logic.

Let us now forget nonnormal worlds for a moment and assume that we have more ordinary possible worlds models, models having only normal

---

[3] It is an improvement of the semantics of epistemic logic in Rantala (1982a; 1982b).

[4] See my (2006).

worlds. (If nonnormal worlds semantics are used, what we say below must be modified according to principles described above.) The propositional (intuitive) interpretation of contexts leads us to look at the familiar model-theoretic facts concerning the intensional operators $\mathbf{S}(\alpha)$ and context operators from a slightly uncommon point of view. If W is the set of all possible worlds of a given model M and $R(\alpha)$ its alternativeness relation, for each α, then to say that

$\mathbf{S}(\alpha)\varphi$ is true at a world $w \in W$ (in M) if and only if φ is true at every world $w' \in W$ such that $wR(\alpha)w'$

equals saying that

$\mathbf{S}(\alpha)\varphi$ is true at a world $w \in W$ (in M) if and only if φ is true at every world $w' \in W$ such that $w' \in R(\alpha)[w]$,

where

$R(\alpha)[w] = \{w' \in W \mid wR(\alpha)w'\}$.

If we follow Hintikka, we may say that whenever $wR(\alpha)w'$, the world w' is visually possible relative to w in the context α, or w' is compatible with what the agent sees at w in the context α. If we accept this terminology, then it is natural to call the set $R(\alpha)[w]$, which is a proposition according to the parlance customary in possible worlds semantics, the *visually possible proposition* (within M) relative to w in the context α, and say that this *proposition is compatible* with what the agent sees at w in that context. If now

$||\varphi|| = \{w \in W \mid \varphi \text{ is true at } w \text{ in } M\}$

is the proposition in M determined by the sentence φ and if $\mathbf{S}(\alpha)\varphi$ is true at w, then $R(\alpha)[w] \subseteq ||\varphi||$, so it is perhaps appropriate to state that this visually possible proposition *implies* (in M) what the agent sees at w in that context.

Let us emphasize the assumption that contexts are intuitively interpreted as propositions by using familiar operators, such as $\neg, \wedge, \vee, \rightarrow$ – hoping that no confusion arises – to form compound contexts $\neg\alpha, \alpha \wedge \beta, \alpha \vee \beta, \alpha \rightarrow \beta$, and reflect their "propositional" character by means of the respective alternativeness relations in models. For example, if M, R, and W are as above, define

$R(\neg\alpha) = (W \times W) \setminus R(\alpha);$
$R(\alpha \wedge \beta) = R(\alpha) \cap R(\beta);$

$R(\alpha \vee \beta) = R(\alpha) \cup R(\beta)$;
$R(\alpha \to \beta) = R(\neg\alpha) \cup R(\beta)$,

and therefore,

$R(\neg\alpha)[w] = W \setminus R(\alpha)[w]$;
$R(\alpha \wedge \beta)[w] = R(\alpha)[w] \cap R(\beta)[w]$;
$R(\alpha \vee \beta)[w] = R(\alpha)[w] \cup R(\beta)[w]$;
$R(\alpha \to \beta)[w] = (W \setminus R(\alpha)[w]) \cup R(\beta)[w]$.

Then it follows, for example, that if a context imitates a tautology, as, e.g. $\alpha \vee \alpha \to \alpha$, then in each model the respective visually possible propositions equal the set of all possible worlds of the model. In other words, if **T** is the symbol denoting such "tautological" contexts, then $R(\mathbf{T}) = W \times W$, so $R(\mathbf{T})[w] = W$ for each $w \in W$. Therefore, such a context could be called the *empty* propositional context.

Thus it seems appropriate to define noncontextual, pure seeing as follows:

Sφ = **S**(**T**)φ.

Then **S**φ logically implies all sentences of the form **S**(α)φ, that is, all sentences of the form **S**φ → **S**(α)φ are logically valid in our semantics. Moreover, if we omit the context symbols from the language and only consider the intensional operator **S** and models M such that the altenativeness relation of M is $W \times W$, where W is the set of all possible worlds of M, we have the well-known semantics for the system S5. Hence, if noncontextual seeing means logically something like what is indicated here, it is logically stronger than contextual seeing, satisfying principles (those of S5) that evidently idealize too much.

University of Tampere

References

Bechtel, W. and A. Abrahamsen 1991: *Connectionism and the Mind*, Blackwell, Oxford–Cambridge, MA.

Bell, J. L. 2002: "Continuity and the Logic of Perception", *Transcendental Philosophy* 3, 1–8.

Churchland, P. S. and T. J. Sejnowski 1996: *The Computational Brain*, The MIT Press, Cambridge, Mass.–London.

Estany, A. 2001: "The Thesis of Theory-Laden Observation in the Light of Cognitive Psychology", *Philosophy of Science* 68, 203–217.
Fodor, J. 1984: "Observation Reconsidered", *Philosophy of Science* 51, 23–43.
Gombrich, E. H. 1960: *Art and Illusion*, Phaidon, London.
Goodman, N. 1968: *Languages of Art*, Bobbs-Merrill, New York.
Hanson, N. R. 1958: *Patterns of Discovery*, Cambridge University Press, Cambridge.
Hatfield, G. 1990: "Gibsonian Representations and Connectionist Symbol Processing: Prospects for Unification", *Psychological Research* 52, 243–252.
Hintikka, J. 1969: "On the Logic of Perception", in J. Hintikka, *Models for Modalities*, D. Reidel Publishing Co., Dordrecht, pp. 151–183.
Hochberg, J. 1972: "The Representation of Things and People", in E. H. Gombrich, J. Hochberg, and M. Black, *Art, Perception, and Reality*, The Johns Hopkins University Press, Baltimore–London, pp. 47–94.
Kuhn, T. S. 1962: *The Structure of Scientific Revolutions*, The University of Chicago Press, Chicago.
Leopold, D. A. and N. K. Logothetis 1996: "Activity Changes in Early Visual Cortex Reflect Monkeys' Percepts during Binocular Rivalry", *Nature* 379, 549.
Marr, D. 1982: *Vision*, W. H. Freeman and Company, New York.
Martindale, C. 1991: *Cognitive Psychology. A Neural-Network Approach*, Brooks/Cole Publishing Company, Pacific Grove, CA.
Niiniluoto, I. 1982: "Remarks on the Logic of Perception", in I. Niiniluoto and E. Saarinen (eds.), *Intensional Logic: Theory and Applications*, Acta Philosophica Fennica, Vol. 35, Helsinki, pp. 116–129.
Norman, J. 2002: "Two Visual Systems and two theories of Perception: An Attempt to Reconcile the Constructivist and Ecological Approaches", *Behavioral and Brain Sciences* 25, 73–96.
Raftopoulos, A. 2001: "Reentrant Neural Pathways and the Theory-Ladennes of Perception", *Philosophy of Science* 68, S187–S199.
Rantala, V. 1982a: "Impossible Worlds Semantics and Logical Omniscience", in I. Niiniluoto and E. Saarinen (eds.), *Intensional Logic: Theory and Applications*, Acta Philosophica Fennica, Vol. 35, Helsinki, pp. 106–115.
Rantala, V. 1982b: "Quantified Modal Logic: Non-Normal Worlds and Propositional Attitudes", *Studia Logica* 41 (1982), 41–65.
Rantala, V. 2006: "Learning and Epistemic Logic", in T. Aho and A.-V. Pietarinen (eds.), *Truth and Games: Essays in Honour of Gabriel Sandu*, Acta Philosophica Fennica 78, pp. 139–157.
Rantala, V. and T. Vadén 1994: "Idealization in Cognitive Science. A Study in Counterfactual Correspondence", in M. Kuokkanen (ed.), *Idealization VII: Structuralism, Idealization and Approximation*, Poznan Studies in the Philosophy of the Sciences and the Humanities, vol. 42, Rodopi, Amsterdam–Atlanta, GA, pp. 179–198.

Suppe, F. 1977: "The Search for Philosophic Understanding of Scientific Theories", in F. Suppe (ed.), *The Structure of Scientific Theories*, University of Illinois Press, Urbana, Chicago, London, pp. 1-232.

Gabriel Sandu

Read on the Liar

Abstract. I analyse Read's solution to the Liar and point out some of its problems relating it to other solutions in the literature.

1. Tarski and the Liar

It is well known that Tarski's requirement on a materially adequate theory of truth, namely that it should entail all instances of the schema

(1) (*T*) *x* is true if and only if *p*

where '*x*' stands for the name of a sentence '*p*', leads to paradoxes. Let us recall how the paradox is derived in Tarski's seminal paper.

We stipulate that *c* denotes the sentence written on the 8th line of this section:

> *c* is false.

Applied to *c* the relevant instance of schema (*T*) is

(2) '*c* is false' is true if and only if *c* is false.

(2) together with the previous stipulation

> *c* = '*c* is false'

entails

(3) *c* is true if and only if *c* is false.

Finally, (3) and the principle of bivalence lead to the contradiction

> *c* is true and *c* is false.

The general mechanism involved should to be clear. On one side, the language contains expressions α ('*c*' in our example) which denote the sentence

'$\neg Tr(\alpha)$'. In addition, the language contains standard names a ('c is false' in our example). The relevant instance of Tarski's T-schema is

$$Tr(a) \leftrightarrow \neg Tr(\alpha)$$

which together with $a = \alpha$ and the laws of identity implies a contradiction.

2. Truth-value Gaps

2.1 Kripke's Fixed Point Construction

One solution is to restrict the T-schema as Tarski did. This solution has been criticized for various reasons that will not be repeated here. One of the the first serious attempts to break with the Tarskian approach was that of Kripke (1975) and Martin-Woodruff (1975).

In more details, we can take Kripke's starting point to a first-order language $\mathcal{L}$ of arithmetic which contains names for its sentences. We add a truth predicate Tr and form the extended language $\mathcal{L}^+ = \mathcal{L} \cup \{Tr\}$. On the interpretational level, we start with an interpretation $M = (U, I, I^+, I^-)$ where U is the universe and I assigns to the nonlogical vocabulary of the language appropriate elements from U in a standard way. The new element is the pair of functions I^+, I^- which interpret the truth predicate in a partial way: $I^+(Tr)$ is the extension of Tr; $I^-(Tr)$ is the counter-extension of Tr, disjoint from $I^+(Tr)$. Thus the universe U may be seen as divided into (a) sentences which belong to the extension $I^+(Tr)$ of the truth-predicate; (b) sentences which belong to its counter-extension; and (c) nonsentences.

The kernel of Kripke's proposal is a fixed point construction resulting in a partial model $M = (U, I, E^+, E^-)$ where E^+ contains exactly the sentences true in M, and E^- contains exactly the sentences false in M. The Liar sentence '$\neg Tr(\alpha)$' is neither true nor false.

The problem with this solution is well known. If '$\neg Tr(\alpha)$' is neither true nor false, then it is not true. But then the sentence $Tr(a)$ which asserts that '$\neg Tr(\alpha)$' is true, is false. But we cannot say that consistently. For if '$Tr(a)$' is false, then by logic, '$\neg Tr(a)$' is true, and thus, by one of the laws of identity, '$\neg Tr(\alpha)$' is true. But if '$\neg Tr(\alpha)$' is true, then '$Tr(a)$' which says of '$\neg Tr(\alpha)$' that it is true, should be true. Applying again the law of identity, we infer that '$Tr(a)$' is true.

Thus from the premise that '$Tr(a)$' is false we ended up with the conclusion that '$Tr(a)$' is true.

Thus the Strong-Kleene proposal of Kripke which says of the Liar that it is neither truth nor false is inadequate for it does not allow one to classify the *Liar* sentence in one's own object language. That can be done only in the metalanguage where one has availbale the notion of contradictory negation.

2.2 IF-logic

Another attempt to overcome Tarski's second impossibility result is given within the so-called *IF*-languages introduced by Hintikka and Sandu (1989) (see Hintikka (1996), Hodges (1997)). These languages express more quantifier dependencies and independencies than ordinary first-order languages whose extensions they are. More concretely, the object language contains sentences of the form

(4) $\quad \forall x_0(\forall x_1/\{x_0\})(\exists x_2/\{x_1\})(\exists x_3/\{x_0,x_2\})R(x_0,x_1,x_2,x_3\})$

which are meant to express the fact that:

$\forall x_1$ is not in the scope of $\forall x_0$;
$\exists x_2$ is not in the scope of $\forall x_1$;
$\exists x_3$ is not in the scope of $\forall x_0$ nor in that of $\exists x_2$.

The slash is thus an outscoping device. The sentence (4) expresses in a linear notation the so-called Henkin or branching quantifier introduced by Henkin (1961):

(5) $\quad \begin{pmatrix} \forall x_0 & \exists x_2 \\ \forall x_1 & \exists x_3 \end{pmatrix} R(x_0, x_1, x_2, x_3).$

The truth-conditions of (5) (and alternatively (4)) are given by a translation in a second-order metalanguage. The basic ideas for this translation are those of game theoretical semantics (GTS). With every formula φ of the object-language *L*, model *M* for *L* and assignment *s* (which is empty in case φ is a sentence) a semantical game is associated, $G(\varphi, M, s)$, opposing *Eloise* (the initial verifier) to *Abelard* (the initial falsifier). In the relevant game associated with (4), the players choose alternatively the elements *a*, *b*, *c* and *d* from the universe of *M* to be the interpretations of the four quantifiers (both players have two choices corresponding to the universal and respectively the existential quantifiers). The play $\langle a, b, c, d \rangle$ is a win for *Eloise* if it belongs to the interpretation R^M of R in M. Otherwise it is a win for *Abelard*. The slash codes the information sets of the players in the semantical games. Thus for her first move *Eloise* knows only *Abelard*'s first choice, and for the second move she knows only *Abelard*'s second choice. The sentence (4) is true in the model M (relatively to the assignmend s) if and only if there is a winning strategy for *Eloise* in the game $G(\varphi, M, g)$, that is, there are two functions f, g defined only on the possible known moves so that $\langle a, b, f(a), g(b) \rangle$ is a win for *Eloise* for any a and b chosen by *Abelard*. And similarly (4) is false in M if and only if there is a winning strategy for *Abelard*, that is, there are elements x and y such that $\langle x, y, c, d \rangle$ is a win for *Abelard* for any choices c and d of *Eloise*. The truth ($M \models^+$) and falsity ($M \models^-$) of (4) is given by the following translations (we abbreviate $\forall x_0 (\forall x_1 / \{x_0\})(\exists x_2 / \{x_1\})(\exists x_3 / \{x_0, x_2\})$ by $H x_0 x_1 x_2 x_3$):

$$M \models^+ H x_0 x_1 x_2 x_3 R(x_0, x_1, x_2, x_3) \iff \exists f \exists g \, \forall x_0 \forall x_1 R(x_0, x_1, f(x_0), g(x_1))$$
$$M \models^- H x_0 x_1 x_2 x_3 R(x_0, x_1, x_2, x_3) \iff \exists x_0 \exists x_1 \, \forall x_2 \forall x_3 \neg R(x_0, x_1, x_2, x_3).$$

There are *IF*-sentences in the pure language of identity (e.g. $\forall x_0 (\exists x_1 / \{x_0\})$) which are neither true nor false in any model which contains at least two elements. It was shown in Sandu (1996; 1998), Hyttinen and Sandu (2000) that there is an *IF*-formula $\Psi(x)$ in the vocabulary of *PA* which defines "true-in-M" for every model M of *PA*; that is

$$M \models^+ \varphi \iff M \models^+ \Psi(\ulcorner \varphi \urcorner)$$

for every *IF*-sentence φ in the vocabulary of *PA*.

As in the previous case, the *Liar* which now takes the form of a sentence β denoting '$\neg \Psi(\beta)$' is neither true nor false (the negation $\neg$ is interpreted as role switching). One is not able to express this fact in the object language but only in an extension containing classical negation. That negation, however, cannot be any longer interpreted game-theoretically, an interesting fact by itself which cannot be discussed here.

Common to both Kripke's partially interpreted languages and *IF*-languages is the fact that $\varphi \rightarrow \varphi$ is defined as $\neg \varphi \vee \varphi$. Accordingly, for sentences φ which are neither true nor false, the implication $\varphi \rightarrow \varphi$ is not valid in Kripke's semantics, neither in *IF*-logic.

In the paper "The Truth Schema and the Liar" Stephen Read offers a thought provoking solution to the *Liar via* a detour through the notion of what a sentence says. According to his view, the *Liar* turns out to be false, and this can be consistently asserted in the object language. In what follows I am going to question this solution. Before doing it, I will give a short presentation of Parsons' solution to the *Liar* which, I think, offers an interesting point of comparison with that of Read.

3. Parsons: Quantifier Shift

One way to get around the problem discussed in connection with truth-valued gaps has been suggested by Charles Parsons (1983) and Tyler Burge (1979). Their idea is that the truth-predicate as it appears in the Liar applies to different entities than the truth-predicate which is used to classify the Liar sentences. There is one essential modification though with respect to Tarski's theory: it is propositions and not sentences which are the truth-bearers. Consider now the reformulation of the Liar sentence in terms of propositions

    (*c*)   *c* expresses a false proposition.

In order to allow for the possibility of a sentence not expressing a proposition at all, Parsons replaces Tarski's *T*-schema by the weaker:

(6) $\forall x(x$ is a proposition and 'p' expresses x, then x is true if and only if p).

Together with the assumption that propositions are bivalent, (6) entails

(7) $\forall x(x$ is a proposition and 'p' expresses x, then x is false if and only if $\neg p$).

Applied to the Liar sentence, (6) and (7) lead to the conclusion that they do not express a proposition at all. Here is the argument:

    Suppose x is a proposition and c expresses x. Then by (6) we get

(8) x is true if and only if c expresses a false proposition.

    Suppose x is not true. By existential generalization we infer

(9) $\exists x(x$ is a a proposition $\wedge \neg(x$ is true$) \wedge c$ expresses $x)$

that is, c expresses a false proposition. But now, from (8) we get that x is true. Thus starting from an abitrary proposition x that c expresses, we landed in the conclusion that x is true.

(10) $\quad \forall x((x \text{ is a proposition} \wedge c \text{ expresses } x) \rightarrow x \text{ is true})$

(10) is equivalent with

(11) $\quad \neg \exists x(x \text{ is a a proposition} \wedge \neg(x \text{ is true}) \wedge c \text{ expresses } x)$.

But then there is no proposition that c expresses, for if c expressed one, say y, then by (10) y would have to be a true proposition, and that together with (8) implies that c expresses a false proposition. But this is in contradiction with (11). A similar argument shows that the Strengthened Liar (d is the sentence: d does not express a proposition) does not express a proposition either.

As Parson notices, there is a difficulty with his proposal: (c) says of a certain sentence which is (c) itself, that it expresses a false proposition. The argument above has shown that (c) does not express any proposition. But then (c) seems to say something false. Aren't we compelled to say that (c) expresses a false proposition after all? If the answer is yes, it can be shown that a contradiction will arise, and we end up in the same predicament in which we found ourselves with the truth-value gaps solution: the Liar sentence cannot be classified within the object language.

Parsons avoids the contradiction by his shifting quantifier domain assumption. According to it, both 'c expresses a false proposition' and 'c does not express any proposition' are true, but the quantifiers range over a domain of propositions which is different from the propositional domain relevant for assessing (c) in the first place. The former is larger than the latter.

The quantifier-shift proposal has been found attractive for the possibility it opens up for narrowing narrow down the gap between set-theoretic and semantic paradoxes introduced by Ramsey. Here is Parsons' argument. Given a predicate 'Fx', the fact that a is its extension is expressed by the condition

(12) $\quad \forall x(x \in a \leftrightarrow Fx)$.

By analogy with (6) above we have

(13) $\quad \forall y(y \text{ is the extension of '}Fx\text{'} \rightarrow \forall x(x \in y \leftrightarrow Fx))$.

If we now take 'Fx' to be '$x \notin x$', we obtain

(14) $\neg\exists y\forall x(x \in y \leftrightarrow x \notin x)$.

But (13) and (14) entail

(15) $\neg\exists y(y$ is the extension of '$x \notin x$').

Parsons adopts here the same solution he proposed for the *Liar*, that is, he takes the two quantifiers in (13) and (15) to range over distinct domains (Parsons, pp. 231–232).

4. Read: The Liar Is False

Read takes seriously what Tarski thought at a certain moment to be the philosophical motivation for his theory of truth, namely a correspondence theory encoded in the principle

(CP)   A sentence is true if and only if things are as the sentence says they are.

Read abbreviates 'x says that p' by '$x : p$' where 'x' designates a sentence.

The notion of 'saying that' is a technical notion, a close relative to Frege's notion of content in the *Begriffsschrift* according to which the content of a sentence comprises all its logical consequences. Read does not explicitly draw the analogy with Frege, but he nevertheless wants his notion of *saying that* to be closed under the principle

(K) $\forall p,q(p \Rightarrow q) \to (x : p \Rightarrow x : q))$

where '$\Rightarrow$' is strict implication and '$\to$' is material implication. It is clear that a sentence says, in this technical sense, more than what it says in the intuitive sense. The crux of Read's proposal is to replace Tarski's *T*-schema by

(A) $T(x) \Leftrightarrow \forall p(x : p \to p)$.

In other words:

(S) x is true if and only if things are wholly as x says they are.

Read is aware that (A) makes true all sentences which say nothing and is thus in need of qualification. The way he qualifies it is to conjoin $\exists p(x : p)$ to the right-hand side of (A). For reasons of simplicity, this proposal is not followed but Read assumes, instead, that each sentence to which (*A*) is applied says something. I shall return to this point later on.

According to Read, the point of replacing the (*T*)-schema with the (*A*)-schema is that, unlike the former, all the instances of the latter are true. In this new setting, the liar sentence *c* turns out to be false without contradiction, and, amazingly, the laws of classical logic still hold. The argument is resumed below.

The liar sentence *c* says that $\neg Tr(c)$. It may say more, say, $\neg Tr(c) \wedge q$. This together with (*A*) entails

(16) $\quad Tr(c) \Leftrightarrow \forall p(c : p \rightarrow p).$

But given that *c* says that $\neg Tr(c) \wedge q$ and that is all that *c* says, we get from (16)

(17) $\quad Tr(c) \Leftrightarrow \neg Tr(c) \wedge q.$

Thus

(18) $\quad \neg Tr(c) \Rightarrow \neg(\neg Tr(c) \wedge q)$

which is equivalent with

(19) $\quad \neg Tr(c) \Rightarrow Tr(c) \vee \neg q.$

(19) and (*K*) entail

(20) $\quad c : Tr(c) \vee \neg q$

which in conjunction with $c : q$ yields

(21) $\quad c : (Tr(c) \vee \neg q) \wedge q$

whence

(22) $\quad c : Tr(c).$

The argument has showed that if *c* says that $\neg Tr(c)$ it also says that $Tr(c)$ as well, i.e. $c : \neg Tr(c) \wedge Tr(c)$. Thus by (*A*)

(23) $\quad Tr(c) \Leftrightarrow (\neg Tr(c) \wedge Tr(c)...)$

whence

(24) $\quad \neg Tr(c).$

The liar sentence is thus not true. Read's conclusion is the following:

[*c*] cannot be true, for to be true, it would have to be both true and not true. Nothing can be both true and not true. So *c* cannot be true. ... The solution is ready to hand. Abandon (T) and realise that the correct theory of truth is given by (A) ... governing all well-formed sentences in a semantically closed language. As applied to *c*, we obtain the correct truth-condition:
$$Tr(c) \Leftrightarrow (\neg Tr(c) \land Tr(c))$$
(Read, p. 10.)

Read's view has some close analogy with Parson's when the later is reformulated to apply to sentences. Parsons himself defines such an explicit truth-predicate by

(*) $Tr(y) \leftrightarrow \exists x(x \text{ is a proposition} \land (x \text{ is true}) \land y \text{ expresses } x)$

and then points out that together with (6) it implies

(T*) $\exists x(x \text{ is a proposition} \land \text{`}p\text{'} \text{ expresses } x) \rightarrow (Tr(\text{`}p\text{'}) \leftrightarrow p)$.

Obviously, for sentences '*p*' which do not express a proposition, we will not be able to assert the consequent of (T*).

Analoguously, for falsity, he ends up with

(F*) $\exists x(x \text{ is a proposition} \land \text{`}p\text{'} \text{ expresses } x) \rightarrow (F(\text{`}p\text{'}) \leftrightarrow \neg p)$.

Let us represent propositions by second-order propositional variables. Then (*) becomes

(25) $Tr(x) \Longleftrightarrow \exists p(p \text{ is a proposition} \land x \text{ expresses } p \land p)$.

Analoguously, we may define falsity by:

(26) $F(x) \Longleftrightarrow \exists p(p \text{ is a proposition} \land x \text{ expresses } p \land \neg p)$.

Making explicit Read's abbreviation '$x : p$', his schema (*A*) becomes:

(27) $Tr(x) \Leftrightarrow \forall p(p \text{ is a proposition} \land x \text{ says that } p \rightarrow p)$.

The analogy between (25) and (27) is now straightforward:

> For Parsons, a sentence is true exactly when it expresses a proposition which is the case.
> For Read, a sentence is true exactly when everything it says is the case.

Unlike Parsons, Read claims that the *Liar* sentences express propositions (i.e., say something) and have truth-values. The key ingredient in his system which allows him to avoid a contradiction is his technical notion of "saying that".

Recall Parsons' analysis of the *Liar*: an argument has shown that the Liar sentence *c* does not express a proposition at all. Hence the Liar sentence asserts something false. But this is what the *Liar* sentence *c* says, hence it is true after all. In the end, the *Liar* sentence which is neither true nor false, receives a determinate truth-value when the domain of the quantifiers shift.

Read's solution is different. He has produced an argument showing that the *Liar* sentence *c* is not true (or false). One may now be tempted to adopt the same line of reasoning as above and continue

"But this is what the *Liar* says, hence the *Liar* is true after all".

If this could be done, a contradiction would be derived. The point of the present solution is that one cannot continue in the way just described. The particularity of Read's approach is that the *Liar* does not only say that it is not true, it also says that it is true. Therefore, in order for the *Liar* to be true, *everything* the *Liar* says must be the case. But as Read puts it, "nothing can be both true and false".

I find two problems with this solution (apart from quantifications over propositions, etc).

The minor problem is the way it deals with falsity. Recalling Read's provision devised to block his schema (*A*) to apply to sentences which say nothing, (27) should be rephrased as:

(28) $Tr(x) \Leftrightarrow \exists p(p$ is a proposition $\wedge\ x$ says that $p) \wedge$
    $\forall p(p$ is a proposition $\wedge\ x$ says that $p \rightarrow p)$.

Reads does not deal with falsity explicitly, but given that he accepts the principle of bivalence, (28) entails:

(29) $F(x) \Leftrightarrow \forall p(p$ is a proposition $\rightarrow \neg(x$ says that $p)) \vee$
    $\exists p(p$ is a proposition $\wedge\ x$ says that $p \wedge \neg p)$

That is, a sentence *x* is false if it either does not say anything or it says something that is not the case. In other words, all sentences which say nothing are false. This conclusion, although philosophically defensible, is absurd in my opinion, but I am not going to dwell upon it.

The feature in Read's treatment that concerns me here is the way his notion of "saying that" is applied in the proof. He starts from the instance

$$Tr(c) \Leftrightarrow \forall p(c : p \to p)$$

and then from the assumption that $\neg Tr(c) \wedge q$ is all that c says, he derives

$$Tr(c) \Leftrightarrow \neg Tr(c) \wedge q.$$

But why should we assume that everything the *Liar* says is expressible by one single proposition? In fact, we should not, given the fact that the *Liar* says of itself both that it is true and that it is false. Hence by the principle (K), for every p, the *Liar* says that p. In other words, there are infinitely many propositions expressed by the *Liar*. Accordingly, (23) cannot be a formula of the relevant object language (unless this language is infinitary): when we explicitate the dots on the right side of the equivalence, we obtain an infinite conjunction. For this reason, Read's argument to the effect that the *Liar* is false but not true, can be properly carried out only in a (n infinitary) metalanguage in which propositions like the ones expressed by the *Liar* can be represented. This is the price he has to pay for the acceptance of the (A)-scheme and of the principle (K). Parsons can attribute a a determinate truth-value to the *Liar* only after shifting the domain of propositions expressible by it. Similarly, Kripke can classify the *Liar* only in the metalanguage in which one has available contradictory negation, and the same goes for *IF*-logic. In Read's case, the proposition expressed by the *Liar* can be shown to receive a determinate truth-value by appeal to an argument which, when properly expressed, requires an infinitary language. The possibility of assigning a determinate truth-value to the Liar while sticking to the rules of classical logic has, in the end, turned out to be illusory.

Institut d'histoire et de philosophie des sciences et des techniques, Paris

References

Hintikka, J. 1996: *The Principles of Mathematics Revisited*, Cambridge University Press.
Hintikka, J. and G. Sandu 1989: "Informational Independence as a Semantical Phenomenon", in J. Fenstad *et al.* (eds.) (1989), *Logic, Methodology and Philosophy of Science* VIII, Elsevier Science Publishers B.V., pp. 571–589.
Hintikka, J. and G. Sandu 1997: "Game-theoretical Semantics", in J. van Benthem and A. ter Meulen (eds.) (1997), *Handbook of Logic and Language*, Elsevier, pp. 361–410.

Hodges, W. 1997: "Compositional Semantics for a Language with Imperfect Information", *Journal of the IGPL* 5, 539–563.
Hyttinen, T. and G. Sandu 2000: "Henkin Quantifiers and the Definability of Truth", *Journal of Philosophical Logic* 29, 507–527.
Kripke, S. 1975: "Outline of a Theory of Truth", *Journal of Philosophy* 72, 690–716, reprinted in Martin (1984), pp. 53–82.
Martin, R. L. (ed.) 1984: *Recent Essays on Truth and the Liar Paradox*, Oxford University Press, Oxford.
Martin, R. L. and P. W. Woodruff 1975: "On Representing 'True-in-L' in L", *Philosophia* 5, 213–217, reprinted in Martin (1984), pp. 47–52.
Parsons, Ch. 1983: *Mathematics in Philosophy*, Cornell University Press.
Read, S.: "The Truth Schema and the Liar", in S. Rahman and T. Tulenheimo (eds.), *Unity, Truth and the Liar: The Modern Relevance of Medieval Solutions to the Liar Paradox*, forthcoming.
Sandu, G. 1996: "IF First-Order Logic, Kripke, and 3-Valued Logic", appendix to Hintikka (1996), pp. 254–270.
Sandu, G. 1998: "IF-logic and Truth-definition", *Journal of Philosophical Logic* 27, 143–164.
Tarski, A. 1983: "The Concept of Truth in Formalized Languages", in A. Tarski, *Logic, Semantics, Metamathematics*, second edition, ed. J. Corcoran, Indianapolis, Hackett.

II
Induction, Truthlikeness, and Scientific Progress

Roberto Festa

Verisimilitude, Qualitative Theories, and Statistical Inferences

Abstract. This paper argues that qualitative theories may have interesting statistical applications. More precisely, it is shown that Q-theories, i.e., qualitative theories stated in monadic languages with two or more families of predicates, can be used in describing the statistical structure of cross classified populations and that their adequacy in this task can be evaluated by measuring their 'statistical verisimilitude'. Firstly, a short survey of some applications of the post-Popperian notions of verisimilitude to the analysis of statistical inferences is provided. Secondly, a new class of measures for the verisimilitude of the universal generalizations stated in monadic languages is introduced. Finally, such measures are applied to Q-theories and used as the starting point for defining appropriate measures of the statistical verisimilitude of Q-theories.

The present contribution is motivated by the idea that the statistical applications of verisimilitude may contribute to showing that verisimilitude "is not only an artificial philosopher's concept, but rather a methodological tool for analysing the real-life success of scientific inquiry" (Niiniluoto 1989, p. 240).

In the last three decades several measures of verisimilitude have been defined for different kinds of qualitative and quantitative theories, and their methodological applications have been thoroughly explored. Among other things it has been shown, w.r.t. quantitative hypotheses, that the notion of verisimilitude can be fruitfully applied in the analysis of some important kinds of statistical inferences. Here we will argue that also qualitative theories may have interesting statistical applications. More precisely, we will show that qualitative theories stated in (first order) monadic languages with two or more families of predicates – henceforth, *Q-theories* – can be used in describing the statistical structure of cross classified populations and that their adequacy in this task can be evaluated by measuring their 'statistical verisimilitude'.

In Section 1 we will outline the basic ideas underlying the post-Popperian approaches to verisimilitude and we will shortly survey some applications of verisimilitude to the analysis of statistical inferences. In Section 2 we will introduce a new class of measures for the verisimilitude of the universal generalizations stated in monadic languages with one or more families of predicates. Finally, in Section 3, the above measures will be applied to Q-theories and used as the starting point for defining appropriate measures of the statistical verisimilitude of Q-theories.

1. Post-Popperian Approaches to Verisimilitude and Statistical Inference

The emergence of post-Popperian approaches to verisimilitude. According to the falsificationist methodology outlined by Popper (1934/1959) the basic steps of a good scientific inquiry are the following: (1) highly informative theories are invented; (2) they are submitted to severe experimental tests; (3) the theories falsified by the results of the tests are rejected; (4) among the non-falsified theories, the most corroborated one, i.e., the theory which passed the most severe tests, is tentatively accepted. In the last part of his life, Popper (1963; 1972) provided an epistemological basis to falsificationism; in fact he stated the view that the main epistemic goal of science is the achievement of a high degree of verisimilitude and claimed that falsificationism is the best methodology to achieve this goal.

At the end of the sixties, the former collaborator of Popper, Imre Lakatos, while sharing with Popper a verisimilitudinarian view of the epistemic goals of science, rejected Popper's claim that falsificationism is a suitable tool to approximate the truth.[1] Accordingly, between the end of the sixties and his premature death, occurred in 1974, Lakatos developed a sophisticated non-falsificationist methodology – known as methodology of scientific research programs, or MSRP – according to which experimental falsification does not necessarily lead to the rejection of scientific theories. Indeed, Lakatos claims that also falsified theories may receive a high degree of corroboration from their successful experimental predictions and, therefore, may be accepted as part of a progressive research program. According to Lakatos the MSRP-rules for the acceptance of scientific theories can be justified by showing that the

---

[1] Below the terms 'approximation (or closeness) to the truth' and 'truthlikeness' will be used as synonyms of 'verisimilitude'.

high degree of corroboration of theories is a sign of their increasing approximation to the truth. To this purpose, the link between corroboration and verisimilitude has to be precisely reconstructed by appropriate inductive principles, providing plausible corroboration-based estimates of the verisimilitude of the scientific hypotheses, *even if already falsified.* Lakatos (1974) was aware that the inductive principles of the requested kind could not be stated in terms of Popper's measure of corroboration.[2] It is not clear whether he realized that also Popper's notion of verisimilitude was problematic. In any case, he died too soon to appreciate the consequences of the theorem discovered by David Miller (1974) and Pavel Tichý (1974), who independently proved that, according to Popper's definition, a false theory can never be closer to the truth than another one.

Lakatos' criticisms of Popper's falsificationism, together with the discovery of serious faults in Popper's notion of verisimilitude, opened the way to the post-Popperian approaches to verisimilitude, emerged since 1975.[3] Two fundamental issues investigated within such approaches are:

(1) The analysis of the *logical problem of verisimilitude*, in order to provide appropriate notions of *verisimilitude* and *distance from the truth*. Such notions should allow us to compare any two theories w.r.t. their closeness to the truth, even in the case where they are false. Post-Popperian theories of verisimilitude generally admit that adequate notions of verisimilitude and distance from the truth should leave room for the possibility that, under certain conditions, a false theory is closer to the truth than a true one.

(2) The analysis of the *epistemic problem of verisimilitude*, in order to define the notions of *estimated verisimilitude* and *estimated distance from the truth*. Such notions should allow us to compare, on the basis of the available data, any two theories w.r.t. their estimated closeness to the truth, even in the case where they have been falsified by such data. It is widely agreed that adequate notions of estimated verisimilitude and estimated distance from the truth should allow that, under certain conditions, the estimated verisimilitude of a falsified theory is higher than that of a non-falsified one.

---

[2] Cf. Lakatos (1974, p. 270, note 122); see also Niiniluoto (1989, p. 236, note 9) and Festa (forthcoming).

[3] An excellent survey of the modern history of verisimilitude, started with Karl Popper's definition of verisimilitude in 1960, is provided by Niiniluoto (1998).

Some scholars involved in the development of post-Popperian theories of verisimilitude explicitly admit the possibility that a hypothesis h is falsified by evidence e but, at the same time, is considered highly verisimilar on e.[4] A possibility of this kind suggests the idea that, *pace* Popper, the achievement of a high degree of verisimilitude requires the adoption of a non-falsificationist methodology.[5]

Verisimilitude and statistical inferences. The post-Popperian notions of verisimilitude and distance from the truth have been applied in the rational reconstruction of a number of widely used statistical procedures. Some of these applications are outlined below.

(1) BAYESIAN POINT AND INTERVAL ESTIMATES. Given a distribution of epistemic probabilities on the relevant set of statements, one can calculate, on the basis the available evidence e, the expected verisimilitude and the expected distance from the truth of any hypothesis h under inquiry.[6] One can construct a theory of inductive acceptance based on the idea that the distance from the truth is the 'cognitive loss' whose expected value should be minimized when a hypothesis is selected for inductive acceptance. This idea has been applied to the rational reconstruction of certain standard results about point estimation obtained in Bayesian statistics. More specifically, Niiniluoto (1982a; 1987, pp. 427–429) shows that, given a suitable loss function, accepting a Bayesian point estimate is equivalent to minimizing the expected distance from the truth. Furthermore, Niiniluoto (1982b; 1986; 1987, pp. 430–441) and others (Festa 1986; Maher 1993, pp. 143–147) extend this perspective to Bayesian interval estimation by approaching interval estimation in decision-theoretic terms, where the loss of accepting an interval is defined by the distance of this interval from the truth.

---

[4] Cf. Niiniluoto (1989, p. 236).

[5] A non-falsificationist methodology which is justified on the basis of a verisimilitudinarian epistemology might be called verisimilitudinarian non-falsificationist methodology or, for short, *VNF-methodology* (for a more detailed account of the emergence of VNF-methodologies, see Festa, forthcoming). The most articulated versions of VNF-methodologies have been developed by Niiniluoto (1987) and Kuipers (2000).

[6] See Niiniluoto (1986; 1987, Ch. 7.2).

(2) BAYESIAN ANALYSIS OF OBSERVATIONAL ERRORS. The expected verisimilitude of a hypothesis h can be seen as a reasonable estimate of the degree of verisimilitude of h, while the probable verisimilitude of h expresses our degree of belief in the possibility that h is highly verisimilar, i.e., that the verisimilitude of h exceeds a fixed threshold (Niiniluoto 1987, Ch. 7.3). The notions of expected and probable verisimilitude have been applied to the analysis of significant statistical problems such as the problem of observational errors. This problem typically arises in the cases where the evidence is obtained by measuring a given quantity: in fact, in such cases there is a positive probability that, due to errors in measurement, the result of measurement deviates from the true value of the quantity. After discussing in general terms the possibility of appraising verisimilitude on the basis of false evidence, Niiniluoto (1987, Ch. 7.4) suggests that the classical Bayesian treatment of observational errors can be understood as an attempt to evaluate the expected and probable verisimilitude of hypotheses on the basis of evidence which is known (with probability one) to be erroneous.

(3) PROBLEMS OF CURVE-FITTING AND REGRESSION ANALYSIS. In many cognitive problems – for instance, whenever the inquiry is based upon counterfactual idealizing conditions – all the relevant hypotheses under inquiry are known to be false.[7] A well known example of this sort of cognitive problems is given by the statistical problems of curve-fitting, where all the relevant quantitative hypotheses, expressed by sufficiently simple linear functions, are known to be false. A verisimilitude interpretation of the typical statistical methods based on the Least Square Difference between a curve and a finite set of data points is provided by Niiniluoto (1987, Ch. 7.5). He suggests also that verisimilitude could be used for the rational reconstruction of the statistical methods of regression analysis which are applied to problems in which all the relevant hypotheses are false (idealizations) and the evidence is known to be erroneous (Niiniluoto 1987, p. 289).

(4) THE CHOICE OF PRIOR DISTRIBUTIONS IN THE BAYESIAN ANALYSIS OF MULTINOMIAL INFERENCES. Multinomial inferences are statistical inferences aiming at establishing the true values of the objective probabilities $\mathbf{P}=(\mathbf{P}_1,...,\mathbf{P}_k)$

---

[7] Cf. Niiniluoto (1987, Ch. 7.5).

of a multivariate Bernoulli process. A major problem for the Bayesian analysis of multinomial inferences – which has been thoroughly investigated in Bayesian statistics and in the theory of inductive probabilities, developed by Rudolf Carnap and his followers – is the choice of an 'optimal' prior distribution F on **P**. The 'verisimilitude solution' to this problem suggested by Festa (1993; 1995) is inspired by the intuitive idea that F should be chosen so that the estimates of $\mathbf{P}_1, ..., \mathbf{P}_k$ based on F approach the truth most effectively. This means that a good reason for selecting F, from a certain class of prior distributions, is that there are grounds to believe that it is 'the optimum tool' for achieving a high degree of verisimilitude.

2. The Verisimilitude of Universal Generalizations

2.1. Constituents, Quasi-constituents and Generalizations

The individuals of a population D can be described by using a monadic (first-order) language L with one or more families $X, Y, ..., Z$ of primitive predicates, where the predicates of each family are mutually exclusive and jointly exhaustive. Any family $X \equiv \{x_1, ..., x_j, ..., x_c\}$ refers to a given character of the individuals of D; hence any predicate $x_j \in X$ will denote one of the possible properties, or states, of the individuals of D w.r.t. the character corresponding to X. For instance, in social sciences X may refer to characters such as gender, religion, occupation, and political party affiliation; hence x_j may denote properties such as 'male', 'catholic', 'manual worker' or 'republican'. Given a monadic language L, one can define the class $\mathbf{Q} \equiv \{Q_1, ..., Q_q\}$ of the Q-predicates of L, where any $Q_h \in \mathbf{Q}$ is a conjunction taking one element from each family of primitive predicates, so that the Q-predicates in $\mathbf{Q}$ will be mutually exclusive and jointly exhaustive. For example, in a language L with two families $X \equiv \{x_1, ..., x_j, ..., x_c\}$ and $Y \equiv \{y_1, ..., y_i, ..., y_r\}$ of primitive predicates, $\mathbf{Q}$ is defined as follows: $\mathbf{Q} \equiv Y \times X \equiv \{(y_1 x_1), (y_1 x_2), ..., (y_2 x_1), ..., (y_i x_j), ..., (y_r x_c)\}$. When L includes one family X of primitive predicates, $\mathbf{Q}$ can be identified with X. $\mathbf{Q}$ is the most specific *classification system* applicable to the individuals of D by using the linguistic resources of L. This means that any *cell* Q_h of $\mathbf{Q}$ corresponds to one of the *kinds of individuals* which can be specified by L. We will say that a Q-predicate, or cell, Q_h of $\mathbf{Q}$ is *instantiated*, or *non-empty*, in the case that Q_h applies to at least an individual in D; otherwise, we will say that Q_h is *empty*.

Constituents. The sentences of L without individual constants are often called *generalizations*.[8] The most informative generalizations of L are the so-called *constituents*. Indeed a constituent C specifies exactly which kinds of individuals exist and which do not exist in the population D under examination, i.e., which cells Q_h of **Q** are empty and which are non-empty. In other words, C determines a partition $\{C^+, C^-\}$ of **Q**, where C^+ and C^- ($= \mathbf{Q} - C^+$) are the sets of the instantiated and empty cells specified by C. C^+ and C^- may be referred to as the *instantiated* (or *non-empty*) and the *empty area of C*, respectively. C can be seen as a complete description of the 'qualitative structure' of a population or, equivalently, as the qualitative description of a 'kind of possible worlds'. One can easily check that there are $t = 2^q - 1$ constituents including the 'trivial' constituent C_T saying that any Q-predicate is instantiated in D.[9] It is easy to see that there is a unique true constituent of L, which can be construed as 'the truth' about D.

A constituent C can be written in the following form:

(1) $(\pm)(\exists u)Q_1(u)$ & ... & $(\pm)(\exists u)Q_q(u)$

where $(\pm)$ is either empty or the negation sign $\neg$.

Below the following notational conventions will be used. Given a set of formulas F, $\vee_{\alpha \in F}\alpha$ and $\wedge_{\alpha \in F}\alpha$ are the disjunction and the conjunction of all the formulas in F. Moreover, given a set of formulas α_i indexed by I, $\vee_{i \in I}\alpha_i$ and $\wedge_{i \in I}\alpha_i$ are the disjunction and the conjunction of all the formulas α_i, $i \in I$.

By applying the above notation, (1) can be rewritten in the following way:

(2) $\wedge_{j \in C^+}(\exists u)Q_j(u)$ & $\wedge_{j \in C^-}\neg(\exists u)Q_j(u)$

where $\wedge_{j \in C^+}(\exists u)Q_j(u)$ expresses the existence claims of C, while $\wedge_{j \in C^-}\neg(\exists u)Q_j(u)$ expresses its non-existence claims. Due to equality $C^+ = \mathbf{Q} - C^-$, $\wedge_{j \in C^-}\neg(\exists u)Q_j(u)$ is logically equivalent to $(u)\vee_{j \in C^+}Q_j(u)$. Hence, (2) can be rewritten as follows:

---

[8] Cf. Niiniluoto (1987, 40).
[9] Here the constituent which claims that no Q-predicate is instantiated in D is excluded from consideration.

(3) $\bigwedge_{j \in C^+} (\exists u) Q_j(u)$ & $(u) \bigvee_{j \in C^+} Q_j(u)$

where $(u) \bigvee_{j \in C^+} Q_j(u)$ represents, as it were, the *universal ingredient* of C.

Quasi-constituents. A simple type of generalization is given by *quasi-constituents* or, for short, *q-constituents*.[10] While a constituent C gives a *complete* list of the instantiated and the empty cells of **Q**, a q-constituent G gives a (possibly) *incomplete* list of such cells. Let G^+ and G^- be the sets of the instantiated and empty cells specified by G; then $G^+ \cup G^-$ may be a proper subset of **Q**. A q-constituent G can be written in the form:

(4) $\bigwedge_{j \in G^+} (\exists u) Q_j(u)$ & $\bigwedge_{j \in G^-} \neg(\exists u) Q_j(u)$

which is strictly similar to the form (2) for C. Since $\bigwedge_{j \in G^-} \neg(\exists u) Q_j(u)$, which states the non-existence claims of (4), is logically equivalent to $(u) \bigvee_{j \in \mathbf{Q}-G^-} Q_j(u)$, (4) can be rewritten as follows:

(5) $\bigwedge_{j \in G^+} (\exists u) Q_j(u)$ & $(u) \bigvee_{j \in \mathbf{Q}-G^-} Q_j(u)$.[11]

A q-constituent G is uniquely characterized by the partition $\{G^+, G^-, G^?\}$, where G^+ and G^- are the *instantiated* (or *non-empty*) and the *empty area of G*, while $G^? \equiv \mathbf{Q} - (G^+ \cup G^-)$ is the *question mark area* of G. Three special types of q-constituents should be mentioned: (i) the *purely non-existential q-constituent*, i.e., the q-constituents with triple $\{\emptyset, G^-, G^?\}$ which do not specify any instantiated cell; (ii) the *purely existential q-constituents*, i.e., the q-constituents with triple $\{G^+, \emptyset, G^?\}$ which do not specify any empty cell; (iii) the *constituents*, i.e., the q-constituents with triple $\{G^+, G^-, \emptyset\}$.

Universal generalizations. Since a purely non-existential q-constituent G satisfies equalities $G^+ = \emptyset$ and $G^? = \mathbf{Q} - G^-$, it is uniquely characterized by the couple $\{G^-, G^?\}$. Moreover the above equalities allow to rewrite (5) as follows:

---

[10] Cf. Oddie (1986, p. 86).

[11] However, note that, since equality $G^+ = \mathbf{Q} - G^-$ is not generally valid, $\bigwedge_{j \in G^-} \neg(\exists u) Q_j(u)$ is not logically equivalent to $(u) \bigvee_{j \in G^+} Q_j(u)$, so that (4) cannot be rewritten in the same form of (3).

(6) $(u)\vee_{j\in G^?} Q_j(u)$.

It appears from (6) that a purely non-existential q-constituent G amounts to a *universal generalization*. There are $t = 2^q - 1$ universal generalizations including the tautological generalization G_T, characterized by the couple $\{\emptyset, \mathbf{Q}\}$, which does not specify any empty cell.[12]

For any $Q_b \in G^-$, G makes the non-existence claim that no individual in D is Q_b, and, thereby, also the *prediction* that no individual sampled from D will be Q_b. Hence G^- might be interpreted as the *predictive content*, or predictive strength, of G. It should be noticed that, given two universal generalizations G and G_*, G is *logically stronger* than G_* - i.e., G logically implies G_* but not vice versa - if and only if $G_*^- \subset G^-$, i.e., if and only if the predictive content of G_* is strictly included in that of G.[13] A simple (normalized) measure of the predictive content of G is:

(7) $cont(G) \equiv \dfrac{|G^-|}{q}$.

Suppose that the constituent C is 'the truth' about D. Then a universal generalization G is true if and only if $G^- \subseteq C^-$, i.e., if and only if all the predictions made by G are true. The *strongest true theory* will be the true universal generalization with maximal predictive content, i.e., the universal generalization G such that $G^- = C^-$. Given a false universal generalization G, we will say that G is *completely false* when $G^- \subseteq C^+$, i.e., when all the predictions made by G are false. It should be pointed out that, if the trivial constituent C_T is 'the truth' about D, then the only true universal generalization is the tautological universal generalization G_T while any non-tautological universal generalization is not only false, but completely false. Finally, w.r.t. false universal generalizations, we can introduce also the fuzzy notion of almost true universal generalization, by saying that the false universal generalization G is *almost true* when almost all the

---

[12] In fact, one can see from (3) and (6) that there is a one-to-one correspondence between universal generalizations and constituents, since for any constituent C characterized by the universal ingredient $(u)\vee_{j\in C^+} Q_j(u)$ one can specify the universal generalization $G \equiv (u)\vee_{j\in G^?} Q_j(u)$, where $G^? = C^+$, and vice versa.

[13] In particular, the tautological universal generalization G_T, with $G_T^- = \emptyset$ and $G_T^? = \mathbf{Q}$, is weaker than any non-tautological theory G.

predictions of G are true, i.e., when almost all the cells in G^- belong to C^-. A quantitative specification of this notion can be given by defining the 'degree of truth' $t(G,C)$ of G w.r.t. C as follows:

$$(8) \quad t(G,C) \equiv \frac{|G^- \cap C^-|}{|G^-|}.$$

Then we could say that G is almost true whenever $t(G,C)$ exceeds a fixed threshold τ, where $\tau \geq \frac{1}{2}$.

Generalizations. Given the set $\Gamma \equiv \{C_1, ..., C_t\}$ of the t constituents of L, the disjunctive closure of Γ - i.e., the set of all the disjunctions $\vee_{i \in J_a} C_i$, where $J_a \subseteq \{1, ..., t\}$ - will be called $D(\Gamma)$. It is well known that any generalization G is logically equivalent to a disjunction of constituents of L, i.e., to a member of $D(\Gamma)$:

$$(9) \quad \vdash G \equiv \vee_{i \in J_G} C_i \qquad J_G \subseteq \{1, ..., t\}.^{14}$$

$\vee_{i \in J_G} C_i$ is called the *distributive normal form* of G - for short: $dnf(G)$. Note that, given a q-constituent G with triple $\{G^+, G^-, G^?\}$, C_i is a disjunct of $dnf(G)$ if and only if $G^+ \subseteq C_i^+ \subseteq \mathbf{Q} - G^-$.[15] In other words, the distributive normal form of a q-constituent G contains all and only the constituents which claim that at least the cells of G^+ are exemplified and at least the cells of G^- are empty. In the special case where G is a universal generalization, so that $G^+ = \emptyset$ and $G^? = \mathbf{Q} - G^-$, C_i is a disjunct of $dnf(G)$ if and only if $C_i^+ \subseteq G^?$.

Although many generalizations make some definite existence and non-existence claims relative to specific kinds of individuals in $\mathbf{Q}$, there also generalizations which do not make any (non-)existence claim. An example of the first type is $G_1 \equiv (\exists u) Q_1(u) \wedge \neg (\exists u) Q_2(u) \wedge ((\exists u) Q_3(u) \vee (\exists u) Q_4(u))$, while $G_2 \equiv (\exists u) Q_1(u) \leftrightarrow (\exists u) Q_2(u)$ is an example of the second type. Note that non-equivalent generalizations may make the same existence and non-existence claims. For instance, although $G_3 \equiv (\exists u) Q_3(u) \leftrightarrow (\exists u) Q_4(u)$ is non-equivalent to G_2, it is characterized by the same empty set of (non-)existence claims.

---

[14] Cf. Niiniluoto (1987, p. 336).
[15] Cf. Niiniluoto (1987, pp. 335–336).

Another couple of non-equivalent generalizations which make the same (non-)existence claims is given by $G_4 \equiv \neg\,(\exists u)\,Q_1(u)$ and $G_5 \equiv \neg\,(\exists u)\,Q_1(u) \land (\neg\,(\exists u)\,Q_2(u) \lor \neg(\exists u)\,Q_3(u))$.

The (non-)existence claims made by a generalization G are expressed by the strongest q-constituent implied by G, which will be called *the q-constituent associated to G*, while the triple $\{G^+, G^-, G^?\}$ of this q-constituent will be called *the triple of G*. Of course two non-equivalent generalizations which make the same (non-)existence claims have the same triple.[16]

A generalization G conveys two sorts of information about the qualitative structure of the population D under inquiry, which will be called *KW-information* and *KI-information*, respectively. KW-information, which is related to the distributive normal form $dnf(G)$ of G, tells us that D is one of the *kinds of possible worlds* described by the constituents C_i in $dnf(G)$. KI-information, which is related to the q-constituent associated to G, tells us that some of the *possible kinds of individuals* listed in $\mathbf{Q}$ are instantiated in D whereas other kinds are not.

2.2. Measuring the Verisimilitude of Universal Generalizations

The verisimilitude of constituents. The distance, or dissimilarity, $d(C_*, C)$ between two constituents C_* and C might be defined in terms of the (lack of) overlap between the corresponding partitions $\{C_*^+, C_*^-\}$ and $\{C^+, C^-\}$. More precisely, $d(C_*, C)$ might be defined as an increasing function of $|C_*^- \cap C^+|$ and $|C_*^+ \cap C^-|$ and as a decreasing function of $|C_*^- \cap C^-|$ and $|C_*^+ \cap C^+|$. This intuition can be developed in different ways. For instance, we could adopt one of the following measures of distance:

$$(10) \quad d_\Delta(C_*, C) \equiv \frac{|C_*^+ \Delta C^+|}{q} = \frac{|C_*^- \Delta C^-|}{q};$$

$$(11) \quad d_e(C_*, C) = \frac{|C_*^+ \Delta C^+|}{|C_*^+ \cup C^+|};$$

---

[16] For instance, the above generalizations G_2 and G_3 share the triple $\{\varnothing, \varnothing, \mathbf{Q}\}$ corresponding to the tautological q-constituent G_T while G_4 and G_5 share the triple $\{\varnothing, Q_1, \mathbf{Q} - Q_1\}$.

$$(12) \quad d_n(C_*,C) \equiv \frac{|C_*^- \Delta C^-|}{|C_*^- \cup C^-|} = \frac{|C_*^+ \Delta C^+|}{|C_*^- \cup C^-|}.$$

The intuitive meaning of the distance measures in (10)–(12) is easily understood by recalling that the symmetrical difference $C_*^+ \Delta C^+$ ($= C_*^+ \cap C_*^-) \cup (C^+ \cap C_*^-) = C_*^- \Delta C^-$) is the set of those Q-predicates which C_* claims to be instantiated and C claims to be empty, or vice versa, i.e., the set of those Q-predicates about which C_* and C make different (non-)existence claims. Hence $d_\Delta(C_*,C)$ can be seen a normalized measure, with $1/q$ as a normalizing factor, of the *absolute (existential and non-existential) disagreement* between C_* and C. Moreover, since $C_*^+ \cup C^+$ is the set of those Q-predicates about which either C_* or C make existence claims and $C_*^- \cup C^-$ is the set of those Q-predicates about which either C_* or C make non-existence claims, $d_e(C_*,C)$ and $d_n(C_*,C)$ can be construed as measures of the *relative existential disagreement* and the *relative non-existential disagreement*, respectively, between C_* and C.

The above distance measures are particular cases of the following parametric class of distance measures:[17]

$$(13) \quad d_{\alpha,\beta,\gamma,\delta}(C_*,C) \equiv$$

$$\equiv \frac{\alpha|C_*^- \cap C^+| + \beta|C_*^+ \cap C^-|}{\alpha|C_*^- \cap C^+| + \beta|C_*^+ \cap C^-| + \gamma|C_*^- \cap C^-| + \delta|C_*^+ \cap C^+|}$$

with $\alpha, \beta, \gamma, \delta \geq 0$.

Given a distance measure $d(C_*,C)$, the proximity, or similarity, $s(C_*,C)$ between C_* and C can be defined as a decreasing function of $d(C_*,C)$, such as $s(C_*,C) \equiv 1 - d(C_*,C)$. If C is 'the truth' about D then $d(C_*,C)$ and $s(C_*,C)$ measure the *distance from the truth* of C_* and its *verisimilitude*, respectively.

The verisimilitude of universal generalizations. Above it has been suggested that a generalization G conveys two sorts of information about the qualitative structure of the population D under examination, i.e., KW-information and KI-information. In order to define the distance $d(G,C)$ between a generalization G and a constituent C one can adopt two types of strategies: (i) the KW-

---

[17] Cf. Festa (1987, p. 158).

strategies, inspired by the idea that $d(G,C)$ should be defined in terms of the KW-information conveyed by G, and (ii) the *KI-strategies*, inspired by the idea that $d(G,C)$ should be defined in terms of the KI-information conveyed by G. Here we will adopt a KI-strategy. More specifically, $d(G,C)$ will be defined in terms of the (lack of) overlap between the partitions $\{G^-, G^?\}$ and $\{C^+, C^-\}$, as an increasing function of $|G^- \cap C^+|$ and $|G^? \cap C^-|$ and as a decreasing function of $|G^- \cap C^-|$ and $|G^? \cap C^+|$.

First of all, let us notice that the empty area G^- of G can be partitioned into the sets $G^- \cap C^+$ and $G^- \cap C^-$ which correspond to the *mistaken* and *correct predictions* of G w.r.t. C, respectively. This suggests that $G^- \cap C^+$ and $G^- \cap C^-$ could be referred to as the *predictive error* and the *predictive success* of G w.r.t. to C. Two simple normalized measures of the predictive error and success of G w.r.t. to C are the following:

$$(14) \quad E_{GC} \equiv \frac{|G^- \cap C^+|}{q}; \quad S_{GC} \equiv \frac{|G^- \cap C^-|}{q},$$

where E_{GC} and S_{GC} are related to the predictive content *cont(G)* (see (7)) by the relation:

$$(15) \quad E_{GC} + S_{GC} = \mathit{cont}(G).$$

One might think that $d(G,C)$ should essentially depend on the overlap between the empty areas of G and C and, more specifically, that it should be a decreasing function of $|G^- \cap C^+|$. This consideration inspires the following very simple definition of $d(G,C)$:

$$(16) \quad d_1 \equiv \frac{|G^- \cap C^+|}{|G^-|}.$$

$d_1(G,C)$ is equal to the ratio between the predictive error E_{GC} and the predictive content *cont(G)*:[18]

---

[18] Note that $d_1(G_T,C)$ is not defined since $\mathit{cont}(G_T) = 0$.

(17) $$d_1(G,C) = \frac{E_{GC}}{cont(G)} = \frac{E_{GC}}{E_{GC} + S_{GC}}.$$

A serious limitation of $d_1(G,C)$ consists in the fact that it depends only on the ratio $E_{GC}/cont(G)$, by neglecting the *absolute* value of the predictive content $cont(G)$ ($= |G^-|/q$). Moreover, although $d_1(G,C)$ is affected by the structure of G^- - i.e., by the way in which G^- is partitioned into the sets $G^- \cap C^-$ and $G^- \cap C^+$ of the correct and mistaken predictions of G - it is *not* affected by the structure of $G^?$, i.e., by the way in which $G^?$ is partitioned into the sets $G^? \cap C^+$ and $G^? \cap C^-$ of the correct and mistaken permissions of G. In other words, $d_1(G,C)$ is not affected by the overlap between the question mark area of G and the instantiated area of C.

Due to the above limitations, d_1 exhibits some problematic features which can be illustrated, for simplicity of exposition and without any loss of generality, by assuming that C is 'the truth' about D.

(i) *Comparing true universal generalizations.* The minimum value of $d_1(G,C)$, which is zero, is reached whenever $G^- \cap C^+ = \emptyset$, i.e., whenever $G^- \subseteq C^-$ or, equivalently, whenever G is true. This means that all the true universal generalizations are maximally close to C, independently of their predictive content. So, for instance, assuming that C^- includes Q_1 and several other Q-predicates, the very weak true universal generalization G_*, with $G_*^- = Q_1$, is as distant from C as the strongest true universal generalization G, with $G^- = C^-$. This feature of d_1 conflicts with the principle that, given two true universal generalizations G and G_*, where G is logically stronger than G_* (in symbols: $G_*^- \subset G^-$), G should be closer to the truth than G_*. The intuitive plausibility of this principle can be easily appreciated by recalling that G and G_* are true in the case where all the predictions associated with G^- and G_*^- are true; hence, saying that $G_*^- \subset G^-$ amounts to saying that G makes all the true predictions made by G_* with the addition of further true predictions.

(ii) *Comparing completely false universal generalizations.* The maximum value of $d_1(G,C)$, which is one, is reached whenever $G^- \cap C^+ = G^-$, i.e., whenever $G^- \subseteq C^+$ or, equivalently, whenever G is completely false. This means that all the completely false universal generalizations are maximally distant from C, independently of their predictive content. This feature of d_1 conflicts with the principle that, given two completely false universal generalizations G and G_*, where $G_*^- \subset G^-$, G should be more distant from the truth than G_*. The plausibility of this principle can be appreciated by recalling that G and G_* are completely false in the case where all the predictions associated with G^- and G_*^- are false; hence, saying that $G_*^- \subset G^-$ amounts to saying that G makes all the false predictions made by G_* plus further false predictions.

(iii) *Comparing true and almost true universal generalizations.* Suppose that C^- includes Q_1 and several other Q-predicates. Then, any true universal generalization, including the very weak true universal generalization G_*, with $G_*^- = Q_1$, will be closer to the truth than any false, but almost true, universal generalization G, independently of the predictive content of G. This conflicts with the intuitively plausible principle that an almost true and highly informative universal generalization G may be closer to the truth than G_*.

The above problematic features of d_1 can be avoided by allowing that $d(G,C)$ depends not only on the structure of G^-, but also on the structure of $G^?$, i.e., on the way in which $G^?$ is partitioned into the sets $G^? \cap C^+$ and $G^? \cap C^-$. Although no genuine mistake can be associated to the members of the question mark area $G^?$, since G does not say anything about the Q-predicates in $G^?$, one cannot neglect the significant differences between the sets $G^? \cap C^-$ and $G^? \cap C^+$, into which $G^?$ is partitioned. Since G leaves room for, i.e., permits the existence of individuals in the question mark cells of $G^?$, we may say that such permissions are mistaken when they conflict with C. More precisely, we may say that, for any Q_h in $G^? \cap C^-$, the existence of Q_h-individuals has been mistakenly permitted by G and that Q_h corresponds to a *mistaken permission* made by G. Analogously, we could say that the permissions made by G are correct when they do not conflict with C. More specifically, we may say that, for any Q_h in $G^? \cap C^+$, the existence of Q_h-individuals has been correctly allowed by G and that Q_h corresponds to a *correct permission* made by G.

The above remarks suggest that we could refer to $G^? \cap C^-$ and $G^? \cap C^+$ as the *permissive error* and the *permissive success* of G w.r.t. to C, respectively. Two simple normalized measures of the permissive error and success of G w.r.t. to C are the following:

(18) $\quad e_{GC} \equiv \dfrac{|G^? \cap C^-|}{q}; \quad\quad s_{GC} \equiv \dfrac{|G^? \cap C^+|}{q}.$

The *total error* TE_{GC} of G w.r.t. to C might be regarded as a combination of the predictive error E_{GC} and the permissive error e_{GC}, where the weight of e_{GC} is not higher than that of E_{GC}, since mistaken permissions cannot be considered more serious than mistaken predictions:

(19) $\quad TE_{GC} \equiv \alpha E_{GC} + \beta e_{GC} \quad\quad$ where $\alpha \geq \beta \geq 0$.

Analogously, assuming that the success related to correct permissions is not more important than that related to correct predictions, the *total success* TS_{GC} of G w.r.t. to C might be defined as follows:

(20) $\quad TS_{GC} \equiv \gamma S_{GC} + \delta s_{GC} \quad\quad$ where $\gamma \geq \delta \geq 0$.

Now we can introduce a parametric class of measures, based on the intuitive idea that the distance between G and C is given by the ratio between the total error TE_{GC} and the sum $TE_{GC} + TS_{GC}$ of the total error and the total success of G w.r.t. C:

(21) $\quad d_{\alpha,\beta,\gamma,\delta}(G,C) \equiv \dfrac{TE_{GC}}{TE_{GC} + TS_{GC}} = \dfrac{\alpha E_{GC} + \beta e_{GC}}{\alpha E_{GC} + \beta e_{GC} + \gamma S_{GC} + \delta s_{GC}}$

$$= \dfrac{\alpha|G^- \cap C^+| + \beta|G^? \cap C^-|}{\alpha|G^- \cap C^+| + \beta|G^? \cap C^-| + \gamma|G^- \cap C^-| + \delta|G^? \cap C^+|}$$

where $\alpha \geq \beta \geq 0$ and $\gamma \geq \delta \geq 0$.[19]

---

[19] Note that $d_{\alpha,\beta,\gamma,\delta}(G,C)$ is similar to the measure $d_{\alpha,\beta,\gamma,\delta}(C_*,C)$ of the distance between constituents. Indeed $d_{\alpha,\beta,\gamma,\delta}(G,C)$ can be obtained from $d_{\alpha,\beta,\gamma,\delta}(C_*,C)$ just by replacing C_*^- and C_*^+ with G^- and $G^?$, respectively, and by adding the restrictions $\alpha \geq \beta$ and $\gamma \geq \delta$ – whose meaning was explained above – on the value of the parameters.

Three interesting special cases of $d_{\alpha,\beta,\gamma,\delta}(G,C)$ are the following: (i) for $\alpha = \beta = \gamma = \delta = 1$, we obtain the distance measure $d_\Delta(G,C) = E_{GC} + e_{GC} = |G^- \Delta C^-|/q$ which can be seen a normalized measure, with $1/q$ as a normalizing factor, of the absolute non-existential disagreement between G and C; (ii) for $\alpha = \beta = \gamma = 1$ and $\delta = 0$, we obtain the distance measure $d_n(G,C) = (E_{GC} + e_{GC})/(E_{GC} + e_{GC} + S_{GC}) = |G^- \Delta C^-|/|G^- \cup C^-|$ which can be construed as a measure of the relative non-existential disagreement between G and C; (iii) for $\alpha = \gamma = 1$ and $\beta = \delta = 0$, we obtain $d_1(G,C) \equiv E_{GC}/(E_{GC} + S_{GC})$ (see (17)).

Given a distance measure $d(G,C)$, the proximity, or similarity, $s(G,C)$ between G and C can be defined as $s(G,C) = 1 - d(G,C)$. For instance, the similarity measures corresponding to d_1 is:

$$(22) \quad s_1(G,C) = \frac{S_{GC}}{S_{GC} + E_{GC}} = \frac{S_{GC}}{cont(G)} = \frac{|G^- \cap C^-|}{|G^-|}.^{20}$$

Suppose that C is 'the truth' about D; then $d(G,C)$ and $s(G,C)$ measure the *distance from the truth* of G and its *verisimilitude*, respectively.

Although for our purposes we only need a measure $d(G,C)$ applicable to universal generalizations, it does not seem too difficult to extend our class of measures $d_{\alpha,\beta,\gamma,\delta}(G,C)$ to any kind of generalization. Indeed, an extended distance measure $d(G,C)$ might be defined in terms of the (lack of) overlap between the partitions $\{G^+, G^-, G^?\}$ and $\{C^-, C^+\}$, by considering, in addition to $G^- \cap C^+$, $G^- \cap C^-$, $G^? \cap C^-$ and $G^? \cap C^+$, also the sets $G^+ \cap C^-$ and $G^+ \cap C^+$, corresponding to the *mistaken* and *correct existential claims* of G, respectively.[21]

---

[20] One can see that $s_1(G,C)$ is identical to the degree of truth $t(G,C)$ of G w.r.t. C, defined in (8).

[21] W.r.t. such sets, one may define normalized measures of the *existential error* and the *existential success* of G w.r.t. to C, which can be incorporated in the extended distance measure $d(G,C)$, along the same lines of definition (21).

2.3. A Comparison with Niiniluoto's and Tuomela's Measures of Verisimilitude

The measures for the verisimilitude of generalizations proposed by Ilkka Niiniluoto and Raimo Tuomela reveal some interesting conceptual relations with our measures $d_{\alpha,\beta,\gamma,\delta}(G,C)$. While Niiniluoto makes use of a KW-strategy, Tuomela adopts a KI-strategy. For this reason we will consider first Tuomela's measures.

Tuomela's measures for the verisimilitude of generalizations. Tuomela (1978) proposes a class of measures of distance applicable to any couple G_1 and G_2 of generalizations of a first-order language. Such measures are defined, to use our terminology, in terms of the relations between the q-constituents associated to G_1 and G_2 or, equivalently, of the relations between the triples $\{G_1^+, G_1^-, G_1^?\}$ and $\{G_2^+, G_2^-, G_2^?\}$. More precisely, Tuomela (ib., 217) claims that $d(G_1, G_2)$ should be proportional to two factors:

(A) The first factor is given by the sizes of the symmetric differences $G_1^+ \Delta G_2^+$ and $G_1^- \Delta G_2^-$ between the instantiated areas as well as between the empty areas of G_1 and G_2: $G_1^+ \Delta G_2^+$ includes those kinds of individuals which G_1 claims to exist and G_2 does not claim to exist (or vice versa) while $G_1^- \Delta G_2^-$ includes the kinds G_1 denies and G_2 does not (or vice versa);

(B) The second factor is given by the amount of errors G_1 and G_2 make with respect to each other. Such errors may belong to one of the two following sorts: (i) the errors related to the Q-predicates in $G_1^+ \cap G_2^-$, i.e., to the cells of **Q** that G_1 claims to be instantiated while G_2 claims to be empty; (ii) the errors related to the Q-predicates in $G_1^- \cap G_2^+$, i.e., to the cells of **Q** that G_1 claims to be empty while G_2 claims to be instantiated.

The above two factors are taken into account by the class of distance measures $d_T(G_1, G_2)$ tentatively suggested by Tuomela (ibid.):

$$d_T(G_1,G_2) =$$

(23)
$$= \gamma \frac{\left|G_1^+ \cap G_2^-\right| \cup \left|G_1^- \cap G_2^+\right|}{q} + (1-\gamma)\left(\beta \frac{\left|G_1^+ \Delta G_2^+\right|}{\left|G_1^+ \cup G_2^+\right|} + (1-\beta)\frac{\left|G_1^- \Delta G_2^-\right|}{\left|G_1^- \cup G_2^-\right|}\right)$$

where $0 \leq \gamma \leq 1$ and $0 \leq \beta \leq 1$.

As far as "the specific form" of $d_T(G_1,G_2)$ is concerned, Tuomela (1978, 218) honestly remarks: "Why not just say that [the distance between G_1 and G_2] is an increasing function of the [above components (A) and (B)] and thus define a comparative notion of distance? Well, why not, but I have ventured a little more and then [(23)] simply gives my proposal." Indeed we feel that the specific form of $d_T(G_1,G_2)$ is unnecessarily baroque. This appears when one considers the special case of the distance $d_T(C_*,C)$ between two constituents which is given by:

(24) $$d_T(C_*,C) = \gamma \frac{\left|C_*^+ \cap C^+\right|}{q} + (1-\gamma)\left(\beta \frac{\left|C_*^+ \Delta C^+\right|}{\left|C_*^+ \cup C^+\right|} + (1-\beta)\frac{\left|C_*^+ \Delta C^+\right|}{\left|C_*^- \cup C^-\right|}\right)$$

$$= \gamma\, d_\Delta(C_*,C) + (1-\gamma)\,[\beta\, d_e(C_*,C) + (1-\beta)\, d_n(C_*,C)].^{22}$$

One can see that $d_T(C_*,C)$ depends in three different ways on the symmetrical difference $C_*^+ \Delta C^+$ and can be expressed as a function of the three distance measures given in (10)–(12), i.e., the measure of absolute disagreement $d_\Delta(C_*,C)$, the measure of relative existential disagreement $d_e(C_*,C)$, and the measure of relative non-existential disagreement $d_n(C_*,C)$.

Given a universal generalization G, equalities $G^- \Delta C^- = (G^- \cap C^+) \cup (C^- \cap G^+)$ and $G^- \cup C^- = (G^- \cap C^+) \cup (C^- \cap G^+) \cup (G^- \cap C^-)$ hold. Due to these equalities it follows from (23) – by recalling the definitions of E_{GC}, e_{GC}, and S_{GC} – that the distance $d_T(G,C)$ between a universal generalization G and a constituent C is given by:

---

[22] Cf. Niiniluoto (1987, p. 320) who kindly attributes equality (24) to Roberto Festa (manuscript, 1982).

(25) $$d_T(G,C) = \gamma \frac{|G^- \cap C^+|}{q} + (1-\gamma)\left(\beta + (1-\beta)\frac{|G^- \Delta C^-|}{|G^- \cup C^-|}\right)$$

$$= \gamma E_{GC} + (1-\gamma)\left(\beta + (1-\beta)\frac{E_{GC} + e_{GC}}{E_{GC} + e_{GC} + S_{GC}}\right)$$

$$= \gamma E_{GC} + (1-\gamma)[\beta + (1-\beta) d_n(G,C)]$$

where $d_n(G,C)$ is a measure of the relative non-existential disagreement between G and C (see below definition (12)).

Although the measures of distance $d_T(G_1, C)$ are qualitatively similar to $d_{\alpha,\beta,\gamma,\delta}(G,C)$, it seems to me that they are intuitively much less perspicuous than my measures of distance.

Niiniluoto's measures for the verisimilitude of generalizations. Niiniluoto's measures $d(G,C)$ for the verisimilitude of generalizations are worked out on the basis of a KW-strategy. More precisely, $d(G,C)$ is defined in terms of the distances $d(C_i,C)$ between the constituents C_i in $dnf(G)$ and C. According to Niiniluoto (1987, pp. 215-216), the following *min-max* and *min-sum* distance functions are plausible enough as a distance measure $d(G,C)$:

(26) $\Delta_{mm}^{\gamma}(G,C) = \gamma \Delta_{min}(G,C) + (1-\gamma) \Delta_{max}(G,C)$

(27) $\Delta_{ms}^{\gamma\gamma'}(G,C) = \gamma \Delta_{min}(G,C) + \gamma' \Delta_{sum}(G,C)$

where $0 \leq \gamma \leq 1$ and $0 \leq \gamma' \leq 1$.

Here $\Delta_{min}(G,C)$ and $\Delta_{max}(G,C)$ are the minimum and the maximum distance between the constituents C_i in $dnf(G)$ and C, while $\Delta_{sum}(G,C)$ is a suitably normalized sum of the distances between such constituents and C.

Although the KW-strategy adopted by Niiniluoto is conceptually very different from the KI-strategy adopted by Tuomela and me, it is interesting to notice that, when G is a q-constituent, Niiniluoto's measures $d(G,C)$ can be restated in terms of the (lack of) overlap between the partitions $\{G^+, G^-, G^?\}$ and $\{C^+, C^-\}$. This possibility depends on the already mentioned circumstance that, in the case of a q-constituent G, C_i is a disjunct of $dnf(G)$ if and only if G^+

$\subseteq C_i^+ \subseteq \mathbf{Q} - G^-$. Let us consider, for instance, the min-max distance measure $\Delta^\gamma_{mm}(G,C)$. For a q-constituent G, $\Delta^\gamma_{mm}(G,C)$ can be rewritten as follows:[23]

$$(28) \quad \Delta^\gamma_{mm}(G,C) = \gamma \frac{|G^+ \cap C^-| + |G^- \cap C^+|}{q} + (1-\gamma)\frac{|G^?|}{q}.$$

In the special case where the q-constituent G is a universal generalization, equalities $|G^+ \cap C^-| = \emptyset$ and $|G^?|/q = 1 - cont(G)$ hold.[24] Hence, by recalling that $E_{GC} = |G^- \cap C^+|/q$ (see (14)), equality (28) can be rewritten as follows:

$$(29) \quad \begin{aligned}\Delta^\gamma_{mm}(G,C) &= \gamma \frac{|G^- \cap C^+|}{q} + (1-\gamma)(1 - cont(G)) \\ &= \gamma E_{GC} + (1-\gamma)(1 - cont(G)).\end{aligned}$$

One can see from (29) that the distance $\Delta^\gamma_{mm}(G,C)$ between a universal generalization G and C can be expressed as an increasing function of the predictive error E_{GC} of G and a decreasing function of its informative content $cont(G)$. This means that $\Delta^\gamma_{mm}(G,C)$ has the same qualitative features and, thereby, the same limitations as our measure $d_1(G,C)$ (see (17)).[25]

3. The Statistical Verisimilitude of Qualitative Theories (Q-theories)

In social sciences the investigated population D is very often described by a monadic language L with two or more families of primitive predicates. In many cases the initially available data show that all the Q-predicates are instantiated in D and, thereby, that 'the truth' about D is given by the trivial constituent C_T. This implies that any (non-tautological) universal generalization is completely false, i.e., that no 'qualitative structure' can be found in D. In such cases, in which evaluating the verisimilitude of universal generalizations is not very

---

[23] See Niiniluoto (1987, p. 337).
[24] This equality follows from $|G^?| = q - |G^-|$ and the definition of $cont(G)$ (see (7)).
[25] This applies also to the min-sum measure $\Delta^{\gamma\gamma'}_{ms}(G,C)$; see Niiniluoto (1987, p. 337).

interesting,[26] a more attractive goal consists in identifying the 'statistical structure' of D. This might mean, for instance, identifying the relative frequency of any Q-predicate in D, or establishing whether the distribution of the relative frequencies of the Q-predicates in D has, or has not, certain significant features. Without denying the importance of such inferences, investigated by the 'traditional' statistical analysis of cross classified populations, we think that the statistical structure of a cross classified population can be captured, in some respect, also by qualitative theories. Below, in Section 3.1, we will see how the qualitative structure of cross classified populations can be described by qualitative laws and qualitative theories (Q-theories). In Section 3.2 we will then recall some basic statistical concepts about cross classification and we will shortly explain the idea, originally introduced within a statistical approach known as prediction logic, that Q-theories can be applied to the analysis of cross classified populations. Finally, in Section 3.3 we will introduce appropriate measures of the statistical verisimilitude of a Q-theory G, i.e., of the similarity between G and the 'statistical truth' about D.

3.1. Q-theories. The Qualitative Structure of Cross Classified Populations

Suppose that the individuals of a population D are described by a monadic language L with two families of predicates $X \equiv \{x_1, ..., x_j, ..., x_c\}$ and $Y \equiv \{y_1, ..., y_i, ..., y_r\}$, where the predicates of each family are mutually exclusive and jointly exhaustive.[27] The family of predicates $\mathbf{Q} \equiv Y \times X \equiv \{(y_1 x_1), (y_1 x_2), ..., (y_2 x_1), ..., (y_i x_j), ..., (y_r x_c)\}$ will include $q \equiv rc$ Q-predicates $y_i x_j$. Below a Q-predicate $y_i x_j$ will be denoted with 'Q_{ij}', so that $\mathbf{Q} = \{Q_{11}, ..., Q_{ij}, ..., Q_{rc}\}$. In an empirical inquiry on the cross classified population D, researchers aim very often to discover the qualitative relations between X and Y. To such a purpose they typically consider *qualitative laws* – or *Q-laws* – L_{y_i} and L_{x_j} of the following form:

---

[26] For instance, it is easy to check that, if C_T is 'the truth' and the similarity function $s_{\alpha,\beta,\gamma,\delta} = 1 - d_{\alpha,\beta,\gamma,\delta}$ is used, then the most verisimilar universal generalization is the tautological universal generalization G_T.

[27] The subscripts 'c' in 'x_c' and 'r' in 'y_r' stay for 'column' and 'row', w.r.t. the tables used to represent cross classifications; see Fig. 1 below.

(30) (a) $L_{y_i} \equiv (u)(y_i(u) \supset D_{y_i}(u))$
(b) $L_{x_j} \equiv (u)(x_j(u) \supset D_{x_j}(u))$

where $D_{y_i}(u)$ ($D_{x_j}(u)$) is a 'disjunctive predicate' formed by the disjunction of (one or) more members of X (Y). For instance, we might have $D_{y_1}(u) \equiv x_2(u) \vee x_3(u)$, in which case L_{y_i}, for $i = 1$, amounts to $(u)(y_1(u) \supset (x_2(u) \vee x_3(u)))$ which tells that all the y_1-individuals have one of the properties x_2 and x_3. A Q-law L_{y_i} (L_{x_j}) of the type shown in (30) will be called L_Y-law (L_X-law), since L_{y_i} (L_{x_j}) predicts the X-state (Y-state) of individuals characterized by the Y-state y_i (x_j).

It should be pointed out that any L_Y-law (L_X-law) is a universal generalization of L, since it only says that certain cells of **Q** are empty, while it does not say anything about the remaining cells. For instance, the L_Y-law L_{y_i} in (30)(a) only says that any Q-cell $Q_{ij} \equiv y_i x_j$, where $x_j(u)$ is not a disjunct of $D_{y_i}(u)$, is empty.

The qualitative relations occurring between X and Y in D can be described by using one or more Q-laws. In particular, a conjunction of r L_Y-laws L_{y_1}, ..., L_{y_r} (c L_X-laws L_{x_1}, ..., L_{x_c}) will be called *qualitative theory*, or *Q-theory*. This definition includes the case in which one or more L_Y-laws (L_X-laws) are tautological. Hence we might say that any conjunction of m L_Y-laws L_{y_1}, ..., L_{y_r}, with $0 \le m \le r$ (n L_X-laws L_{x_1}, ..., L_{x_c}, with $0 \le n \le c$) is a Q-theory, so that also a single L_Y-law (L_X-law) is a Q-theory. Given a Q-theory $G \equiv L_{y_1} \wedge ... \wedge L_{y_r}$, we will say that $\equiv L_{y_1} \wedge ... \wedge L_{y_r}$ is the *conjunctive L_Y-form* of G. It is easy to see that any Q-theory G can be stated both in a conjunctive L_Y-form and in a *conjunctive L_X-form* $L_{x_1} \wedge ... \wedge L_{x_c}$. Finally, it is important to notice that a Q-theory G is a universal generalization of L, where the empty cells in G^- are given by the union of the sets of empty cells specified by the Q-laws occurring in the conjunctive L_Y-form (L_X-form) of G. This means that all the measures $d(G,C)$ for the distance between a universal generalization G and a constituent C can be applied also to Q-theories.

The description of a population D generated by a monadic language L with two families of predicates $X \equiv \{x_1, ..., x_p, ..., x_c\}$ and $Y \equiv \{y_1, ..., y_i, ..., y_r\}$ leads, via the set of Q-predicates $\mathbf{Q} \equiv Y \times X$, to a cross classification of the individuals of D which can be graphically represented by a $r \times c$ table. For instance, given the characters $X \equiv \{x_1, x_2, x_3\}$ and $Y \equiv \{y_1, y_2, y_3\}$, the family **Q** $\equiv Y \times X = \{Q_{11}, ..., Q_{33}\}$ can be represented by the following 3×3 table where any cell corresponds to a Q-predicate:

	x_1	x_2	x_3
y_1	Q_{11}	Q_{12}	Q_{13}
y_2	Q_{21}	Q_{22}	Q_{23}
y_3	Q_{31}	Q_{32}	Q_{33}

Fig. 1. **Q** ≡ {Q_{11}, ..., Q_{33}}

A constituent C, which specifies which cells are empty and which are not, can be graphically represented by shading the empty cells, and by putting a sign '+' into the instantiated ones. For instance, the constituent represented in Fig. 2 tells that the empty cells of **Q** are Q_{11}, Q_{12}, and Q_{21}.

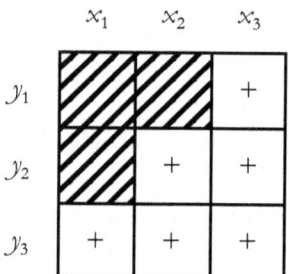

Fig. 2. The constituent C

We can also give a graphical representation of Q-laws and Q-theories. For instance, the Q-theory $G \equiv L_{y_1} \wedge L_{y_2} \wedge L_{y_3}$, where $L_{y_1} \equiv (u)(y_1(u) \supset x_3(u))$, $L_{y_2} \equiv (u)(y_2(u) \supset (x_2(u) \vee x_3(u)))$, and $L_{y_3} \equiv (u)(y_3(u) \supset (x_1(u) \vee x_2(u) \vee x_3(u)))$, is represented in the following figure:

Verisimilitude, Qualitative Theories, and Statistical Inferences

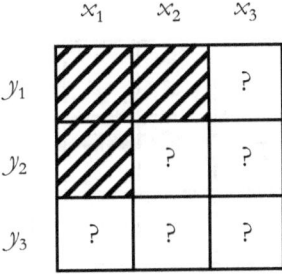

Fig. 3. The Q-theory G

where the empty cells specified by each Q-law are shaded, and a question mark is put in the remaining cells.[28]

3.2. The Statistical Structure of Cross Classified Populations

Some basic statistical concepts about cross classification. Given a population D and the family of Q-predicates $\mathbf{Q} \equiv Y \times X \equiv \{Q_{11}, ..., Q_{ij}, ..., Q_{rc}\}$, the statistical structure of D is expressed by the probability vector $\mathbf{P} \equiv (\mathbf{P}_{11}, ..., \mathbf{P}_{ij}, ..., \mathbf{P}_{rc})$, where $\mathbf{P}_{ij}$ is the relative frequency of Q_{ij} in D, i.e., the probability that an individual randomly drawn from D has the property Q_{ij}. A value $P \equiv (P_{11}, ..., P_{ij}, ..., P_{rc})$ of $\mathbf{P}$ may be represented by a so-called $r \times c$ contingency table where any cell Q_{ij} is filled with P_{ij}. For instance, given the family $\mathbf{Q} \equiv \{Q_{11}, ..., Q_{33}\}$, represented in Fig. 1, the 3×3 contingency table for $P \equiv (P_{11}, ..., P_{33})$ is the following:

---

[28] It should be pointed out that L_{y_3} is a tautological law, since the disjunctive predicate $x_1(u) \lor x_2(u) \lor x_3(u)$ applies to any individual; hence in the row for L_{y_3} of Fig. 3 one finds only question marks.

	x_1	x_2	x_3	
y_1	P_{11}	P_{12}	P_{13}	$P_{1.}$
y_2	P_{21}	P_{22}	P_{23}	$P_{2.}$
y_3	P_{31}	P_{32}	P_{33}	$P_{3.}$
	$P_{.1}$	$P_{.2}$	$P_{.3}$	

Fig. 4. The contingency table for $P \equiv (P_{11}, ..., P_{33})$

$P_{11}, ..., P_{33}$ are the *cell probabilities* of P while the probabilities $P_{1.}, P_{2.}, P_{3.}$ and $P_{.1}, P_{.2}, P_{.3}$, shown in the margins of the table, are called *marginal probabilities*. Any marginal probability indicates the (unconditional) probability that an individual in D lies in the respective row or column; for example, $P_{1.}$ is the probability that an individual of D is in the state y_1;[29] more generally, $P_{i.}$ ($P_{.j}$) is the probability that an individual randomly drawn from D will be y_i (x_j). The probability vectors $P_X \equiv (P_{.1}, P_{.2}, P_{.3})$ and $P_Y \equiv (P_{1.}, P_{2.}, P_{3.})$, corresponding to the states X and Y, will be called *marginal vectors*. $P_{i/j}$ ($P_{j/i}$) is the *conditional probability* that the next x_j-individual (y_i-individual) randomly drawn from D will be y_i (x_j). The following relations between cell probabilities P_{ij}, marginal probabilities $P_{i.}$ and $P_{.j}$, and conditional probabilities $P_{i/j}$ and $P_{j/i}$ should be recalled: $P_{i/j} = P_{ij}/P_{.j}$, $P_{j/i} = P_{ij}/P_{i.}$ and $P_{ij} = P_{.j} P_{i/j} = P_{i.} P_{j/i}$.

We will say that Y and X are *statistically independent* w.r.t. P – or, equivalently, that P is an *independence distribution* – in the case that, for any y_i and x_j, $P_{ij} = P_{.j} P_{i.}$. Given a statistical distribution P, with marginal vectors P_Y and P_X, the *independence counterpart* of P, i.e., the independence distribution with the same marginal vectors of P, will be denoted with P^*. This means that, for any y_i and x_j, $P^*_{ij} = P^*_{i.} P^*_{.j} = P_{i.} P_{.j}$. Given a statistical distribution $P \equiv (P_{11}, ..., P_{ij}, ..., P_{rc})$, we will say that Q_{ij} is *P-positive* in the case where $P_{ij} > 0$. Moreover, we will say that P is *typical* in the case where any Q_{ij} is P-positive.

---

[29] The dot in the symbol $P_{1.}$ indicates that the variable X whose index j is omitted has been ignored.

The true, usually unknown, value $P \equiv (P_{11}, ..., P_{ij}, ..., P_{rc})$ of **P** can be seen as 'the statistical truth' about D. The aim of the statistical analysis of a cross classified population D consists in discovering the 'statistical structure' of D, by making appropriate inductive inferences about the true value of **P**. Some widely used types of inductive inferences about **P** are the following: (i) one might estimate **P**, which amounts to making reasonable guesses about the true value of **P**; (ii) given a specific value P of **P** – which, for some reason, is considered especially interesting – one could test the hypothesis '**P** = P'; (iii) one might evaluate the hypothesis that Y and X are statistically correlated or also estimate their 'degree of correlation'. In particular, inferences of the last type concern the *statistical relations* between the characters Y and X.

Prediction logic. Using Q-theories for the statistical analysis of cross classified populations. The so-called *prediction logic* was developed by the statisticians and social scientists David K. Hildebrand, James D. Laing, and Howard Rosenthal – henceforth H-L-R – in a series of papers published between 1974 and 1977, which culminated in their not yet adequately appreciated book *Prediction Analysis of Cross Classification* (1976).[30] H-L-R argue that the 'traditional' measures of association between qualitative variables are far from being a completely appropriate conceptual tool for the statistical analysis of cross classified populations. In this respect, they remark:

> The basic tools for analyzing relations among qualitative variables have consisted of various measures and tests of statistical 'association'. [...] [T]he concept is best described by a double negative: *two or more variables are 'associated' if they are 'not unrelated.'* [...] Knowing that two variables are not unrelated and that they depart from unrelatedness to some specified 'degree' says nothing about the *form* of the relation that does obtain. Ordinary chi-square, [...] for example, is used only to determine whether or not there is *any* relation between the variables. (Hildebrand et al. 1976, pp. 43–44)

This means that the chi-square test and other measures and tests of statistical association do not allow us to evaluate the specific qualitative theories, or predictions, that a researcher could state about the population

---

[30] For an outline of the basic features of prediction logic and its conceptual relations with VNF-methodologies see Festa (forthcoming).

under inquiry. The constructive purpose pursued by H-L-R with their prediction logic is to "develop a general model that yields a specific measure of the success achieved by a particular [Q-theory]" (ib., p. 17). In the terminology of prediction logic, Q-theories are indicated by terms such as 'predictions for qualitative variables', or just 'predictions'. In conflict with the widespread view according to which the methodological analysis of Q-theories is an exercise in armchair methodology with scarce relations to scientific practice, Hildebrand et al. (ib., pp. 3–4) point out that Q-theories play an important methodological role in many sciences, starting from social sciences, and complain that a satisfying statistical analysis of the predictive success, or adequacy, of Q-theories is still missing.

According to H-L-R social scientists are usually convinced that all the (non-tautological) Q-theories proposed in an empirical inquiry are false. However, scientists may accept also presumably false – and even empirically falsified – Q-theories, whenever such theories seem to give a predictively adequate description of the statistical structure of the population under inquiry. In the terminology of the present paper, this amounts to saying that scientists may accept falsified Q-theories in the cases where their estimated statistical verisimilitude is high enough. H-L-R argue (ibid.) that appropriate prediction analysis methods for qualitative variables should fulfill two related tasks: (1) to provide a precise definition of what they call the 'predictive success' of any Q-theory G in the population D under inquiry, as described by the true statistical distribution about D; (2) to establish plausible criteria to estimate the predictive success of any G on the basis of empirical samples randomly drawn from D. While fulfilling task (1) amounts to defining appropriate measures for the statistical verisimilitude of Q-theories, i.e., to solving the logical problem of the statistical verisimilitude of Q-theories, fulfilling task (2) amounts to working out appropriate methods to estimate the statistical verisimilitude of Q-theories, i.e., to solving the epistemic problem of the statistical verisimilitude of Q-theories (see Section 1).

The statistical predictive content of Q-theories. Recalling that the qualitative predictive content of a Q-theory G is defined as $cont(G) \equiv |G^-|/q$ (see (7)), one realizes that $cont(G)$ amounts to the weighted cardinality of G^-, where the weight of any Q_{ij} in G^- is $1/q$:

(31) $$cont(G) \equiv \frac{|G^-|}{q} = \sum_{Q_{ij} \in G^-} \frac{1}{q}.$$

This means that any Q_{ij} in G^- brings the same contribution to $cont(G)$.

A measure of the *statistical predictive content* of G can be obtained by modifying equality (31) so that the weight of any Q_{ij} in G^- depends, as it were, on the 'statistical importance' of Q_{ij}. To this purpose, imagine a situation where the available prior information in an inquiry on D – although not including the knowledge of the true value P of $\mathbf{P}$ – does includes the knowledge of the marginal vectors P_X and P_Y. In lack of any information about the statistical relations between X and Y, it seems reasonable to consider $P_i P_j$ as an appropriate prior estimate of the (unknown) relative frequency P_{ij} of Q_{ij} in D. Now, recalling that, for any $Q_{ij} \in G^-$, G claims that no individual in D is Q_{ij} and, consequently, predicts that no individual sampled from D will be Q_{ij}, it would seem reasonable to assume that the informative value of this prediction increases with the prior estimate $P_i P_j$ of the relative frequency of Q_{ij} in D. Such considerations suggest that we should define the statistical predictive content $cont_P(G)$ of G as the weighted cardinality of G^-, where the weight of any Q_{ij} in G^- is $P_i P_j$:

(32) $$cont_P(G) \equiv \sum_{Q_{ij} \in G^-} P_i P_j .$$ [31]

It appears from (32) that the contribution of any Q_{ij} in G^- to $cont_P(G)$ is equal to the prior estimate $P_i P_j$ of the relative frequency of Q_{ij} in D.[32]

---

[31] One might object that the informative content of a statement should be an 'intrinsic feature' of that statement, and not depend on the actual, usually unknown, state of the world. Hence, the term 'content', applied to $cont_P(G)$, would be inappropriate since $cont_P(G)$ *does depend* on the actual state of the world, as described by the statistical (usually unknown) truth P, via the marginal probabilities P_i, P_j. While we recognize that this objection is meaningful, we think that using 'content' w.r.t. $cont_P(G)$ is not entirely inappropriate since it can be assumed that, before starting an empirical inquiry on the statistical relations between X and Y, researchers, while not knowing the statistical truth P, typically know the actual marginal probabilities P_i and P_j. In any case, one should recognize at least that $cont_P(G)$ can be defined without knowing exactly P but only some, relatively unimportant, features of it.

[32] A nice feature of $cont_P(G)$ is that, in the special case of a uniform distribution P, where $P_{ij} = 1/q$ for any Q_{ij}, $cont_P(G) = cont(G)$.

Statistically common Q-predicates. Below we will show that any statistical distribution $P \equiv (P_{11}, ..., P_{ij}, ..., P_{rc})$ determines a partition $\{P^+, P^-\}$ of Q where P^+ can be interpreted as the set of the *statistically common* Q-predicates and $P^- \equiv \mathbf{Q} - P^+$ as the set of the *statistically rare* Q-predicates. For short, the Q-predicates in P^+ and P^- will be called *P-common* and *P-rare*.

Suppose that an appropriate definition of the degree of commonness $com(Q_{ij})$ of Q_{ij} w.r.t. P is available. For instance, as will be seen below, $com(Q_{ij})$ might be identified with P_{ij}, or with some increasing function of P_{ij}. Then P^+ and P^- are defined as follows:

(33) $P^+ \equiv \{Q_{ij} | com(Q_{ij}) > 0\}; \qquad P^- \equiv \{Q_{ij} | com(Q_{ij}) \leq 0\}.$

The degree of rareness $rar(Q_{ij})$ of Q_{ij} can be quite naturally defined as:

(34) $rar(Q_{ij}) \equiv - com(Q_{ij}).$

From (33) and (34) it follows that:

(35) $P^- = \{Q_{ij} | rar(Q_{ij}) \geq 0\}.$

Given a statistical distribution P, the partition $\{P^+, P^-\}$, defined on the basis of the commonness function *com*, will be called the P_{com}-partition. In the case where $P^- = \emptyset$ and $P^+ = \mathbf{Q}$, i.e, in the case where all the Q-predicates are classified as *P*-common, we will say that the P_{com}-partition $\{P^+, P^-\}$ is *trivial* and that P is trivial w.r.t. *com*, or *com-trivial*. It would be obviously useless to distinguish between *P*-common and *P*-rare Q-predicates on the basis of a commonness function *com* such that *any* typical probability distribution P is com-trivial, so that any Q-predicate is classified as *P*-common. We will say that a commonness function *com* of this kind is trivial.

Above it was suggested that $com(Q_{ij})$ might be identified with P_{ij} or with some increasing function of P_{ij}. However, it should be noticed that the first possibility leads to unpalatable consequences. Indeed, if one adopts the commonness function com_t defined as follows:

(36) $com_t(Q_{ij}) \equiv P_{ij}$

then any *P*-positive Q-predicate belongs to the set P^+ of the *P*-common Q-predicates. This implies that any typical distribution P is com_t-trivial and, therefore, that com_t is trivial.

Fortunately, this unpleasant feature of com_t can be avoided by adopting a commonness function in the following parametric class:

(37) $com_\sigma(Q_{ij}) \equiv P_{ij} - \sigma P_{i.} P_{.j}$ where $0 \leq \sigma \leq 1$.

From (34) and (37) it follows that $rar_\sigma(Q_{ij}) = \sigma P_{i.} P_{.j} - P_{ij}$. Hence, the P_σ-partition $\{P^+, P^-\}$, defined on the basis of com_σ, is the following:

(38) $P^+ \equiv \{Q_{ij} | P_{ij} > \sigma P_{i.} P_{.j}\}$; $P^- \equiv \{Q_{ij} | P_{ij} \leq \sigma P_{i.} P_{.j}\}$.

One sees that the P_σ-partition $\{P^+, P^-\}$ defined in (24) can be specified also in one of the following two equivalent ways:

(39) (a) $P^+ \equiv \{Q_{ij} | P_{i/j} > \sigma P_{i.}\}$; $P^- \equiv \{Q_{ij} | P_{i/j} \leq \sigma P_{i.}\}$;
(b) $P^+ \equiv \{Q_{ij} | P_{j/i} > \sigma P_{.j}\}$; $P^- \equiv \{Q_{ij} | P_{j/i} \leq \sigma P_{.j}\}$.[33]

That amounts to saying that Q_{ij} is P-rare, w.r.t. com_σ, if and only if the probability $P_{i/j}$ ($P_{j/i}$) that an x_j-individual (y_i-individual) randomly drawn from D is y_i (x_j) is not higher than a fraction σ (with $0 \leq \sigma \leq 1$) of the probability $P_{i.}$ ($P_{.j}$) that an individual randomly drawn from D is y_i (x_j), i.e., if and only if the X-state x_j (the Y-state y_i) reveals a sufficient amount of negative relevance for the Y-state y_i (the X-state x_j), where the threshold of negative relevance is fixed by σ.

The trivial commonness function com_t is equal to the extreme member com_σ^0, with $\sigma = 0$, of com_σ. The other extreme member of com_σ is com_σ^1, with $\sigma = 1$,

(40) $com_\sigma^1(Q_{ij}) = P_{ij} - P_{i.} P_{.j}$

Statistically true Q-theories. Above it was pointed out that, given the true constituent C, a Q-theory G is true if and only if $G^- \subseteq C^-$. The notion of statistically true Q-theory can be defined in a similar way. Indeed, given the statistical truth P and the corresponding P_{com}-partition $\{P^+, P^-\}$, defined on the basis of the commonness function *com*, we will say that G is *statistically true* - for short: *s-true* - if and only if $G^- \subseteq P^-$; otherwise, we will say that G is *statistically false*, or *s-false*.[34]

---

[33] This immediately follows from the equalities $P_{i/j} = P_{ij}/P_{.j}$, $P_{j/i} = P_{ij}/P_{i.}$ and $P_{ij} = P_{.j} P_{i/j} = P_{i.} P_{j/i}$.
[34] It should be noted that any true theory is s-true, while the reverse does not hold.

The above definition of the notion of an s-true Q-theory can be justified as follows. Recalling that, for any $Q_{ij} \in G^-$, G makes the (non-existence) claim, or prediction, that Q_{ij} *is not* instantiated in D, one might say that such a claim is statistically true (s-true) in case Q_{ij} is statistically rare in D, and statistically false (s-false) otherwise. Then the above definition of 's-true Q-theory' amounts to saying that G is s-true if and only if all its claims are s-true.[35]

Given a s-false Q-theory G, we will say that G is *completely s-false* when $G^- \subseteq P^+$, i.e., when all its claims are s-false.[36] Finally, we will use also the fuzzy notion of an almost s-true Q-theory, by saying that the s-false Q-theory G is *almost s-true* when almost all its claims are s-true, i.e., when almost all the cells of G^- belong to P^-.

The statistical verisimilitude of Q-theories. Suppose that P is 'the statistical truth' about D. Then, given a Q-theory G, its *distance from the statistical truth* can be identified with a suitable measure $d(G,P)$ of the distance of G from P. Similarly, the *statistical verisimilitude* of G can be identified with a suitable measure $s(G,P)$ of the similarity between G and P. Below, we will see how the measures $d(G,P)$ and $s(G,P)$ can be defined.

The distance $d(G,P)$ between a Q-theory G and the statistical distribution P could be defined along the same lines followed above in defining the distance $d(G,C)$ between G and a constituent C. More precisely, $d(G,P)$ could be defined in terms of the sets $G^- \cap P^-$, $G^- \cap P^+$, $G^? \cap P^-$, and $G^? \cap P^+$. However, one should not hope that the definition of $d(G,P)$ can be obtained from that of $d_{\alpha,\beta,\gamma,\delta}(G,C)$ in (21) just by replacing everywhere C^+ and C^- with P^+ and P^-, respectively. This depends on the fact that, as will be seen below, the 'statistical analogues' E_{GP}, S_{GP}, e_{GP}, and s_{GP} of the measures $E_{GC} \equiv |G^- \cap C^+|/q$, $S_{GC} \equiv |G^- \cap C^-|/q$, $e_{GC} \equiv |G^? \cap C^-|/q$ and $s_{GC} \equiv |G^? \cap C^+|/q$ of the errors and success of G cannot be obtained just by replacing C^+ and C^- with P^+ and P^-.

---

[35] If one does not like the notion of statistically true Q-theory and finds that 'G is s-true' is a problematic statement, one could introduce the 'statistical' statement $G° \equiv$ 'The Q-predicates in G^- are P-rare', and interpret 'G is s-true' as a handful reformulation of the '$G°$ is true'.

[36] It should be noticed that, if the statistical truth P is trivial, then the only s-true Q-theory is the tautological Q-theory G_T while all the Q-theories different from G_T are completely s-false.

Let us consider, first of all, a basic requirement which should be satisfied by any adequate measure E_{GP} of the *statistical predictive error* of G w.r.t. P. Recalling that, for any $Q_{ij} \in G^- \cap P^+$, G makes the s-false claim that Q_{ij} is not instantiated in D, it seems reasonable to assume that the *seriousness* of the error made by such claim increases with the commonness $com(Q_{ij})$ of Q_{ij}. This requirement is satisfied by the following definition:

(41) $\quad E_{GP} \equiv \sum_{Q_{ij} \in G^- \cap P^+} com(Q_{ij}).$

Secondly, consider the measure S_{GP} of the *statistical predictive success* of G w.r.t. P. Recalling that, for any $Q_{ij} \in G^- \cap P^-$, G makes the s-true claim that Q_{ij} is not instantiated in D, it seems reasonable to assume that the *importance* of the success related to such claim increases with the rareness $rar(Q_{ij})$ of Q_{ij}. This requirement is satisfied by the following definition:

(42) $\quad S_{GP} \equiv \sum_{Q_{ij} \in G^- \cap P^-} rar(Q_{ij}) = - \sum_{Q_{ij} \in G^- \cap P^-} com(Q_{ij}).$

Thirdly, consider the measure e_{GP} of the *statistical permissive error* of G w.r.t. P. Recalling that, for any $Q_{ij} \in G^? \cap P^-$, G does not make any claim about the emptiness of Q_{ij} in D, it seems reasonable to assume that the *seriousness* of this mistaken permission increases with the rareness $rar(Q_{ij})$ of Q_{ij}. This requirement is satisfied by the following definition:

(43) $\quad e_{GP} \equiv \sum_{Q_{ij} \in G^? \cap P^-} rar(Q_{ij}) = - \sum_{Q_{ij} \in G^? \cap P^-} com(Q_{ij})$

Finally, consider the measure s_{GP} of the *statistical permissive success* of G w.r.t. P. Recalling that, for any $Q_{ij} \in G^? \cap P^+$, G permits the existence of Q_{ij}-individuals in D, it seems reasonable to assume that the *importance* of the success related to such a correct permission increases with the commonness $com(Q_{ij})$ of Q_{ij}. This requirement is satisfied by the following definition:

(44) $\quad s_{GP} \equiv \sum_{Q_{ij} \in G^? \cap P^+} com(Q_{ij}).$

The statistical total error TE_{GP} and the statistical total success TS_{GP} of G w.r.t. P can be defined in terms of E_{GP}, S_{GP}, e_{GP}, and s_{GP}, along the same lines as their 'qualitative counterparts' TE_{GC} and TS_{GC} (see (19) and (20)):

(45) $\quad TE_{GP} \equiv \alpha E_{GP} + \beta e_{GP}; \qquad TS_{GP} \equiv \gamma S_{GP} + \delta s_{GP}$

$$\text{where } \alpha \geq \beta \geq 0 \text{ and } \gamma \geq \delta \geq 0.$$

Now we can introduce a parametric class of measures which represents the statistical analogues of the class $d_{\alpha,\beta,\gamma,\delta}(G,C)$ (see (21)):

(46) $\quad d_{\alpha,\beta,\gamma,\delta}(G,P) = \dfrac{TE_{GP}}{TE_{GP} + TS_{GP}} = \dfrac{\alpha E_{GP} + \beta e_{GP}}{\alpha E_{GP} + \beta e_{GP} + \gamma S_{GP} + \delta s_{GP}}$

$$= \dfrac{\alpha \sum_{Q_{ij} \in G^? \cap P^+} com(Q_{ij}) - \beta \sum_{Q_{ij} \in G^? \cap P^-} com(Q_{ij})}{\alpha \sum_{Q_{ij} \in G^- \cap P^+} com(Q_{ij}) - \beta \sum_{Q_{ij} \in G^? \cap P^-} com(Q_{ij}) - \gamma \sum_{Q_{ij} \in G^- \cap P^-} com(Q_{ij}) + \delta \sum_{Q_{ij} \in G^? \cap P^+} com(Q_{ij})}.$$

It is easy, although rather tedious, to calculate the value of $d_{\alpha,\beta,\gamma,\delta}(G,P)$ on the basis of any specific com_σ, in accordance with the definition (37) of com_σ.

Given a distance measure $d(G,P)$, the proximity, or similarity, $s(G,P)$ between G and P can be defined as $s(G,P) \equiv 1 - d(G,P)$. Suppose that P is 'the statistical truth' about D; then $d(G,P)$ and $s(G,P)$ measure the *distance from the statistical truth* and the *statistical verisimilitude* of G, respectively.

Concluding remarks. The above measures $d(G,P)$ and $s(G,P)$ represent a possible solution of the logical problem of the statistical verisimilitude of Q-theories. They provide the starting point for the analysis of the related epistemic problem, that is, for the design of appropriate methods for estimating the distance from the statistical truth and the statistical verisimilitude of Q-theories. For instance, a plausible Bayesian method to estimate the statistical verisimilitude of a Q-theory G is the following. Given a prior distribution of epistemic probabilities F on the probability vector $\mathbf{P} \equiv (\mathbf{P}_{11}, ..., \mathbf{P}_{ij}, ..., \mathbf{P}_{rc})$ associated to $\mathbf{Q} \equiv \{Q_{11}, ..., Q_{ij}, ..., Q_{rc}\}$, the expected statistical verisimilitude $ES(G|e)$ of G – i.e., the expected value of the statistical verisimilitude $s(G,P)$ of G, calculated w.r.t. the available evidence e – may be considered a reasonable estimate of the statistical verisimilitude of G. In spite of the conceptual plausibility of this method, we have to admit that the calculation of $ES(G|e)$ for our measures of statistical verisimilitude does not seem a easy task. However, a relatively more simple procedure to estimate the statistical verisimilitude of G can be tentatively suggested. Let $P^e \equiv (P^e_{11}, ..., P^e_{ij}, ..., P^e_{rc})$ be the estimated value of **P** – for instance P^e_{ij} might be the expected value $E(\mathbf{P}_{ij}|e)$ of $\mathbf{P}_{ij}$. Then the estimated statistical verisimilitude $EstS(G|e)$ of G might be identified with the similarity between G and the estimated value of **P**: $EstS(G|e) \equiv s(G,P^e)$.

References

Festa, R. 1983: "Epistemic Utilities, Verisimilitude, and Inductive Acceptance of Interval Hypotheses", in *Abstracts of the 7th International Congress of Logic, Methodology and Philosophy of Science*, Vol. I, Saltzburg, pp. 212–215.

Festa, R. 1986: "A Measure for the Distance Between an Interval Hypothesis and the Truth", *Synthese* 67, 273–320.

Festa, R. 1987: "Theory of Similarity, Similarity of Theories, and Verisimilitude", in T. A. F. Kuipers (ed.) (1987), pp. 145–176.

Festa, R. 1993: *Optimum Inductive Methods. A Study in Inductive Probabilities, Bayesian Statistics, and Verisimilitude*, Kluwer, Dordrecht.

Festa, R. 1995: "Verisimilitude, Disorder, and Optimum Prior Probabilities", in T. A. F. Kuipers and A. R. Mackor (eds.), *Cognitive Patterns in Science and Common Sense*, Rodopi, Amsterdam, pp. 299–320.

Festa R. forthcoming: "Verisimilitude, Cross Classification, and Prediction Logic. Approaching the Statistical Truth by Falsified Qualitative Theories", *Mind and Society*.

Hildebrand, D. K., J. D. Laing, H. Rosenthal 1976: *Prediction Analysis of Cross Classification*, John Wiley and Sons, New York – London.

Kuipers, T. A. F. (ed.) 1987: *What is Closer-to-the-truth?*, Rodopi, Amsterdam.

Kuipers, T. A. F. 2000: *From Instrumentalism to Constructive Realism*, Kluwer, Dordrecht.

Lakatos, I. 1974: "Popper on Demarcation and Induction", in P. A. Schilpp (ed.), *The Philosophy of Karl Popper*, Open Court, La Salle, Illinois, pp. 241–273.

Maher, P. 1993: *Betting on Theories*, Cambridge University Press, Cambridge.

Miller, D. 1974: "Popper's Qualitative Theory of Verisimilitude", *The British Journal for the Philosophy of Science* 25, 166–177.

Niiniluoto I. 1982a: "What Shall We Do with Verisimilitude?", *Philosophy of Science* 49, 181–197.

Niiniluoto I. 1982b: "Truthlikeness for Quantitative Statements", in P. D. Asquith and T. Nickles (eds.), *PSA 1982*, Vol. 1, Philosophy of Science Association, East Lansing, pp. 208–216.

Niiniluoto, I. 1986: "Truthlikeness and Bayesian Estimation", *Synthese* 67, 321–346.

Niiniluoto, I. 1987: *Truthlikeness*, Reidel, Dordrecht.
Niiniluoto, I. 1989, "Corroboration, Verisimilitude, and the Success of Science", in K. Gavroglu, Y. Goudaroulis, and P. Nicolacopoulos (eds.), *Imre Lakatos and Theories of Scientific Change*, Reidel, Dordrecht, pp. 229–243.
Niiniluoto, I. 1998: "Verisimilitude: The Third Period", *The British Journal for the Philosophy of Science* 49, 11–29.
Oddie, G. 1986: *Likeness to the Truth*, Reidel, Dordrecht.
Popper, K. R. 1934/1959: *Logik der Forschung*, Vienna. Translated as: *The Logic of Scientific Discovery*, Hutchinson, London.
Popper, K. R. 1963: *Conjectures and Refutations*, Routledge and Kegan Paul, London.
Popper, K. R. 1972: *Objective Knowledge*, Clarendon Press, Oxford.
Tichý, P. 1974: "On Popper's Definition of Verisimilitude", *The British Journal for the Philosophy of Science* 25, 155–160.
Tuomela, R. 1978: "Verisimilitude and Theory-Distance", *Synthese* 38, 213–246.

Theo A. F. Kuipers

The Hypothetico-Probabilistic (HP)-Method as a Concretization of the HD-Method

Abstract. As Ilkka Niiniluoto happens to have suggested already in 1973, analogous to the Hypothetico-Deductive (HD-) method of testing hypotheses, a Hypothetico-Probabilistic (HP-) method can be developed, with the HD-method as an extreme special case. It amounts to 'deriving' probabilistic test implications I from a hypothesis H, in the sense of (relatively) observational statements I satisfying 'positive relevance' for H, i.e., $p(I/H) > p(I)$. When such probabilistic predictions come true, H is (said to be) probabilistically confirmed, otherwise H is probabilistically disconfirmed. In this paper I will first elaborate the HP-method of testing as a concretized version of the HD-method.

Elsewhere, I have argued that the comparative evaluation of two hypotheses, based on a repeated application of the HD-method, even when both have already been falsified, is functional for achieving empirical progress, with articulated perspectives on truth approximation. In this paper I argue along the same lines that the HP-method is functional for achieving (a concretized version of) empirical progress, at least as far as general probabilistic test implications of deterministic theories are concerned. Analysis of the perspectives on truth approximation has to be postponed to another occasion.

1. Introduction

By mid 1999 I completed a chain of argumentation (Kuipers 2000) for the claim that a sophisticated form of the Hypothetico-Deductive Method, the HD-method, was functional for achieving empirical progress and even truth approximation. The HD-method was put at work as a method of separate and comparative evaluation of hypotheses, rather than as merely a method of testing for their truth-value. In view of my earlier work on probabilistic confirmation, the natural question was whether this highly idealized HD-story could be 'concretized' to a HP-story, that is, a story in terms of something like a Hypothetico-Probabilistic Method, a HP-method. Since mid 2003 I have been working on it from time to time, with many problems, in particular with respect to the final stages of the argument. However, up to describing HP-testing and separate and comparative HP-evaluation, the main lines were soon clear to me.

It was a great surprise to (re-)discover recently, just by seeing a reference in a manuscript of Ilkka Niiniluoto (2007), that he has already introduced in the early 1970s the idea of a Hypothetico-Inductive Method, the HI-method. In fact, Niiniluoto and Tuomela (1973) is devoted to its crucial notion of a hypothetico-inductive inference. According to the *Preface*, the last chapter, entitled "Towards a non-inductivist logic of induction", is written by Niiniluoto, apart from the 5th, and final, section 'Conjectures', being a joint piece of work. I quote from p. 209 (section 2): "Thus, one can develop a hypothetico-inductive method of testing hypotheses which is based, for example, upon some symmetrical explicate of inductive inference (such as positive relevance)." Here 'positive relevance' of evidence E for a hypothesis H refers to the probabilistic condition '$p(H/E) > p(H)$', which is almost[1] equivalent to '$p(E/H) > p(E)$', exemplifying the symmetry Niiniluoto is imposing for 'inductive inference' in general. In his (Niiniluoto 1999, p. 176) he briefly refers to that idea of 1973, and recently confirmed, in conversation, my impression on that basis that he had never elaborated the idea further. But the idea is very much present in 1973. I must have read that chapter in the late 1970s, at least diagonally, and I certainly must have read the brief reference in the 1999-book. But between mid 2003 and mid 2005 I took it to be my own idea to explicate, repeating the crucial phrases above, "a hypothetico-inductive method of testing hypotheses which is based, for example, upon (…) positive relevance". That is, apart from my additional agenda about further evaluation, and except that I prefer to speak of a hypothetico-*probabilistic* method, for relatively marginal reasons that I will make clear soon. My crucial point of departure, explicating a probabilistic version of a test implication of a hypothesis along the lines of positive relevance, which occurred to me as a brain wave in 2003, may well be a pure case of cryptomnesia, that is, a case of unconscious plagiarism, I am afraid.

---

[1] That is, assuming $p(E)$ and $p(H)$ to be non-zero, for otherwise $p(H/E)$ and $p(E/H)$, respectively, are not defined. That is, at least not by the standard definition of conditional probability, according to which, for example, $p(E/H) = p(E\&H)/p(H)$. Of course, there may be good reasons to add a specific definition in the case that $p(H) = 0$. In particular, when H logically entails E, $p(E/H) =_{df} 1$ is highly plausible and will be assumed throughout.

I am nevertheless glad for the occasion to elaborate the method for testing and evaluation up to its functionality for achieving empirical progress in probabilistic terms. I still have to study its possible functionality for truth approximation more intensely, before that can be put on paper. But some provisional findings will be formulated. They lead to new challenges to whom it may concern and appeal, notably to myself and, I hope, to Niiniluoto.

After a methodological section (2) on concept explication by idealization and concretization, the basic concretization of a deductive consequence to a probabilistic consequence is introduced in section 3, followed by an intermezzo with some comparisons with related approaches and an impression of its formal properties in section 4. Section 5 presents the core transition of this paper, viz., from the HD-method of testing by deriving and checking deductive test implications to the HP-method of testing by deriving and checking probabilistic test implications, and section 6 the corresponding transition from deductive to probabilistic confirmation. After an intermezzo (section 7) about deductive- and probabilistic-nomological explanation and -prediction, some further challenges with respect to hypothesis evaluation are formulated in section 8. In section 9, HD- and HP-evaluation are presented as continuations of separate and comparative hypothesis evaluation even when HD- or HP-testing have resulted in their falsification. In section 10, I argue along similar lines as in the case of the HD-method that repeated application of the HP-method is functional for achieving empirical progress, at least as far as general probabilistic test implications of deterministic theories are concerned. In the concluding remarks (section 11) I note that it is more difficult to argue for the HP-method along the same lines as for the HD-method that empirical progress provides good reasons for having achieved truth approximation. Hence, presenting the additional perspectives of the HP-method on truth approximation relative to the HD-method, if they exist at all, has to be postponed to another occasion.

The title of Niiniluoto's 1973-chapter, recall "Towards a non-inductivist logic of induction", may sound strange in light of the title of the book of Niiniluoto and Tuomela, *Theoretical concepts and hypothetico-inductive inference*, and the suggestion (on p. 209) of the possibility of developing 'a hypothetico-inductive method of testing'. From section 1 ("Deductivism and Inductivism") it becomes already clear that he does not want to be an inductivist in any of the five senses of 'inductivism' he distinguishes, to be sure, nor a deductivist in any of the three senses he describes. "Thus, the idea of hypothetico-inductive inference stands in contrast to both inductivism and deductivism" (p. 204). On p. 208 we find the crucial characterization of the type of inference: "Thus, [ ] the situations in which a hypothetical theory occurs among the premises of an inductive argument represent *hypothetico-inductive inference*. This notion is, therefore, an obvious generalization of the hypothetico-deductive inference." Here an inductive argument can be any non-deductive argument, but it is clear that Niiniluoto prefers arguments in which 'positive relevance' is crucial. The main technical aim (and result) of the whole book is to "extend Hintikka's system of inductive logic to apply to situations in which new concepts are introduced to the original language." (*Preface*, p. IX). On p. 197 the authors say to do so in the spirit of Bunge: "Consequently, one can agree with Mario Bunge's wish that a 'non-inductivist logic of induction should be welcome' (cf. Bunge 1963, p. 152). Our development of Hintikka's system of inductive logic is a small but resolute step towards this goal."

This brings me to my reason for speaking of the hypothetico-probabilistic method[2], instead of the hypothetico-inductive method. As is well known Hintikka has, building upon Carnap's continuum of inductive methods, developed in the 1960s an impressive two-dimensional continuum of inductive methods, taking, in contrast to Carnap, universal generalizations into account.[3] As I have elaborated elsewhere (Kuipers 2001, and 2006), Hintikka's probability function is inductive even in two ways. The same applies to the functions presented in Niiniluoto and Tuomela (1973). Though impressive as such 'double-inductive' probability functions are, in the present project I would also like to take non-inductive probability functions into account.

---

[2] Creating in this way another confusion. On p. 8 Niiniluoto and Tuomela state that they want to deal with "nomological, non-probabilistic inductive systematization of general statements". But here 'non-probabilistic' refers to the deterministic (as I would say) nature of the nomological statements involved, not to the relation between such statements and observation statements. The restriction to deterministic hypotheses will become relevant in section 9, and it is inherent in my qualitative theory of truth approximation, and so in my attempts at proving that the HP-method serves that purpose.

[3] Later, Hintikka and Niiniluoto presented an alternative continuum for essentially the same purpose. I greatly enjoyed the study in the 1970s of, among other things, the relation between the old and the new system (Kuipers 1978).

Unfortunately, I was unable to completely (re)read the book of Niiniluoto and Tuomela, but my partial reading strongly suggests that I, and all other formally inclined philosophers of science, should reread this book. It contains many other useful ideas and technical results. But this may seem an *ad hoc* expression of my admiration for the work of Niiniluoto (and of Raimo Tuomela, to be sure). So let me close this introduction by quoting from my *Foreword* to Kuipers (2001), in order to make clear that my admiration for the work of Ilkka Niiniluoto is not something invented for the present occasion:

> I like to mention Ilkka Niiniluoto's *Critical Scientific Realism* (1999) as, as far as I know, the most learned recent exposition of some of the main themes in the philosophy of science in the form of an advanced debate-book, that is, a critical exposition and assessment of the recent literature, including his own major contribution, viz. *Truthlikeness* of 1987. Despite our major differences regarding the topic of truth approximation, I like to express my affinity to, in particular, his rare type of constructive-critical attitude in the philosophy of science.

2. Concept Explication by Idealization and Concretization

I will present this paper to some extent explicitly in terms of 'idealization and concretization'. In my view, elaborated in (Kuipers 2007), this is not only an important methodology in the empirical sciences, but also in philosophy, at least as far as philosophy is engaged in 'concept explication'. In concept explication one aims at the construction of a simple, precise and useful concept, which is, in addition, similar to a given informal concept, used in everyday life, science or philosophy. According to the standard strategy of concept explication one tries to derive from the informal concept to be explicated and relevant empirical findings, if any, conditions of adequacy that the explicated concept will have to satisfy, and evident examples and counter-examples that the explicated concept has to include and exclude, respectively.

As in the empirical case, it may be very useful to start with an idealized way of catching cases and conditions, in order to make it gradually more realistic. However, the concretization should not only be more realistic than the idealized point of departure, the latter should also appear as an extreme special case of the former. This I will call here the C(oncretization)-test. In a way, concretization is just generalization and so a way of aiming at some continuity.

However, speaking of 'explication by idealization and concretization' highlights in typical cases the fact that 1) the informal usages suggest both versions of the relevant concept, 2) the idealized version is an *extreme* (special) case of the concretized version, which 3) frequently has no real life applications, that is, at most toy applications. Although the HD-method is nowadays not very popular among philosophers of science, I would not dare to say the last thing about the HD-method, on the contrary. In my view, both informal usages of 'test implication' belong to 'the language of scientific common sense', i.e., a definite and a more or less probable consequence, and, hence, instead of merely being a generalization of the HD-method, the HP-method should also be an explication of an informal notion, with the HD-method as an extreme special case.

It should be noted that in this respect the Bayesian method, as standardly conceived in terms of the prior and posterior probabilities of hypotheses in the face of evidence, is not very adequate. Although it certainly appeals to informal ways of speaking and although it can deal with deductive relations, it is, first, not very plausible to see the HD-method as an *extreme* special case of it. And, second, it is not very realistic as long as it puts all falsified hypotheses on the scrapheap of hypotheses with zero posterior probability.

3. The Basic Concretization: from d-consequences to p-consequences

The idealized starting point is that a statement S is a deductive (d-)consequence of the hypothesis H; formally:

(1) $H \models S$ S is a d-consequence of H.

There are at least two plausible ways to concretize this in a probabilistic way. The one is by imposing the condition of '(weak) positive relevance'; formally:

(2) $p(S/H) \geq p(S)$ S is a p-consequence of H.

Hence, S is made at least as likely by H as it is *a priori*. The other way requires that S is made at least as likely as its negation; formally:

(2*) $p(S/H) \geq p(\neg S/H)$ S is a p*-consequence of H.
which is equivalent to:
$p(S/H) \geq \frac{1}{2}$.

Both definitions satisfy the necessary, but by no means sufficient condition for being a concretization. According to the C-test, the concretized version should reduce to the idealized version as an extreme special case. If S is a d-consequence of H in the sense of (1) we have, or impose, for whatever probability function p, that p(S/H) = 1, which can't be lower than p(S), let alone than p(¬S/H), which is 0. Hence, a d-consequence is an extreme special case of a p- as well as of a p*-consequence.

The choice between (2) and (2*) is not an easy one, both have attractive and problematic features. Whereas (2) may be intuitively more appealing, (2*) has the advantage of not needing to bother about unconditional probabilities. We will meet other (dis-)advantages in due course. However this may be, as we will see, as soon as we introduce the comparative perspective of hypothesis evaluation the choice becomes irrelevant.

From the methodological perspective of hypothesis evaluation, we should in fact have started with the notion of a conditional deductive (cd-consequence), for a prediction usually needs at least one 'initial condition' for its derivation of a hypothesis. Many more types of conditions may be involved, for example, background knowledge. But this makes no difference for the general idea. Hence, we introduce just a condition C in the following concretizations of d- and p- and p*-consequences.

(1C) H&C ⊨ S   S is a cd-consequence of H, assuming C.
(2C) p(S/H&C) ≥ p(S/C)   S is a cp-consequence of H, assuming C.
(2C*) p(S/H&C) ≥ p(¬S/H&C)   S is a cp*-consequence of H, assuming C.
      which is equivalent to:
      p(S/H&C) ≥ ½.

Note that all three 'conditional concretizations' reduce to the unconditional versions by taking a tautology as an extreme special case of C.

From a purely formal perspective it is interesting to study the class of p- or p*-consequences of a hypothesis, and to see which interesting properties of the class of d-consequences remain intact, which properties get lost, and which new properties become interesting. However, our concern is primarily hypothesis evaluation. In the next section, by way of an optional digression, we will only give some indications of these formal concerns and refer to a selection of the literature devoted to several other approaches to 'consequences from a probabilistic perspective'.

4. Intermezzo. Some Comparisons and Formal Properties

There is quite some literature on issues that are at least to some extent related to our notion of a p-consequence. One dominant issue is the question of the assertability or acceptability of a conditional in probabilistic terms. The proposals of Lewis (1976) and Jackson (1979) were shown by the latter (Jackson 1987) to amount to: 'P→Q' is assertable/acceptable if and only if $p(Q/P)$ is high. Douven (*manuscript*), from whom I borrow this information, proposes to sharpen this by imposing an explicit threshold value (close but unequal to 1) for $p(Q/P)$ and the condition of positive relevance, $p(Q/P) > p(Q)$. In a further refinement, he takes measures to escape from the lottery paradox on the basis of the background knowledge. Although '(weak) positive relevance' is also crucial in my approach, a threshold in Douven's style seems to me a rather arbitrary restriction when one ultimately aims at a general explication of a 'probabilistic consequence'.

According to Adams (1966; 1975), the assertability/acceptability of 'P→Q' should just be measured by $p(Q/P)$, without some threshold. He adds a rather strong definition of 'probabilistic validity', viz., when, for all (rational!) probability functions, the uncertainty of the conclusion Q (i.e., $1-p(Q)$) does not exceed the sum of the uncertainties of the premises. Bradley and Swartz (1979, p. 196) define 'Q follows probably from P' if and only if most models of P are models of Q. In a note (7) to section 6 we will see when and how this definition fits in the definition of 'p*-consequence', i.e., (2*). Note, by the way, that the expression 'follows probably from' is also frequently used for informally characterizing an inductive argument in general.

Hailperin (1996) and Wagner (2004) have studied probabilistic versions of valid arguments, notably *modus ponens* and *modus tollens*. Assuming probabilities for the relevant premises leads to boundaries of the probability of the conclusion. Whereas Hailperin focused on the probabilities of a conditional premise, Wagner replaces this by conditional probabilities. For *modus ponens*, for example, the first and the second approach yield, respectively:

If $p(Q\to P) = a$ and $p(Q) = b$ then $a+b - 1 \le p(P) \le a$.
If $p(P/Q) = a$ and $p(Q) = b$ then $ab \le p(P) \le ab + 1 - b$.

For *modus tollens* the first approach is also easy, but not so the second. Interesting as this is as such, it does not seem to be directly relevant for our purposes.

Malinowski (*manuscript*) certainly is directly relevant for our purposes. Restricting himself to finite propositional languages, he defines 'Q (probabilistically) supports P' by the condition of weak positive relevance of Q for P, $p(Q/P) \geq p(Q)$, *for all possible probability functions*. A very strong extra condition indeed, but with some plausibility, for he interprets a probability function as a generalization of a valuation function. But he does not want to speak of 'probabilistic entailment' or, like Adams does for a related definition (see above), of 'probabilistic validity', because the defined notion is not truth preserving. He reports (Theorem 7) that Makinson (2005) has proved that the definition is equivalent to: "either P logically entails Q or vice versa".

Let me first list some trivial properties of p-consequences ('nd' indicates that a property, or rule, does not hold for d-consequences):

(reflexivity)	P is p-consequence of P.
(or-elimination)[nd]	P is a p-consequence of PvQ.
(and-introduction)[nd]	P&Q is a p-consequence of P.
(tautology)	if $p(Q) = 1$, then Q is a p-consequence of any P.
(contradiction)[nd]	if $p(Q) = 0$, then Q is a p-consequence of any P.

Following Kraus, Lehmann and Madigor (1990), Malinowski starts with listing (Theorem 5) Gentzen style rules of 'weak positive relevance', that is, of p-consequences:

(symmetry)[nd]	if Q is a p-consequence of P, then *vice versa.*
(left logical equivalence)	if P and Q are logically equivalent and R is a p-consequence of P then also of Q.
(right logical equivalence)	if P and Q are logically equivalent and P is a p-consequence of R then so is Q.
(classic logic)	if Q is a d-consequence of P then it is a p-consequence of P. (Our C-test condition!)
(contraposition)	if Q is a p-consequence of P then ¬P is a p-consequence of ¬Q.
(proof by cases)	if R is p-consequence of P&Q and of P&¬Q then R is a p-consequence of P.

They are all easy to prove, even for any extension of a propositional language.

Malinowski continues by reporting several rules that the strong notion just can take over from 'deductive entailment', and which are easy to prove on the basis of Theorem 7, see above, and shows that they fail to hold in general for an arbitrary probability function, that is, for p-consequences. I rephrase five of them as, partially related, 'non-properties' of p-consequences:

(non-monotonicity)   if R is a p-consequence of P, then not necessarily of P&Q.
(non-and)   if R is a p-consequence of P as well as of Q, then R need not be a p-consequence of P&Q.
(non-transitivity)   if Q is a p-consequence of P and R of Q, then R need not be a p-consequence of P.
(non-equivalence)   if Q is a p-consequence of P, and vice versa (trivially due to symmetry), and R is a p-consequence of P, then R need not be a p-consequence of Q.
(non-right weakening)   if P→Q is a tautology and P a d-consequence of R, then Q need not be a p-consequence of R.

However, Malinowski also notes that three other ones have restricted validity for p-consequences (Theorem 11). I reformulate these properties in p-terms, where 'r' refers to their being restricted relative to the properties of the strong notion:

(cut-r)   if R is a p-consequence of P&Q and Q is a d-consequence of P then R is a p-consequence of P.
(cautious monotonicity-r)   if R is a p-consequence of P and Q is a d-consequence of P then R is a p-consequence of P&Q.
(or-introduction-r)   for incompatible P and Q, if R is a p-consequence of P as well as of Q then R is a p-consequence of PvQ.

These three rules can be transformed into equivalent versions by reversing (one or more of) the p-consequence relations (due to symmetry), e.g., (or-introduction-r) is equivalent to

> for incompatible P and Q, if P and Q are both p-consequences of R then PvQ is p-consequence of R.

All proofs are easy, and it is again easy to check that they hold for any extension of the propositional language. One gets the corresponding unrestricted properties of the strong notion by 1) weakening in the first two cases the logical entailment premise to a p-consequence premise (i.e., weak positive relevance) and in the third case by skipping the incompatibility premise and 2) by adding to all resulting p-consequence claims 'for all probability functions p'.

As suggested before, for our purposes, it would also be interesting to further study more specifically the formal properties of the class of p-consequences of a given hypothesis P, p-Cn(P) $=_{df}$ {Q | p(Q/P) ≥ p(Q)}, of course, being a superset of the class of its d-consequences, d-Cn(P) $=_{df}$ {Q | P $\models$ Q}. A crucial difference is that the latter is consistent, as soon as P is non-contradictory, whereas the former is, as a rule, inconsistent, as is easy to check.[4] For this and other reasons it may also be worthwhile to restrict the above classes to contingent hypotheses and consequences, and also to assume that they are (not weak but) positively relevant for each other and non-equivalent to each other, respectively.

5. From the HD-method of Testing to the HP-method

The crucial idea of the HD-method of testing is the derivation of genuine test implications of a hypothesis. We define:

(3) I is a d-test implication of H iff
    I is a non-tautological, observational d-consequence of H.

Hence, I must be phrased in 'observational terms' and it may not be the case that "$\models$ I" holds. Of course, observational terms are here understood in the sophisticated theory-relative sense, that is, I may not be laden by H, but may well be laden by other, 'underlying' theories.

The plausible concretizations of (3) are:

(4) I is a p-test implication of H iff
    I is an observational p-consequence of H even such that p(I/H) > p(I).

---

[4] For example, for a fair die, the hypothesis of an even outcome is positively relevant for all three mutually incompatible outcomes 4, 5 and 6.

(4*)   I is a p*-test implication of H iff
       I is an observational p*-consequence of H even such that $p(I/H) > p(\neg I/H)$ (or, equivalently, $p(I/H) > \frac{1}{2}$).

Both have plausible 'conditional concretizations', where a condition only needs to be observational when it pertains to an 'initial condition'.

For checking that (4) and (4*) are concretizations of (3) in the minimal sense of the C-test, we have to assume that "$H \models I$" and "not $\models I$". In this case, the condition in (4), viz., $1 = p(I/H) > p(I)$, holds, except for an 'almost tautological' test implication, that is, a non-tautological test implication which has nevertheless probability 1. Although such implications do exist, they always have some infinite aspect, which we might want to exclude from being a genuine test implication. However, without further proviso, this would also exclude an observational universal generalization from being a test implication, which we would consider as too high a price. It is interesting to note that (4*) does not have this problem, for "$p(I/H) > \frac{1}{2}$" straightforwardly holds for an almost tautological implication. Hence, we note a second technical advantage of (4*). In sum, a d-test implication is (an extreme special case of) a p-/p*-test implication, with a marginal exception for the unstarred case.

As a method of testing a hypothesis H, the HD-method can be said to aim at d(eductive)-confirmation or falsification of H by checking d-test implications of H. That is, it amounts to deducing test implications and checking by (experiment followed by) observation whether they are (or become) true or false. If false, H has been falsified as well, if true, H has been (deductively) confirmed. Note that an HD-test never can lead to a neutral observation.

In light of 'probabilistic testing', the HD-method of testing is an idealized point of departure. Assuming that we can concretize (deductive) confirmation and falsification to some kind of 'p(robabilistic)-confirmation and -falsification',[5] or p*-versions of them, it is plausible to submit that the Hypothetico-Probabilistic (HP-)method of testing aims at p-/p*-confirmation or -falsification by checking p-/p*-test implications. If this assumption is correct, we may conclude that the HP-method of testing is a concretization of the HD-

---

[5] The phrase 'probabilistic disconfirmation' is not only usually used instead of 'probabilistic falsification', it is also to be preferred for reasons which will be indicated below.

method of testing. That is, an explication of the title of this paper as far as testing is concerned.

6. From Deductive to Probabilistic Confirmation

In Kuipers (2000, Part I) we have argued that there is a coherent landscape of confirmation notions, allowing different languages of (degrees of) confirmation. This has been presented more systematically in Kuipers (2001, Section 7.1.2, and 2006). Here we just represent the main lines, but add the *-versions.

Starting from the idealized notion of deductive (d-)confirmation of a hypothesis H by some evidence E:

(5) $H \models E$ E d-confirms H,

the general form of the probabilistic concretization in line with that of a p-consequence focuses on the so-called likelihood of H in the light of E: $p(E/H)$, viz.,

(6) $p(E/H) > p(E)$ E p-confirms H.

Although this deviates from the standard formulation, $p(H/E) > p(H)$, it is almost equivalent with the latter, viz., if both $p(E)$ and $p(H)$ are non-zero. The *-version:

(6*) $p(E/H) > \frac{1}{2}$ E p*-confirms H.

is rather unusual, but has some plausibility.

Analogously we get two concretizations of 'falsification', i.e., E (deductively) falsifies H iff

(7) $H \models \neg E$ E (d-)falsifies H,

viz.,

(8) $p(E/H) < p(E)$ E p-disconfirms (p-falsifies)[6] H,

---

[6] Speaking of 'probabilistic falsification' in case $p(E/H) < p(E)$ may have some plausibility, but in view of the probabilistic 'Confirmation Square', building upon the 'Deductive Confirmation Matrix', it is much more plausible to speak of 'probabilistic disconfirmation' with deductive disconfirmation as extreme special case, i.e., the case that E is logically entailed by the negation of H. See Kuipers (2001, p. 22, p. 46).

(8*) $p(E/H) < \frac{1}{2}$ E p*-disconfirms (p*-falsifies) H.

All these definitions have plausible conditional versions, straightforwardly in line with the definitions of conditional consequences.

Whatever definition of p-confirmation one prefers, they all presuppose that we choose a probability (p-)function. Now it is plausible to make a distinction between non-inductive and inductive p-functions. Inductive p-functions have some ampliative effect on the probability that "the future resembles the past", to use Hume's well known phrase. Assuming that the p-function can be decomposed over a partition of hypotheses and corresponding likelihoods, an ampliative effect can be generated in three ways, advocated by three famous philosophers, viz., by inductive likelihoods (Carnap), by inductive priors (Bayes), and by a combination of both (Hintikka). As a matter of fact most, if not all, probability functions presented in Niiniluoto and Tuomela (1973) are of this 'double-inductive' nature.

In order to get some more idea of the crucial types of inductive p-functions, that is, inductive likelihoods and inductive priors, we will specify the idea of non-inductive p-functions: whatever their precise form, they do not "learn from the past" by themselves. They nevertheless enable straightforward cases of confirmation, which I prefer to call 'structural confirmation', as opposed to 'inductive confirmation'. So let me illustrate structural confirmation by my favorite example dealing with a fair die. Let E indicate the even (elementary) outcomes 2, 4, 6, and H the 'high' outcomes 4, 5, 6. Then (the evidence of) an even outcome confirms the hypothesis of a high outcome according to all three criteria (standard, (6), (6*)), since $p(E/H) = p(H/E) = 2/3 > 1/2 = p(H) = p(E)$. By speaking of a fair die, we indicate that our p-function is an objective p - dealing with, or representing an objective probability process. However, by leaving out this information, and simply assuming a language of a family of 6 monadic predicates, with two defined disjunctive predicates, we would be able to generate the case of confirmation by consistently assuming the principle of indifference for the 6 possibilities. In this way we get an illustration of the 'logical probability function' or the 'logical measure function', introduced by Kemeny (1953), from now on indicated by 'm'.

Structural confirmation is defined as confirmation on the basis of an objective probability function or on the basis of m.[7] Moreover, inductive confirmation is now defined as confirmation based on an inductive p-function, that is, restricted to monadic predicates, a p-function that has the property of 'instantial confirmation':

$p(Fa/E\&Fb) > p(Fa/E)$ instantial confirmation,

where 'a' and 'b' represent distinct individuals, 'F' an arbitrary monadic property and 'E' any kind of contingent evidence compatible with Fa. Note that this definition can be generalized to n-tuples and n-ary properties. As said, inductive (probability) functions can be obtained in two ways, which can now be specified, viz., by:

– 'inductive priors', i.e., positive prior p-values p(H) for m-zero hypotheses, and/or
– 'inductive likelihoods', i.e., likelihood functions p(E/H) having the property of instantial confirmation.

Note first that standard confirmation of m-zero hypotheses requires inductive priors, whereas confirmation of such hypotheses according to (6) and (6*) is always possible, assuming that p(E/H) can be interpreted.[8]

Popper rejected both kinds of inductive confirmation, for roughly three reasons: two problematic ones and a defensible one. The first problematic one (Popper 1934/1959) is that he tried to argue, not convincingly (see e.g., Earman 1992; Howson and Urbach 1989; Kuipers 1978), that p(H) could not be positive for universal hypotheses. The second one is that any probability function has the property "$p(E{\rightarrow}H/E) < p(E{\rightarrow}H)$" (Popper and Miller 1983). Although the claimed property is undisputed, the argument that a proper inductive probability function should have the reverse property, since 'E→H' is the 'inductive conjunct' in the equivalence "$H \Leftrightarrow (E \vee H)\&(E{\rightarrow}H)$", is not

---

[7] Note that if 'most' in the definition of Bradley and Swartz (1979, p. 196) (recall from section 4: 'Q follows probably from P' if and only if most models of P are models of Q) is interpreted as 'more than not', this definition amounts to (2*), (4*) and (6*), i.e., the *-definitions, specifically based on the logical probability function.

[8] Confirmation according to an inductive probability function is, as a rule, a mixture of structural and inductive confirmation. A precise definition of (degree of) inductive confirmation abstracts from the structural component. See (Kuipers 2001, 2006).

convincing. The indicated reverse property may well be conceived as an unlucky first attempt to explicate the core of (probabilistic) inductive intuitions, which should be replaced by the property of instantial confirmation.

The defensible reason is that the property of instantial confirmation merely reflects a subjective attitude and, usually, not an objective feature of the underlying probability process, if there is such a process at all.

However this may be, the logical measure function perfectly permits structural confirmation, in particular in the form of (6) or (6*), for then the m-zero character of many interesting hypotheses need not prevent confirmation, assuming the likelihoods are well-defined. Hence, taking Popper's reserves regarding inductive probabilities into account, and assuming that he would like the idea of a non-inductive probabilistic version of the hypothetical method, I prefer to speak of the HP-method, instead of speaking of the Hypothetico-Inductive (HI-) method, as Niiniluoto did in (1973) and (1999).

One argument against the very idea of speaking of confirmation, albeit structural confirmation', in view of Popperian considerations, is that Popper did not like the term 'confirmation'. This is for at least two reasons. First, 'confirmation' suggests judging the 'future performance' of a hypothesis, whereas hypotheses should be evaluated on the basis of their 'past performance'. Second, aiming at confirmation, whether deductive or probabilistic, becomes an easy task of searching for probable confirmation. However, the first argument may apply to the standard definition of p-confirmation for p-non-zero hypotheses[9], it does not apply to (6) and (6*) for p-zero hypotheses, assuming the likelihoods are well-defined. Moreover, explicitly according to (4), (6) and (8), and implicitly according to (4*), (6*) and (8*), the lower the prior probability of a test implication, the more one expects disconfirmation and hence the more surprising it is when confirmation follows. Hence, these definitions leave a lot of room for Popperian motives.

Which of the four kinds of p-functions to choose, is no fact of the matter. Hence, as soon as one uses the probability calculus, it does not matter very much which 'confirmation language' one chooses, for that calculus provides the crucial means for updating the likelihood of a hypothesis in the light of

---

[9] But even then it is a matter of probabilistic consistency, for then, assuming p-non-zero evidence, it is just equivalent to (6) or (6*), which evaluate past performance.

evidence. Hence, the only important point which then remains is always to make it clear which confirmation language one has chosen.

In all cases, it is practical to have a degree of confirmation (or degree of success). For various reasons, explained in (Kuipers 2000), I prefer the ratio measure.

$p(E/H)/p(E)$

It is not only directly in line with the unstarred versions (4), (6) and (8), but it also explicitly accounts for the degree of surprise involved: the lower $p(E)$ and the higher $p(E/H)$, the more this ratio exceeds 1, and the more happily surprised we will be with this performance of H. Conversely, the more the ratio is below 1, the more disappointed we will be with its performance. Since the ratio degree of confirmation is not so obvious starting from the starred versions (4*), (6*) and (8*), I prefer the unstarred versions, and will assume them from now on.

Anticipating later sections, note already now that as soon as we come to compare the performance of two hypotheses, H1 and H2, relative to the same evidence, it becomes highly plausible to compare the respective likelihoods and the plausible comparison measure becomes the so-called 'likelihood ratio', $p(E/H2)/p(E/H1)$, in which a role for $p(E)$ disappears. Hence, by the way, the comparative perspective relativizes all reasons to prefer the starred or the unstarred versions.

7. Intermezzo. Deductive- and Probabilistic-Nomological Explanation and -Prediction

Not only probabilistic confirmation of hypotheses by evidence has been studied intensely. Also several forms of probabilistic explanation and prediction as concretization of their deductive forms have been studied. Here we will just indicate the main points of Probabilistic-Nomological (PN-) explanation and -prediction as concretization of the their Deductive-Nomological forms. They nicely fit in the basic probabilistic concretization of a deductive consequence. The basic idea of DN-explanation of some evidence E by a well established nomological statement N, assuming some further well established conditions C, is

N&C $\models$ E, and not C $\models$ E,

with the probabilistic form, PN-explanation, satisfying the C-test:

$p(E/N\&C) > p(E/C)$.

Moreover, it is now plausible to say that, assuming C, N2 explains E better than N1 iff

$p(E/N2\&C) > p(E/N1\&C)$.

Note that this comparative claim has no straightforward deductive version, except when one brings non-empirical differences between N1 and N2 into account.

As a matter of fact, the whole idea of DN- and PN-*prediction* is even more analogous to deriving HD- and HP-test implications. The only difference is that the implications no longer have the status of *test* implications, for, as in the case of DN- and PN-explanation, N as well as C are again supposed to be well established. Hence, to obtain DN- and PN-prediction, 'E' can simply be re-interpreted as 'predicted evidence' in the first and the second formula above, respectively.

Analogous to the case of explanation, we now also have a comparative version of saying that, assuming C, E2 is a more surprising, or a more risky, prediction of N than E1 iff

$p(E2/C) < p(E1/C)$ and $p(E2/N\&C) \geq p(E1/N\&C)$

Here there is a deductive version, assuming that E1 and E2 are logically entailed by N&C and that there is some qualitative way of arguing that E2 is less plausible than E1. Of course, speaking of 'more risky' may suggest that we are back at the topic of hypothesis testing and evaluation, to which we return now.

8. Further Challenges

In Kuipers (2000) we have given a coherent explication of the HD-method, not only as a method of testing hypotheses, in search for an answer to 'the truth question', but also as a method of separate and comparative empirical success evaluation, and, finally, as a method of evaluating truth approximation claims. Whereas testing becomes superfluous after identifying a (convincing) counterexample, i.e., after falsifying a hypothesis, the indicated types of evaluation of hypotheses remain highly meaningful.

The general challenge of this project may now be described as the task of giving a coherent probabilistic concretization of the HD-method for testing and for the various types of evaluation. So far we have described how the first task can be fulfilled, heavily leaning on (Kuipers 2000 and 2001). We will now describe for the first time how the probabilistic concretization of separate and comparative success evaluation naturally emerges from their deductive point of departure.

9. HD- and HP-Testing and -Evaluation

Following the HD-method, the result of checking a deductive test implication I of H is either a case of d-confirmation or a case of falsification. Since 'confirmation' has the connotation of 'not yet falsified', we jump for describing the results of continued (separate) evaluation of H by checking new deductive test implications to the terminology of positive and negative evaluation d-results. This naturally leads to an 'evaluation report' of H, listing the positive d-results at one side and the negative d-results at the other.

Analogously we have that the result of checking a probabilistic test implication I of H is either a case of p-confirmation or p-disconfirmation. And again we may better describe such results for the general evaluation of H as positive and negative p-results, leading to an evaluation report of H listing these results.

Note that the terminology of positive and negative (d- and p-)results is a variant of Larry Laudan's instrumentalist terminology of (empirical) successes and problems.

For the comparative evaluation of two hypotheses H1 and H2 we of course compare the two evaluation reports. In the deductive case we now have to take into account that a deductive test implication of one hypothesis, nor its negation, need to be a test implication of the other, in which case the result of the test becomes a 'neutral d-result' for the latter. Hence, we get in total 9 possible combinations of being a positive, negative or neutral d-result. Of course, we say that a test result favors H2 over H1 if it is a positive d-result of H2 and a neutral or even negative d-result for H1 or if it is a neutral d-result for H2 and a negative one for H1. A test result is indifferent between H1 and H2 when it is of the same kind for both. Note that a result neutral for both can only come into play due to a third hypothesis.

For the probabilistic concretization it is highly plausible to generalize the definition of a test result E favoring H2 over H1 by comparing the likelihoods, as follows,

E favors H2 over H1   iff $p(E/H2) > p(E/H1)$,

that is, if and only if the likelihood ratio of H2 relative to H1 exceeds 1. It is easy to check that this definition covers the three cases constituting the deductive definition of "E favors H2 over H1" as extreme special cases. Hence, the definition can be seen as a probabilistic concretization.

A further differentiation is plausible. So far we included in our definition of, for example, a positive p-result a positive d-result as an extreme special case. We may speak of a proper positive p-result, or a positive pp-result, when it is a positive p-result but not a positive d-result. Note that this classifies as positive pp-result 'almost positive d-results', that is, cases where $p(E/H) = 1$, but H does not logically entail E. A similar differentiation for negative pp-results can be made. Finally, a neutral p-result is defined by $p(E/H) = p(E)$, and hence this is a very special case of a neutral d-result, that is, when neither E nor $\neg$E is entailed by H. Of course, a neutral d-result is either a positive or a negative pp-result, or a neutral p-result.

In total this generates 25 possible combinations of comparative results. Table 1 gives a survey of the basic types of the results of HD- and HP-testing and -evaluation.

Test implications	Test results	Evaluation results	
		separate	comparative
Deductive $H \models I$	d-confirmation or falsification	positive or negative d-results	d-results favoring H2 over H1, or H1 over H2, or being indifferent (9 possible kinds)
Probabilistic $p(I/H) \geq p(I)$	p-confirmation or p-disconfirmation	positive or negative p-results	p-results favoring H2 over H1, or H1 over H2, or being indifferent (25 possible kinds)

Table 1. Basic types of results of HD- and HP-testing and -evaluation.

Table 2 differentiates the comparative results into the indicated 9 and 25 possible combinatory results for HD- and HP-comparative evaluation. In line with (Kuipers 2000, p. 117; 2001, p. 235), in which I called the table restricted to deductive results the 'deductive evaluation matrix', I like to call the probabilistic extension, the probabilistic evaluation matrix.[10] In Table 2 I have indicated the 9 'deductive (elementary or unified) boxes' DB1-9 that result by interpreting p-results as d-results. Four DB's just coincide with the 'probabilistic boxes' (PB's) on the four corners (DB1, 4, 6, 9), four DB's are the union of three PB's (DB2, 3, 7, 8) and one is the disjunction of the remaining 9 PB's (DB5). Hence, the deductive version of Table 2 is a coarse version of its probabilistic version, with the consequence that the deductive second row of Table 1 is a coarse version of the probabilistic third row. This shows that the C-test of both tables amounts to the conclusion that HP-testing and -evaluation form a concretization of their HD-variants in the form of a refinement.

		H1				
		negative d-result	negative pp-result	neutral p-result	positive pp-result	positive d-result
H2	negative d-result	DB4 0	DB2 −	DB2 −	DB2 −	DB1 −
	negative pp-result	DB8 +	DB5 0	DB5 −	DB5 −	DB3 −
	neutral p-result	DB8 +	DB5 +	DB5 0	DB5 −	DB3 −
	positive pp-result	DB8 +	DB5 +	DB5 +	DB5 0	DB3 −
	positive d-result	DB9 +	DB7 +	DB7 +	DB7 +	DB6 0

Table 2. The deductive and probabilistic evaluation matrix.

It is also easy to check that for HP- and HD-notions alike, separate evaluation is a concretization of testing, and comparative evaluation is a concretization of separate evaluation. For the first C-test, assume the extreme (!)

---

[10] I also introduced a quantitative version (l.c. p. 119; p. 237, respectively). From the present point of view, just inserting the likelihood ratios seems at first sight the most plausible thing to do, but I did not investigate this in detail.

special case that H has not yet been falsified. For the second C-test, assume that H1 is a tautology.

10. Empirical Progress

The ultimate aim of comparative HD-evaluation in terms of positive and negative d-results is to come to conclusions about whether one hypothesis is in the light of the obtained empirical evidence really better than the other. The main line of reasoning developed in Kuipers (2000, Part II, reproduced in 2001, Part IV) is as follows. First we need a definition of "more successfulness", which amounts in the present formulation to:

> More Successful (MSF)
> H2 is more successful than H1 in light of the available evidence iff
> (1) all positive d-results of H1 are positive d-results of H2
> (2) all negative d-results of H2 are negative d-results of H1
> (3) H2 has some more positive d-results or some fewer negative d-results than H1.

Note that this definition does not refer to neutral results. Assuming three possible results, they are handled implicitly. For one might expect, for example, "all neutral d-results of H1 are neutral or positive d-results of H2". But a neutral d-result of H1 being a negative d-result of H2 is excluded by clause (2). Note also that MSF is rather strict; more liberal versions are easy to imagine, but for our purposes we want to focus on clear-cut cases. They are sufficient to tell the principal story.

Of course, H2 may be more successful than H1 by sheer luck or prejudice regarding the choice of test implications so far. Hence, this situation only suggests the

> Comparative Success Hypothesis (CSH)
> H2 (is and) remains more successful than H1.

CSH is an interesting hypothesis, even if H2 has already been falsified (by negative d-results). It can be tested by a comparative application of the HD-method of testing: regarding clause (1), derive and test potential positive d-results of H1 that cannot be derived from H2; and similarly for clause (2).

Only after various tests of CSH may we come to the conclusion, for the time being, that H2 is really better than H1. We like to state this in the form of the

Instrumentalist Rule of Success (IRS)
When H2 has so far proven to be more successful than H1, i.e., when CSH has been 'sufficiently confirmed' to be accepted as true, discard H1 in favor of H2, at least for the time being.

Of course, the phrase 'sufficiently confirmed' hides a lot of problems dealing with inductive generalization. However this may be, the acceptance of CSH and the consequent application of IRS I would claim to be the core idea of 'empirical progress', a new hypothesis that is better than an old one. IRS may even be considered the (fallible) criterion and hallmark of scientific rationality, acceptable for the empiricist as well as for the realist. Of course, IRS is called instrumentalistic because it remains to take hypotheses seriously when they have been falsified. By IRS such a hypothesis is only discarded or eliminated, for the time being, when there is a really better, but perhaps also already falsified, hypothesis.

Finally, it is possible to show (Kuipers 2000, Part III and IV) that IRS is functional (or instrumental) for truth approximation, in precise senses depending on the particular type of truth approximation (basic or refined, and observational, referential, or theoretical). For the moment, however, I restrict myself to empirical progress by the HD-method as specified along the lines of d-MSF, d-CSH and d-IRS, where I introduce the prefix 'd-', to distinguish it from their possible p-versions. The question is whether this story can be concretized in probabilistic terms.

The crucial point of departure is the concretization of d-MSF. Of course, we just might replace 'd-results' by 'p-results', only guaranteeing that the scores (positive, neutral, negative) in the report of separate evaluation for H2 are never 'lower' than for H1 and sometimes 'higher'. We may even differentiate this in terms of (positive and negative) d- and proper p-results (pp-results). This would only amount to a translation of the deductive story into probabilistic terms. A genuine form of more successfulness in probabilistic terms is, of course, in terms of 'p-results favoring one hypothesis over another'. Recall:

E favors H2 over H1   iff $p(E/H2) > p(E/H1)$.

This suggests:

> Probabilistically More Successful (p-MSF)
> H2 is probabilistically more successful than H1 in light of the available evidence iff
> (1) no p-results favor H1 over H2
> (2) some p-results favor H2 over H1

Note that p-MSF confronts each p-result with H1 and H2, independently of the other results. This could of course be done otherwise, notably with successive concatenation. This would not only be more complicated but would also lead to a less strict version of 'more successfulness', though perhaps a no less interesting one, in particular when one favors an inductive rather than a structural probability function.

Before we apply the C-test, to check whether d-MSF is indeed an extreme special case of p-MSF, we will discuss the fact that the present form of p-MSF may be criticized for being not very realistic. Not in the sense of being too strict to have only few, if any, real life applications, that is, the same reason why d-MSF can also be said to be very strict. In both cases this is what we may expect for a genuinely superior hypothesis H2 relative to H1, that is, being an exception rather than a rule in actual science.

But even if H2 is superior to H1 in any intuitive sense, there is something strange to the definition of p-MSF. For even in this case, a particular p-result may well be the negation of the result 'predicted' by a p-test implication of H2, providing a higher likelihood of H1, just by bad luck. This is particularly the case for so-called 'individual test implications', that is, test implications that probabilistically predict an outcome of an individual experiment in the face of the hypothesis and some initial conditions. However, if we look at 'general test implications',[11] the definition of p-MSF becomes more plausible. Suppose that for some general statement I holds $p(I/H2) > p(I)$ and $p(I/H2) > p(I/H1)$. Hence it is a more probable p-test implication of H2 than of H1, if it is a p-test implication of H1 at all, that is, if $p(I/H1) > p(I)$, which need not be the case. If such an I becomes established as evidence, no sheer luck can be involved and it should of course be counted as a p-success favoring H2 over H1, as

---

[11] For details about the distinction between individual and general test implications, see Kuipers (2000; 2001).

prescribed by clause (2). On the other hand, if H1 scores such a success relative to H2, clause (1) is rightly violated, for H2 apparently is not overall superior to H1. Hence we conclude that the definition of p-MSF is adequate, provided we restrict the attention to general test implications, which we will indicate by pg-MSF, etc.

Let us now turn to the C-test. In Kuipers (2000, Section 5.1.5) we have argued that there are in fact three interesting models of HD-evaluation, viz. in terms of 'individual successes' and 'individual problems (counterexamples)', or in terms of 'general successes' and 'individual problems', or in terms of 'general successes' and 'general problems (falsifying general facts). The definition of d-MSF can then be given in one of these combinations. The second, 'asymmetric' model follows most naturally when we focus HD-evaluation on general test implications, either leading to general successes or to individual problems. In this model,[12] d-MSF reads:

dg-More Successful (dg-MSF)
H2 is more successful than H1 in light of the available evidence iff
(1) all general successes of H1 are general successes of H2
(2) all individual problems of H2 are individual problems of H1
(3) H2 has some extra general successes or some fewer individual problems than H1.

For the C-test of dg-MSF relative to pg-MSF we now first assume that we have a general success of H1, that is, some (general) I logically entailed by H1, and established as correct. According to dg-MSF-(1), I is also entailed by H2. In probabilistic terms it follows that $p(I/H1) = 1$ and if I is established $p(I/H2)$ should, according to pg-MSF-(1), not be lower than $p(I/H1)$, hence it also equals 1. Hence, pg-MSF-(1) covers dg-MSF-(1) as an extreme special case. Let us now, second, assume that we have an individual problem of H2, that is, some (general) I logically entailed by H2 was established as incorrect, by a counterexample of I and hence of H2. According to dg-MSF-(2) it should also be a counterexample to H1. In probabilistic terms we have that $p(E/H2) = 0$.

---

[12] As suggested, the first, symmetric model makes no sense in probabilistic terms, due to the possibility of bad luck. Of course, also the possibility of sheer luck interferes. However, the third, also symmetric model may well make sense in probabilistic terms, but I did not yet investigate this in enough detail.

Again according to pg-MSF-(1), $p(E/H1)$ may not be higher than 0, hence $p(E/H1) = 0$ (and E amounts to a counterexample of H1 in probabilistic terms). Hence, pg-MSF-(1) also covers dg-MSF-(2) as an extreme special case. By two similar arguments it is easy to show that pg-MSF-(2) covers both alternatives in dg-MSF-(3). Q.e.d.

Having restricted the notion of being probabilistically more successful to general test implications, the rest of the concretization follows easily. The probabilistic versions of the Comparative Success Hypothesis (pg-CSH) and the Instrumentalist Rule of Success (pg-IRS) result from simply substituting pg-MSF for 'more successful', instead of dg-MSF. Finally, again the acceptance of pg-CSH and consequent application of pg-IRS is the core idea of 'empirical progress' in probabilistic terms, a new hypothesis that is better than an old one, now also taking probabilistic general test implications into account.

It is important to note that the expression 'probabilistic general test implications' should not be taken in the sense of an implication being probabilistic in nature. In that expression, 'probabilistic' only refers to the non-deductive nature of the relation between the relevant hypothesis and the implication. Hence, the restriction to general test implications of the HP-method is a restriction to deterministic, non-probabilistic hypotheses. For the HD-method this restriction is even presupposed for individual test implications. However, I would not exclude the possibility to extend the whole analysis of the HD- and the HP-method to probabilistic hypotheses.

11. Concluding Remarks

We may conclude that there is an HP-method of testing and separate and comparative evaluation, which is functional for empirical progress in probabilistic terms and which is a straightforward concretization of the HD-method. The relevant probability function and corresponding 'confirmation language' may be of a Popperian, Carnapian, Bayesian or Hintikkian nature. DN-explanation and -prediction can similarly be concretized to PN-explanation and -prediction. All relevant C-tests are successfully passed, where 1 and 0 represent the extreme special cases of $p(I/H)$ or $p(E/H)$. Recall also that for HP- and HD-notions alike, separate evaluation is a concretization of testing, and comparative evaluation is a concretization of separate evaluation.

In sum, a coherent story can be told up to and including the functionality of the HD- and the HP-method for achieving empirical progress in deductive and probabilistic terms, respectively. The remaining challenge is the question whether the HP-method is, like the HD-method, functional for truth approximation.

In my *From Instrumentalism to Constructive Realism. On some relations between confirmation, empirical progress and truth approximation*, to quote the title of Kuipers (2000) in full, I have argued that the comparative evaluation of two hypotheses, based on repeated application of the HD-method, to be continued if both have already been falsified, not only results in a plausible explication of the notion of empirical progress, but also leads to articulated perspectives on (observational, referential and theoretical) truth approximation.

Hence, it is plausible to try to argue along similar lines that the comparative evaluation of theories based on the repeated application of the HP-method, also results in positive perspectives on truth approximation. However, although the explication part of the project has been successfully completed, that is, the HP-method also suggests a probabilistic concretization of the (qualitative, 'deductive') explication of 'truth approximation', so far the rest of the project only turned into restricted but very surprising grist to the mill of dogmatic Popperians, or so it seems. For, as far as equally probable theories are concerned, it seems provable that the HP-method is, relative to the HD-method, redundant: empirical progress and truth approximation by the HD-method, regarding deterministic hypotheses, seem to entail these respective objectives of the HP-notions in probabilistic terms, hence including an important relativization of the analysis of probabilistic empirical progress in the last section.

Whether or not this project can nevertheless be finished successfully in both cases it is also a challenge for Ilkka Niiniluoto to investigate to what extent his quantitative theory of truth approximation can be adapted to, the above or his own future explication of, the HP-method such that the latter is functional for quantitative probabilistic empirical progress and truth approximation, at least as far as deterministic theories are concerned.

12. Acknowledgements

I would like to thank for all the comments I got after presenting the above ideas in Bielefeld (2003), Rotterdam (2004), Konstanz (2004), and in the PCCP research group in Groningen (2004). In particular I want to thank Igor Douven, Dale Jacquette, Jacek Malinowski and Elliott Sober for their detailed comments in various stages, and Malinowski also for even writing a useful paper on the formal properties of (a strong version of) probabilistic consequence. I like to thank the Netherlands Institute of Advanced Study (NIAS) in Wassenaar for the privilege of allowing me to return for two weeks to this paradise for thinking and writing. Finally, I like to thank Ilkka Niiniluoto for his stimulating friendship, starting in June 1974 in Warsaw on a conference on Formal Methods in the Methodology of Empirical Sciences. I hope he will forgive me my, not unlikely, unconscious plagiarism with respect to the basic idea of this paper.

University of Groningen

References

Adams, E. W. 1966: "Probability and the Logic of Conditionals", in J. Hintikka and P. Suppes (eds.), *Aspects of Inductive Logic*, North-Holland, Amsterdam, pp. 265–316.

Adams, E. W. 1975: *The Logic of Conditionals*, Reidel, Dordrecht.

Bradley, R. and N. Swartz 1979: *Possible Worlds*, Basil Blackwell, Oxford.

Bunge, M. 1963: *The Myth of Simplicity*, Prentice Hall, Englewood Cliffs.

Douven, I. *manuscript*: "The Evidential Support Theory of Conditionals", to appear in *Synthese*.

Earman, J. 1992: *Bayes or Bust. A Critical Examination of Bayesian Confirmation Theory*, MIT-press, Cambridge.

Hailperin, T. 1996: *Sentential Probability Logic*, Lehigh University Press, Bethlehem.

Howson, C. and P. Urbach 1989: *Scientific Reasoning: the Bayesian Approach*, Open Court, La Salle.

Jackson, F. 1979: "On Assertion and Indicative Conditionals", *Philosophical Review* 88, 565–589.

Jackson, F. 1987: *Conditionals*, Blackwell, Oxford.

Kemeny, J. 1953: "A Logical Measure Function", *The Journal of Symbolic Logic* 18.4, 289–308.

Kraus, S., D. Lehmann and M. Magidor 1990: "Nonmonotonic Reasoning, Preferential Models and Cumulative Logics", *Artificial Intelligence* 44, 167–207.

Kuipers, T. 1978: *Studies in Inductive Probability and Rational Expectation*, Synthese Library 123, Reidel, Dordrecht.

Kuipers, T. 2000: *From Instrumentalism to Constructive Realism*, Synthese Library 287, Kluwer Academic Publishers, Dordrecht.

Kuipers, T. 2001: *Structures in Science*, Synthese Library 301, Kluwer Academic Publishers, Dordrecht.
Kuipers, T. 2006: "Inductive Aspects of Confirmation, Information, and Content", forthcoming in R. E. Auxier and L. E. Hahn (eds.), *The Philosophy of Jaakko Hintikka*, Library of Living Philosophers, Vol. 30, Open Court, LaSalle, pp. 855-883.
Kuipers, T. 2007: "On Two Types of Idealization and Concretization. The Case of Truth Approximation", in J. Brzezinski et al., *The Courage of Doing Philosophy: Essays dedicated to Leszek Nowak*, Rodopi, Amsterdam/New York.
Lewis, D. K. 1976: "Probabilities of Conditionals and Conditional Probabilities", *Philosophical Review* 85, 297–315.
Makinson, D. 2005: *Bridges from Classical to Nonmonotonic Logic, Texts in Computing*, Kings College, London.
Malinowski, J. *manuscript*:, Bayesian Propositional Logic.
Niiniluoto, I. 1987: *Truthlikeness*, Reidel, Dordrecht.
Niiniluoto, I. 1999: *Critical Scientific Realism*, Oxford University Press, Oxford.
Niiniluoto, I. 2007: "Evaluation of Theories", in T. Kuipers (ed.), *General Philosophy of Science: Focal Issues, Handbook of the Philosophy of Science*, Vol. 1, Elsevier, Amsterdam, pp. 175-217.
Niiniluoto, I. and R. Tuomela 1973: *Theoretical Concepts and Hypothetico-Inductive Inference*, Synthese Library, Vol. 53, Reidel, Dordrecht.
Popper, K. 1934/1959: *Logik der Forschung*, Vienna, 1934; translated as *The Logic of Scientific Discovery*, Hutchinson, London, 1959.
Popper, K. and D. Miller 1983: "A Proof of the Impossibility of Inductive Probability", *Nature* 302, 687-688.
Wagner, C. G. 2004: "*Modus Tollens* Probabilized", *The British Journal for the Philosophy of Science* 55, 747-753.

Isaac Levi[*]

Is a Miss as Good as a Mile?

Ilkka Niiniluoto has been an effective champion of the cause of making some notion of verisimilitude important to scientific methodology.

Niiniluoto defends a conception of scientific progress as progress towards the truth and maintains that a well-crafted conception of verisimilitude contributes to our understanding of this conception of progress. In this setting, verisimilitude is taken to be relevant to the *ultimate* aims of scientific inquiry.

Verisimilitude is also important, according to Niiniluoto, in understanding the common features of the *proximate aims of specific inquiries*. Niiniluoto shares with me the view that sometimes scientific inquirer's should change their states of belief in a way that best promotes proximate cognitive goals that are autonomous in the sense that they are not readily reducible to non cognitive economic, moral, political, aesthetic or prudential aims.

Niiniluoto and I disagree, however, concerning our understanding of progress in inquiry, of the proximate cognitive goals of specific inquiries and of the range of activities that are regulated by these goals. I have discussed our disagreements in Levi (1986). Niiniluoto returned the favor in Niiniluoto (1987). The following remarks are a response mainly to his discussion in chapter 12 of that classical treatment of verisimilitude.

At the outset, I should emphasize a point that is, I believe, overlooked in Niiniluoto's own comparisons between his and my proposals. Although central elements of the model of inductive inference developed in *Gambling with Truth* (Levi 1967a) are preserved in my subsequent accounts of inductive inference or inductive expansion, in the early 1970's my perspective underwent some important changes.[1]

---

[*] John Dewey Professor of Philosophy Emeritus, Columbia University, New York, NY.
[1] During the period from 1970–75, I read various versions of a paper "Truth, Fallibility and the Growth of Knowledge" at Rutgers University, Case Western Reserve

Inductive inference is a species of change of belief. What kind of change and in what kind of belief? Around 1970, I became convinced that the change in question is a change in state K of what I now call "full belief" or "state of absolute certainty".[2] Most importantly I insisted that this state of absolute certainty is subject to modification not only by expansion where new information is added to the standard for serious possibility but by contraction where information initially judged to be certain is legitimately dislodged from this

---

University, Rockefeller University, Cambridge University, The London School of Economics, the British Society for the Philosophy of Science, the University of Michigan, the University of Pittsburgh and Boston University. It was published in Cohen and Wartofsky (1983), pp. 153–174 and reissued as ch. 8 of Levi (1984) that gave the first expression of my change of view. Other early expressions are found in Levi (1974) reprinted in Levi (1997) and Levi (1976). The mature expression of this view appeared in Levi (1980). The revisions of Levi (1967a) found in these references combined with the revision of the model of epistemic utility that appeared in Levi (1967b) (reprinted in Levi 1984, ch. 5) together with the modifications required when indeterminacy in utility; and probability is allowed constitute the major departures from the model of inductive inference proposed in Levi (1967a).

[2] I also called it "X's state of knowledge". That is because I think that the condition "X at t knows that h if and only if X at t fully believes that h and it is true that h" is a perfectly acceptable characterization of knowledge insofar as definitions of knowledge express our epistemic ideals. From X's point of view K prior to making the inductive leap, X fully believes that all propositions in K are true and that there is no serious possibility that they are false. As far as X is concerned, X has true full belief that all propositions in K are true.

The received view is that additional conditions need to be met in order that the members of K can qualify as knowledge. One such condition is that X have a warrant for X's full belief that h. Following Peirce, however, I suggest that X is not under any rational requirement to have full beliefs that are justified. X should justify *changes* in X's state of full belief from K to K* but need not justify the contents of X's current state of full belief K. Thus, the terminological dispute concerning the use of the verb "to know" is an expression of a dispute over ideals. I continue to embrace my anti foundationalists ideals but when this issue is not the one being addressed, I tend to avoid calling K a state of knowledge as being gratuitously inflammatory.

Perhaps, it also may seem inflammatory to call this state a state of "absolute certainty". My excuse for persisting in this usage is that it is widely acknowledged that assigning probability 1 may be necessary but not sufficient for being certain in the sense I have in mind. And one of the many ways the difference between cases where probability corresponds to full belief and cases where it is not has been expressed is by contrasting absolute with moral certainty or absolute certainty with almost certainty.

status.³ I insisted that X at t may coherently rule out the serious possibility that any consequence of X's state of full belief is false while maintaining that it is seriously possible that information entailed by K will be subject to legitimate scrutiny in the future – scrutiny that requires its removal from the standard for serious possibility.

In adopting this view of inductive inference as change in state of full belief or state of knowledge, I was turning my back on the dogma of fallibilism – the thesis that for every extralogical h,⁴ if X fully believes that h, there is, nonetheless, a serious possibility that the belief that h is false. I did, however, retain corrigibilism – the thesis that if X believes that h, there is a serious possibility according to X that X will have good reason to give up that belief. Thus, I abandoned the thesis that fallibilism and corrigibilism are equivalent – a claim taken for granted by foundationalists and antifoundationalists (including authors like William James) alike.

In Levi (1967a), I had not as yet made a clear break with the notion that fallibilism and corrigibilism are equivalent. Indeed, the main project of that book was to identify a notion of belief that (a) satisfied this equivalence and (b) recognized the legitimacy of inductive expansion of a state of full belief where belief was understood in a sense that is relevant to both practical deliberation and scientific inquiry while preserving the autonomy of the cognitive interests of scientific inquiry.

In Levi (1967a), I reviewed and rejected three approaches: (1) naïve cognitivism according to which belief that h is preparedness to act as if h were true relative to every practical goal, (2) a behavioralism according to which belief that h is preparedness to act relative to some specific goal selected by context, (3) probabilism according to which belief is replaced by degree of belief spelled out in terms of betting rates. I elaborated a fourth view, critical

---

³ Formally deductively closed theories that represent potential states of full belief may be changed in other ways as well: by replacing K containing h with K' containing $\sim h$ and by "residual shifts that are neither expansions, contractions nor replacements. But replacements and residual shifts may be decomposed into sequences of expansions and contractions. I have required *justified* changes in states of belief to be sequences of such expansions and contractions in some order or other each step of which can be justified. (See Levi 1991.)

⁴ I mean also to include views that add to the logical truths, various allegedly conceptual or *a priori* but extralogical truths.

cognitivism according to which change of belief is evaluated relative to autonomous cognitive goals. This elaboration took the form of a model of cognitive decision-making that was the centerpiece of the book. However, at the close of the book, I acknowledged that belief so conceived is not sufficient to serve as evidence in subsequent inquiries. Whether or not X is justified in coming to believe that h according to X's cognitive goals, if these goals do not justify X in accepting h as evidence and thereby ruling out the logical possibility that h is false, the justification has not served its purpose.

I sought to make repairs but, as I read my own proposals, I do not see how the repairs remove the difficulty. If X is justified in accepting h as evidence, X must rule out the falsity of h as a serious possibility. But if beliefs that qualify as evidence are subject to revision as I thought they should be, they might be false according to the equivalence of corrigibilism and fallibilism. If serious possibility of being false (fallibility) and revisability (corrigibility) are equivalent, we cannot ever be justified in coming to accept the truth of h as evidence.

By 1970, I was prepared to bite the bullet and abandon the equivalence. I began reworking my epistemological outlook. The only option to abandoning the equivalence seemed to me to be to embrace some variant or other of R. C. Jeffrey's "radical probabilism" (Jeffrey 1965). But I could not in good conscience do so. Credal or subjective probability judgment is an expression of fine-grained discriminations among propositions that count as seriously possible. One could not eliminate full belief, certainty or knowledge in favor of judgments of probability. Judgments of probability *presuppose* judgments of serious possibility. These judgments of serious possibility are derivable from states of full belief.[5] Propositions judged seriously possible may carry any probability value from 0 to 1 *inclusive*. To rule h out as a serious possibility requires assigning it 0 probability. But propositions assigned 0 probability may be judged to be serious possibilities coherently. To be absolutely certain that h

---

[5] It is worth noting that B. De Finetti (1974, ch. 2) developed a notion of possibility very close to close to serious possibility as I understand it. The chief differences seem to be found in De Finetti's ambivalence concerning how to address the question of logical omniscience. Modulo that issue, it is clear that De Finetti is committed to a view in opposition to radical probabilism.

requires assigning *h* probability 1. But one may assign probability 1 to *h* without being absolutely certain that *h*.[6]

Some fallibilists maintain that all logical possibilities are serious possibilities. If X's state of full belief is X's standard for serious possibility, X should restrict X's full beliefs to logical truths as many skeptics insist. Other fallibilists allow a somewhat stronger standard for serious possibility. But the tradition required that full belief should be immune to revision. If and only if there is an important sense in which belief is fallible, such belief must be corrigible.

By distinguishing between the fallibilism of full belief (which I reject) and the corrigibilism of full belief (which I endorse), I was also able to embrace more whole-heartedly Peirce's model of inquiry as focused on justifying changes in standards for serious possibility. I endorsed Peircean fallibilism insofar as it amounted to corrigibilism while rejecting this fallibilism insofar as it entailed commitment to a fixed single standard of serious possibility. These important modifications of my ideas took place in 1970 although I have spent much of my career since that time in spelling out their ramifications.

In inductive (or ampliative) inference, X adds to K some sentence *h* representing a potential answer to X's problem, as X recognizes it, and forms

---

[6] Radical probabilists may undertake to draw the distinction between serious possibility and impossibility with the aid of conditional probability. Following de Finetti (1974) and H. Jeffreys (1948), I took conditional credal probability to be the primitive notion in a rational reconstruction. And the set of seriously possible propositions according to X at t are those propositions y for which the credal probability $Q(x/y)$ are well defined whether or not $Q(y)=0$. But this approach remains unsatisfactory. Agent X's credal state B is intended to be a function of X's state of full belief at time t. X is supposed to be committed to a rule for adopting credal states given states of full belief. I call such a function X's *confirmational commitment* $C: K \rightarrow B$ at time t (Levi 1980). Keynes (1921), Jeffreys (1948) and Carnap (1962) all hoped to identify a standard confirmational commitment dictated by probability logic obligatory on all rational agents. Had this program been achievable, it is entertainable that full belief could be characterized in terms of this standardized confirmational commitment and credal probability so that the spirit if not the letter of radical probabilism could be satisfied. For those of us who doubt the feasibility of a program for identifying a standard confirmational commitment, there can be different credal states associated with the same state of full belief according to different confirmational commitments. In principle, a rational inquirer can change confirmational commitments independently of changes in state of full belief. From this perspective, radical probabilism seems utterly untenable.

the deductive closure of K and h. This is the expansion of K by adding h (Levi 1974) and following Alchourrón, Gärdenfors and Makinson (1985) is written K^+_h. X also expands X's "corpus of knowledge" in response to sensory inputs and to testimony of witnesses X trusts. This type of expansion is "routine expansion" (Levi 1980) and differs from "deliberate expansion" (Levi 1991) in being the product of following a program to which X is precommitted rather than being deliberately chosen as the best answer to the given question according to the standards of evaluation X endorsed in state K. I focus on inductive expansion here.[7]

There are three other important modifications of the model from (Levi 1967a) worth mentioning. One modification was proposed at about the time (Levi 1967a) was published in Levi (1967b) written in response to some challenging work of J. Hintikka and J. Pietarinen (1966) that came to my attention after I had completed the manuscript for Levi (1967a). Another is a development of the idea of iterating the application of inductive expansion rules suggested but not developed in Levi (1967a) and a comparison of this

---

[7] The importance of this point cannot be overrated. Let us suppose that X begins with corpus B and expands the corpus routinely by adding the beliefs formed in response to observation of the outcome of an experiment. Call the result $K = B^+_e$. X then assesses on the basis of the information in K which of a roster of potential answers in K to add to K in order to answer some specific question.

According to an alternative scenario, X precommits him (or her) self to a program before observing the outcome of the experiment for determining which potential answer to adopt afterwards. X evaluates the program on the information B and chooses a program. X makes the observation and in respeonse to sensory stimulation forms not only the belief e but the belief h and adds e∧h to B. In this scenario, X does not infer the conclusion h from background B and new evidence e. X uses, instead, the "data" as input and not as "evidence" or as premise in the implementation of a program. Often enough the program is implemented because the agent has unreflectively embraced it because of nature or nurture. But if the agent X chose it as best among a set of rival programs, X chose it in ignorance of what the outcome of the experiment would be. The output of the program and stimulus is belief that e∧h added to B. There is no inference from e (relative to background B) to h. In inductive expansion there is.

This distinction would not be terribly important if it were always true that the precommitment to the routine reproduced the results that legitimate inductive inference warrant. In the typical case this is not true. One of the most obvious differences is that routine expansion can be flawlessly implemented and result in inconsistency. Deliberate or inductive expansion into inconsistency is never legitimate.

iterated inductive expansion with models of non monotonic entailment (Levi 1997; 2001). The third is the generalization of these ideas that emerges when probabilities and utilities go indeterminate. (See Levi 1979 republished as ch. 7 of Levi 1984 and Levi 1980, ch. 2.)

Let a roster of potential answers to a given question relative to initial state of belief K be given. I require the set of potential answers relative to K to be representable by a set U_K of hypotheses such that K entails that exactly one member of U_K be true and that each member of U_K is consistent with K. (I restrict attention to finite U_K.) Given this ultimate partition U_K, a potential answer is representable as the *rejection* of some subset of U_K and the expansion of K by taking the disjunction of the unrejected members of U_K together with all the logical consequences of K and that disjunction.

The set U_K is the set of potential answers that provide complete answers to the question under investigation as is demanded by the inquirer. The elements of U_K are not maximally consistent expansions of K or representations of possible worlds in any other sense than this.[8] During the course of inquiry, the inquirer may change his or her demands for information and modify U_K and the notion of a complete answer accordingly. Niiniluoto does not appear to disagree with this notion of a complete answer.

In Levi (1967b), as in Levi (1967a), attention was focused on two-dimensional epistemic goals. The inquirer is concerned to avoid error and to obtain valuable information. I constructed epistemic utility functions that are weighted averages of a utility function $T(h,x)$ that would be exclusively operative if the inquirer were concerned solely with avoiding error in adding h to K and a utility function $C(h)$ that represented the value of new information obtained by adding h to K if X were concerned solely with adding new information.

The concern to avoid error may be represented by the utility function $T(h,t) = 1$ and $T(h,f) = 0$.[9]

Let $M(x)$ be a probability distribution over U_K. For any potential answer $M(h) = \sum_{|h|} M(x)$ where $|h|$ is the set of elements of U_K that entail h given K. $Cont(h) = 1-M(h)$ (or any positive affine transformation $aCont(h) + b$ where

---

[8] I would go further and deny that there is any acceptable theoretical reason for recognizing a true or privileged partition into possible worlds.

[9] Or any positive affine transformation $aT + b$ where $a>0$.

a > 0) represents the informational value of expanding K by adding h. The probability $M(h)$ is the informational value determining probability and is not the probability $Q(h)$ used to determine expectations.

The epistemic utility of adding h to K in an effort to obtain new error free and valuable information when h carries the truth-value x is $\alpha T(h,x) + (1-\alpha)C(h)$ where $0 \le \alpha \le 1$. The expected epistemic utility of adding h is $\alpha Q(h) + (1-\alpha)C(h)$. When α is restricted to non negative values less than or equal to 1 so that no error is preferred to an answer that avoids error, $Q(h)-qM(h)$ is a positive affine transformation of this expected epistemic utility where $0 \le q \le 1$. Maximizing this index maximizes expected epistemic utility. A maximum is reached when all elements x of U_K such that $Q(h) < qM(h)$ are rejected and none where $Q(h) > qM(h)$ are rejected. When $Q(h) = qM(h)$, rejection is optional. I have argued that adopting the weakest optimal answer is to be recommended. This means that x in U_K is rejected if and only if $Q(x) < qM(x)$.

In Levi (1967a), I assumed that for elements of U_K, the value of $M(x) = 1/n$ where n is the number elements of U_K. Otherwise the account proposed is the same as the one I subsequently endorsed. In particular, all four conditions (L1)–(L4) correctly attributed to me on p. 414 of Niiniluoto (1987) are satisfied in my later work as long as the M-function is any probability distribution over U_K assigning elements of U_K positive values. This includes (L3) according to which the epistemic utility of adding h to K when h is false increases with *Cont(h)*. Niiniluoto (1987) concludes from this that all "complete" potential answers in U_K must carry the same content – i.e., the same informational value and, hence the same epistemic utility.

It is true that given the *extra* assumption found in Levi (1967a) that complete potential answers carry equal informational value, L3 entails that all false complete potential answers carry equal epistemic utility. But I gave up the extra assumption a long time ago and it is not true that L3 (which I continue to hold in case assessments of informational value are determinate) implies this form of equal treatment for erroneous complete answers. Nor does my view imply the analogous claim for correct complete potential answers. L2 entails that all error free complete potential answers carry epistemic utility that increases with their informational value or *Cont*-value. But these values need not be the same.

Niiniluoto maintains that his non-triviality condition M3 for assessments of truthlikeness (Niiniluoto 1987, p. 233) conflicts with L2 and with L3. I am not sure I understand M3 so I quote it:

> (*Non-triviality*) All true statements do not have the same of degree of truthlikeness; all false statements do not have the same degree of truthlikeness.

Niiniluto cannot intend to suggest that no true (false) statements have the same degree of truthlikeness. And to insist that some true (false) statements do not have the same degree of truthlikeness is to trivialize non-triviality. The trivial reading entails M3. Neither L2 nor L3 is ruled out. Indeed, Niiniluoto's own approach must allow for the possibility that in some inquiries all false complete answers are equally far from the truth even if it is not the most interesting case. This is quite parallel to my view according to which all false complete answers may (but may not) carry equal informational value.

The differences in outlook that exist between Niiniluoto's approach with its emphasis on distance from the truth and the epistemic utility model I favor cannot be found in the circumstance that the model proposed in Levi (1967a), required elements of U_K to carry equal informational value and, hence, treated all cases of importing complete but false answers the same. I abandoned this feature of the model almost immediately after the book was published some thirty-eight years ago. Both Niiniluoto's approach to evaluating verisimilitude and my approach to assessing informational value in inductive expansion have two components: (1) An assignment of informational value or verisimilitude to "complete" potential answers (i.e., elements of U_K) and (2) a method of deriving assignments of informational value (verisimilitude) for Boolean combinations of these. Both approaches allow for the possibility that false elements of U_K can be assigned equal value in some contexts.

Niiniluoto's condition M4 maintains that stronger true answers are closer to the truth than weaker ones. It is a weaker version of L2 that the difference in epistemic utility between two true answers h and g is linear in the difference in their *Cont-values*, and is, therefore, also satisfied by my approach. But when condition M5 that denies that the truthlikeness of false answers covary with logical strength is converted into a condition on epistemic utility, it conflicts with L3. Both Niiniluto and I agree that the values of false answers need not be the same. For me, the variation is a variation in the *Cont*-value or the value of information. For Niiniluoto the variation is due to distance from the truth. Is there more than a verbal difference here?

The answer is "Yes". Given any two potential answers carrying the same truth-values and the same informational values, conditions L2 and L3 entail that the epistemic values assigned these answers should be the same. Truth-likeness, according to Niiniluoto, can violate this implication of L2 and L3. Thus, two complete false answers can (although they need not) carry the same *Cont*-value and yet differ with respect to closeness to the truth. Indeed, a similar remark holds for any pair of false answers carrying the same *Cont*-value whether they are complete or not.

There is another point of difference of relevance here. In inductive expansion, I claim that an erroneous answer should never be preferred to a true one – no matter how close to the truth in Niiniluoto's or anybody else's sense it may be.

Both of these differences give some rhetorical substance to the slogan I introduced many years ago in a previous comment on Niiniluoto's notion of verisimilitude: A miss is as good as a mile (Levi 1986, p. 357). That is the attitude I favor in evaluating rival potential inductive expansions. Niiniluoto clearly disagrees.

I conceded that I can offer no demonstration of the superiority of my approach to Niiniluoto's. But there is one point that I do think weighs heavily in favor of the model of the aims of inductive expansion I favor.

From the point of view of an inquirer X with initial state of full belief K and ultimate partition U_K, for every x in U_K, it is a serious possibility that x is false. X is contemplating rejection of members of some subset R of U_K as serious possibilities. To reject x as a serious possibility is to change to the view that the falsity of ~x is not a serious possibility from the view that it is.

In contemplating the change, X could register complete indifference as to whether x is true or false without incoherence. Nonetheless, one might well wonder why X should be concerned with change in the standard for serious possibility at all if avoiding error in ruling out x as a serious possibility were of no concern. And if X does care whether x is true or false, it seems at least perverse to prefer that x be ruled out as a serious possibility when x is true rather than when it is false.

But perhaps x might be ruled out as a serious possibility when it is true as long as a set of elements of U_K sufficiently similar with respect to truth to x is accepted into evidence as the new standard for serious possibility. But that is just a variant of preferring to rule out x as a serious possibility when it is true rather than when it is false. As just indicated, that seems just plain perverse.

Keep firmly in mind the problem under consideration. This is the problem of justifying inductive expansion understood as justifying a change in standard for serious possibility by eliminating propositions initially judged to be seriously possible from that status. One can consider other contexts where a deductively closed theory is formally expanded *but where no change in standard for serious possibility is contemplated.* This happens in suppositional reasoning where one asks what would be the case were h true. In answering such a question, one supposes that h is true without changing one's view as to whether h is true or false. Whether X is convinced that h is true or that h is false or is in suspense, supposing that h is true does not alter X's view. X is engaged in a form of fantasizing. The argument I have just offered does not apply to such transformations.

Niiniluoto declares himself a fallibilist but nonetheless seeks to address my concerns. He writes:

> The fallibilism defended in this book agrees with Levi that hypotheses are often tested and accepted in order that they can be used as background knowledge "in subsequent inquiries in science" (Levi 1980, p. 24). I also agree with Levi, contra Popper, that this expansion is a result of *inductive* inference. Furthermore, even if my epistemic utility assignment does not satisfy Levi's L1, avoidance of error is a serious concern of my measure of truthlikeness.
>
> However, it is important to acknowledge also that in many inquiries it is not rational to accept *any* hypothesis *as true* in Levi's strong sense. There are inquiries where the cognitive problem is defined relative to a false presupposition.... For many theoretical and practical purposes, the best we can do is to study some problem with counterfactual idealizing assumptions. In such situations, avoidance of error cannot be an absolute demand: it will be replaced by the desire to find the least absolute demand. Inquiries of this sort are even more 'myopic' than Levi recommends: their results are not intended to be preserved in the corpus of knowledge in the long run since they can be replaced in the short run by more 'concrete' results relative to more realistic background assumptions (Niiniluoto 1987, pp. 421–422).

As I understand the use of background information (and empirical data) as evidence, that use is as a standard for serious possibility in the sense briefly explained at the beginning of the discussion. In the first paragraph of the citation from Niiniluoto, there is, therefore, nearly complete agreement. I say "nearly" because the final sentence of the paragraph continues to defend the use of assessments of truthlikeness in the setting of efforts to change standards for serious possibility by induction.

Niiniluoto maintains that "it is not rational" to accept any proposition as true in my strong sense in contexts where inquiry procedes on suppositions that the inquirer believes are false.

I entirely agree with Niiniluoto that in many inquiries (or in many stages of inquiry) one asks what would or might be the case were a given supposition true where the supposition may, in some contexts be ruled out as a serious possibility by the inquirer, in other contexts may be judged to be certainly true and in others held in suspense. (See Levi 1997, ch. 1.) I agree also that in contexts where the supposition is judged to be certainly false, the answer sought to the question raised cannot, from the inquirer's point of view, be judged to be certainly true.

Consider the kind of issue with which we are dealing in cases of belief-contravening supposition. Inquirer X starts with state of full belief K. X supposes that hypothesis e is true purely for the sake of the argument even though X is convinced that e is false. Supposing that e is true requires a transformation of K to some alternative potential state of belief K' to the current state – a potential state of belief that does not represent X's current point of view. However, K' does represent a state X is supposing to be true in the sense that supposition is a form of fantasy. The fantasy could be formed for X's amusement. But it may have a more earnest application. Thus, the behavior of balls moving down frictionless inclined planes is often instructive when we seek to understand the motions of bodies under conditions of daily life.

There is some controversy as to the nature of the transformation from K to K'.

According to a widely held view, X should consider those possible worlds in which e is true closest to the actual world and ask what is true in all these possible worlds. Let h be true in all the closest e-worlds to the actual world. X can then claim "if e were true, h would be true".

According to the view I favor, if X is convinced that e is false, K' should be the product of first "contracting" K by removing ~e in a manner that minimizes loss of valuable information and then adding e and forming the deductive closure. There are several competing accounts of how to assess loss of valuable information. But as long as K entails ~e, the net product of contraction followed by expansion should satisfy the requirements of AGM revision K*e.

For the purposes of this discussion, it makes little difference whether the Lewis approach (Lewis 1975) which yields K#e or the AGM (1985) approach yielding K*e is adopted. The question before us concerns the inductive expansion of the revision. We may call it the inductively extended revision of K by adding e according to Lewis $(K\#e)^i$ or according to AGM $(K*e)^i$. (See Levi 1996, ch. 6.) Consider, for example, the belief contravening supposition that a coin in my pocket is tossed a billion times. Presumably, the number of heads would on that supposition be any integer from 0 to a billion. By inductively extending this suppositionally grounded thesis, we may conclude that the coin would land heads approximately a half billion times.

Whether the inductively extended suppositional reason is $(K\#e)^i$ or $(K*e)^i$, the inductive expansion is based on suppositions that are believed false. X cannot be concerned to avoid error when error is judged according to X's evolving doctrine K.

Niiniluoto's point is indisputable. But it is not clear what its relevance is. In inductively extended suppositional reasoning, the effort to obtain new valuable error free information is based on judgments of truth and falsity controlled by the suppositional doctrine K#e or K*e. X is certain that the suppositional doctrine is false; but X supposes for the sake of the argument that it is true and is the standard for serious possibility.

In any case, appealing to closeness to the truth faces the same difficulties (if they are, indeed, difficulties) that avoiding error does when engaged in inductively extended reasoning from false suppositions. Verisimilitude is based in such cases on distance from claims that are supposed to be true even though they are believed to be false.[10]

---

[10] Niiniluoto points out rightly that we often resort to suppositional reasoning for practical purposes and not for purely theoretical purposes. But we also resort to suppositional reasoning for theoretical purposes – that is to say, for cognitive purposes.

Suppositional reasoning is an important feature of scientific inquiry. It is sometimes belief contravening. When explanations of events are offered, suppositional reasoning is belief conforming. In practical deliberation aimed at making decisions, the suppositions are typically "open". The deliberating agent is in suspense. In all such cases, the reasoning may sometimes involve inductively extended argumentation. In such cases, the reasoning simulates efforts to avoid error while acquiring new error free information.

Of course, we cannot take the simulation literally. Confusing fantasy with reality is a form of madness. Niiniluoto seeks to convert this important truth into a case for verisimilitude. In my judgment, the importance of appreciating the difference ought to remind us that after all a miss *is* as good as a mile.

References

Alchourrón, C., P. Gärdenfors, and D. Makinson, D. 1985: "On the Logic of Theory Change: Partial Meet Functions for Contraction and Revision", *Journal of Symbolic Logic* 50, 510–530.

Carnap, R. 1962: *The Logical Foundations of Probability*, University of Chicago Press, Chicago.

De Finetti, B. 1974: *Theory of Probability*, v. 1, Wiley, New York.

Hintikka, J. and J. Pietarinen 1966: "Semantic Information and Inductive Logic", in J. Hintikka and P. Suppes (eds.), *Aspects of Inductive Logic*, North Holland, pp. 96–112.

Jeffrey, R. C. 1965: *The Logic of Decision*, McGraw-Hill, New York.

Jeffreys, H. 1948: *Theory of Probability*, 2nd ed., Clarendon Press, Oxford.

Keynes, J. M. 1921: *A Treatise on Probability*, Macmillan, London.

Levi, I. 1967a: *Gambling with Truth*, Knopf, New York. Reprinted in paper in 1983 by MIT Press.

When the suppositional reasoning is inductively extended, the concern is to avoid error while acquiring valuable new information on the supposition that the premises of the suppositional reasoning are true.

Niiniluoto goes on to point out that invoking a double standard for serious possibility – one for theoretical and one for practical purposes "implies an untenable dualism between theory and practice" (Levi 1980, p. 72). Niiniluoto contends that this particular dualism was "an important message" of Levi (1967a). I did not claim that belief is a standard for serious possibility in that essay but it is easy to give that reading to what I did say. Given that reading, I was advocating such a double standard. This observation is another variant of the claim I accept that in the early 1970's I revised my earlier view. I would say that I *corrected* my earlier view.

Levi, I. 1967b: "Information and Inference," *Synthese* 17, 3659–3691.
Levi, I. 1974: "On Indeterminate Probabilities," *The Journal of Philosophy* 71, 391–418.
Levi, I. 1976: "Acceptance Revisited", in R. Bogdan (ed.), *Local Induction*, Reidel, Dordrecht, pp. 1–71.
Levi, I. 1979: "Abduction and Demands for Information", in I. Niiniluoto and R. Tuomela (eds.), *The Logic and Epistemology of Scientific Change*, Proceedings of a Philosophical Colloquium, Helsinki, December 12–14, 1977 (Acta Philosophica Fennica 30), North-Holland, Amsterdam, pp. 405–429.
Levi, I. 1980: *The Enterprise of Knowledge*, MIT Press, Cambridge, Mass.
Levi, I. 1983: "Truth Fallibility and the Growth of Knowledge", in R. S. Cohen and M. Wartofsky (eds.), *Language, Logic and Method*, Reidel, Dordrecht, pp. 153–174.
Levi, I. 1984: *Decisions and Revisions*, Cambridge University Press, Cambridge.
Levi. I. 1986: "Estimation and Error Free Information", *Synthese* 67, 347–360.
Levi, I. 1991: *The Fixation of Belief and Its Undoing*, Cambridge University Press, Cambridge.
Levi, I. 1996: *For the Sake of the Argument*, Cambridge University Press, Cambridge.
Levi, I. 1997: *The Covenant of Reason*, Cambridge University Press, Cambridge.
Levi, I. 2001: "Inductive Expansion and Nonmonotonic Reasoning", in M.-A. Williams and H. Rott (eds.), *Frontiers in Belief Revision*, Kluwer, Dordrecht, pp. 7–56.
Niiniluoto, I. 1976: "Inquiries, Problems and Questions: Remarks on Local Induction", in R. J. Bogdan (ed.), *Local Induction*, Reidel, Dordrecht, pp. 263–296.
Niiniluoto, I. 1979: "Verisimilitude, Theory-Change and Scientific Progess", in I. Niiniluoto and R. Tuomela (eds.), *The Logic and Epistemology of Scientific Change*, Proceedings of a Philosophical Colloquium, Helsinki, December 12–14, 1977 (Acta Philosophica Fennica 30), North-Holland, Amsterdam, pp. 243–264.
Niiniluoto, I. 1980: "Scientific Progress", *Synthese* 45, 427–462.
Niiniluoto, I. 1982: "What Shall We Do with Verisimilitude", *Philosophy of Science* 49, 181–197.
Niiniluoto, I. 1986: "Truthlikeness and Bayesian Estimation", *Synthese* 67, 321–346.
Niiniluoto, I. 1987: *Truthlikeness*, Reidel, Dordrecht.

Graham Oddie

Truthlikeness and Value

In Scandinavian Universities it is customary for there to be a Chair of Practical Philosophy and a Chair of Theoretical Philosophy. And it is the Chair of Theoretical Philosophy at the University of Helsinki that Professor Ilkka Niiniluoto has occupied with distinction for over a quarter of a century. In his book *Is Science Progressive?* Ilkka (Niiniluoto 1984) characterizes the theoretical-practical divide thus: whereas the central question of theoretical philosophy is *what should be believed?* the central question of practical philosophy is *what should be done?* In the course of his career Ilkka has made significant and lasting contributions in answer to the core question in theoretical philosophy – within the sub disciplines of epistemology, philosophy of science and logic. In particular he has made seminal and original contributions both to the theory of induction and to the theory of verisimilitude, and to illuminating their central and intertwining roles in any adequate realist, fallibilist account of science.

I have to confess that I am innately disposed to admire all things Scandinavian. Perhaps this is because some Oddie ancestors were Vikings, who, after they had grown tired of their notorious and somewhat disruptive adventures abroad, settled down and became peaceful, productive and (I like to think) reflective farmers in the Ribble Valley in Lancashire, England. In spite of this innate, possibly inherited, disposition, over the years I have become increasingly dissatisfied with this one Scandinavian tradition – the neat and institutionally enshrined Practical/Theoretical divide. Things are even worse in the United States, of course, where the institutional expectation is that one specialize in some much narrower sub-discipline, like philosophy of physics, or biomedical ethics, or philosophy of music. This seems wrong. Philosophy is, or ideally ought to be, a single garment woven from a seamless cloth.

I have come to agree with Plato that the unity of philosophy derives from a central question, one which informs all the others (including the two mentioned above), namely *what is the Good?* And the point of doing philosophy is to try to become better acquainted with the Good. If Plato is right, then all problems in philosophy are, or at the very least involve, problems of value. I do not say that Theoretical Philosophy is, at bottom, Practical Philosophy, but rather that both are, at bottom, branches of *Axiology*, the theory of value.

Like Ilkka I am too thoroughly a product of the Analytic tradition to rest content with making sweeping unsubstantiated claims. I have had drilled into me Russell's famous quip about the relative virtues of theft and honest toil. Thus I feel compelled to toil, at least a little, to make this particular grandiose claim of mine seem a bit more plausible. And I propose to do this by taking a problem (one which both Ilkka and I have both beavered away at over the years) which apparently lies at the very purest end of the theoretical spectrum – the problem of truthlikeness – and show that it is a problem, or bunch of problems, in value theory. In the first part of this paper I discuss the general problem of the value of truthlikeness. In the second part I show in some detail that the theory of the structure of truthlikeness is an aspect of the theory of the logical structure of value.

1 At its simplest the problem of truthlikeness is the problem of determining what it takes for one proposition to be closer to the truth than another. That way of framing the problem presupposes that there is indeed a non-trivial ordering on propositions of closeness to truth, and this presupposition may well, on closer philosophical examination, turn out to be false. But I am interested in another presupposition, one that lurks in the posing of the problem – namely, the presupposition that we should care about the problem in the first place. Why should we *bother* expending time, energy, printer ink (and Ilkka and I have expended lots of all of these over the last thirty odd years) wrestling with this problem? There are, surely, many other ways of spending one's time, quite a few of them more pleasant or more profitable, than this.

There are a range of possible answers, and a rather glib answer is that truth is *valuable*. Further, if truth is valuable then presumably the more truth the better. Thus of two propositions that are false, one may still be more valuable than the other – the one which contains more truth, is closer to the truth, is more truthlike. If this is so then it would be good to know in what closeness to truth, or truthlikeness, consists, and consequently good for some philosophers to beaver away at the problem of truthlikeness.

This is glib because it pushes the problem back onto an essentially equivalent problem – the value of truth. But it is also glib because it assumes that the fundamental value is being true, that the fundamental disvalue is being false, and the derived value is being truthlike. However it may turn out that the dependence is the other way around. If being close to the truth is what is valuable, then being false may not be such a bad thing in itself, and hence being true may not be such a good thing in itself.

Still, the basic idea that truthlikeness is valuable has something going for it, and I suspect many would endorse it. But it is still puzzling for a couple of reasons. Propositions are, I assume, eternal necessary items. The existence of the proposition that *it is hot, rainy and still*, for example, does not depend on the weather, or on any other contingent facts. The truth, on the other hand, is a highly contingent matter, as is closeness to the truth. The proposition that *it is hot and rainy and windy* is quite close to the truth if the weather happens to be hot and rainy and still, but it is rather far from the truth if it happens to be cold, dry and are still. If truthlikeness is valuable, then the value of a proposition is as fragile and fickle as transient truth itself. Now the *intrinsic* value of a necessary entity presumably cannot fluctuate with the weather, because intrinsic value is the value that an item possesses *in itself*, without reference to any external items or their properties. So if truthlikeness is valuable then it is not an intrinsic value of a proposition – rather, it must be extrinsic, a value which a proposition possesses or fails to possess depending on where the contingent truth happens to lie. Further, if propositions are necessarily existent entities, all of them are already out there with all their values, extrinsic or intrinsic, already in tact. How does the project of inquiry then add to value?

It might be useful to compare the propositional value of truthlikeness with the value of a work of art, like a musical work. Works of music are clearly repeatable items. There have been many different performances of, say, Bach's Goldberg Variations, and hopefully there will be many more. These different performances are all instances of the *same* work. A musical work like this one is a universal, a *sound structure*, one which can be instantiated in different places at different times through different performances or the playing of different recordings. Note that the written score or a CD of the Goldberg Variations is not an instantiation of the work, even though they both share some of the structure of the work, since neither the score nor the CD is a structure of *sound*. Rather, both are devices for encoding the sound structure, which they do by means of an isomorphic structure, in a different medium, together with a contingent code connecting the two structures. The written score and the CD certainly have extrinsic value – given the code, one can use them to access the sound structure, which does indeed have intrinsic value.

As a realist about universals I take these sound structures (universals) to be, like propositions, necessary and eternal entities, although where and when they are instantiated is clearly a contingent affair. Furthermore, it is pretty clear that at least some of the more significant values of a musical work are intrinsic to it. The beauty and richness of the Goldberg variations, for example, is at least in part a necessary product of the relations between the elements of the sound structure that it has, and it has those internal structural features of necessity. So Bach's genius cannot have consisted in *creating* the Goldberg Variations *ex nihilo* – since it was already there amongst the plethora of necessary beings, lurking amongst the myriad of sound structures – but rather in *plucking* such a singularly beautiful, rich sound structure out of an infinite sea of sound structures, nearly all of which are complete rubbish. (I dare say that on any reasonable measure of the set of all sound structures, the set of musical works as beautiful, complex, unified and rich as the Goldberg Variations has measure close to zero. Imagine a random generator of sound structures churning them out *seriatum*. Set 2^{10000} such devices going, working in parallel. I suspect that a sound structure as good as the Goldberg variations would turn up very rarely indeed. Anyone who hears the Goldberg Variations and knows them for what they are must conclude that their discovery was due not merely to the work of an intelligent being, but to the sensibility of an extraordinarily perceptive being.)

Of course, different performances of a musical work may embody quite different values. Is this compatible with the view that the work possesses its values necessarily? The work is a sound structure but it is a sound structure which leaves considerable latitude in performance – not every aspect of the sound is perfectly determined by the work – and so two different performances (or interpretations) may realize different values of the work. One performance might maximize the beauty at some cost to the unified complexity or richness of the work. Another might do the reverse. Whether or not a particular sound sequence counts as a performance of the work depends partly on whether it is a good enough realization of (approximation to) the sound structure, and whether or not there is an intention to make it an instance of that sound structure. A performance need not be a perfect realization of the work for it to be a performance of that work. There is such a thing as a bad performance and it may miss the mark by a long shot. It may be that only a performance of the

Goldberg Variations on a harpsichord would perfectly realize that work, but still a performance on a piano forte might be close enough to count. For all that, the beauty and richness of the *work* resides in the internal structure of sound, something *that* work possesses essentially, of necessity.

If Bach's contribution to value did not consist in *creating* the Goldberg Variations *ex nihilo*, or in creating something of musical value then in what did it consist? His contribution to value consisted in his *selecting* this sound structure for consideration through *recognizing* that this particular sound structure (as it gradually unfolds through the process of selecting its parts from amongst the possible alternatives) is amongst the most beautiful and intriguing of sound structures – something which *we* would never have done left to our own devices – and thereby allowing us also to become acquainted with that singular example of rich beauty. In other words, *tracking* value and *recognizing* value, and thereby *realizing* instances of value to enable others to so recognize it, is itself valuable. Bach was not only the first to attend to that structure, but the first to recognize the beauty and richness of it, the first to become thoroughly acquainted with its value, and the one who thereby enabled the rest of us to become acquainted with that work and its value.

To sum up, then, a particular kind of interaction with a necessary and abstract existent, one which possesses its value intrinsically and of necessity, can itself enhance the value of the world because it consists in the recognition of that value, becoming intimately acquainted with that value, and making it possible for others to also become acquainted with that value. Thus the intrinsic value of a necessary entity can impart value to a grasper of that entity, given that the grasping has the requisite features.

Return now to the value of truthlikeness. Can anything similar be said for the grasping of a valuable proposition? The proposition that *it is hot and rainy and windy* is, like the Goldberg Variations, an eternal and necessary entity, one of many such, but one we are all too familiar with already. It is a rather simple proposition, and it requires no special talent to pluck it out of the set of propositions. It is not much of achievement to focus one's attention on it, or to bring it to the attention of others. Other propositions, like Newton's theory of Gravitation, or String Theory, are, admittedly, rather more difficult to pluck out of the set of all propositions. It takes considerable ingenuity to think of them first, and to single them out for the attention of others. Is the value of singling out String Theory for the attention of others comparable to the contribution to value that Bach made by singling out the Goldberg Variations? It is true that String Theory, as abstract structure, has a certain intrinsic value, one which can be appreciated independently of whether or not it happens to be true or close to the truth. Purely mathematical structures, or possible models of the physical world, can be quite rich and beautiful in themselves. (And if the world has a comparably rich structure then the world also shares that value.) But that is not where the bulk of the value of String Theory, if it indeed it has much value, really lies. The bulk of the supposed value of grasping such theories does not reside in the intrinsic structural features of the theories. Most of the value of a scientific theory resides in its being a good approximation to the truth. The larger portion of the value of these theories, considered as propositions is thus extrinsic. Is there, then, some kind of interaction with such entities the value of which is mostly extrinsic which might transfer the extrinsic value of the structure to the grasper of such?

Suppose that the two of us are locked in that famous concrete bunker, without any access to the external weather conditions and, to pass the time, we amuse ourselves by thinking of and adverting to various propositions about the weather. You say: *I am thinking of the proposition that it is hot, rainy and still.* I say: *I am thinking of the proposition that it is cold, dry and still.* Outside, unbeknownst to us, it is hot, rainy and windy. So the proposition of which you are thinking is closer to the truth than the one of which I am thinking. But your thinking of your proposition doesn't really seem any more valuable than my thinking of mine. Plucking a proposition out of the set of all propositions, grasping it and adverting to it – even a proposition which happens to be extrinsically valuable in virtue of being close to the truth – does not in itself seem to be a particularly valuable thing to do. It isn't at all like Bach's discerning, in the vast ocean of sound structures, the intrinsically valuable Goldberg Variations. What do we have to add to an interaction between bunker residents and weather propositions, to make the interaction itself valuable?

Note that in Bach's case he does not merely pluck a valuable object out of a sea of sound structures randomly. He plucks the Goldberg Variations out because, as they unfold in his mind's ear, he experiences them as rich and beautiful, and Bach's experiences of musical beauty do indeed reliably track musical beauty. Bach selects that sound structure because he is *acquainted* with its value. And his drawing our attention to the Goldberg Variations is valuable because we also have the capacity to become acquainted with that value, even though our becoming so acquainted is much clunkier and less reliable than Bach's. In the bunker you advert to a proposition which happens to be valuable (it is close to the contingent weather truth), but not *because* you recognize it as valuable (you have no reason to think it close to the truth) and your random propositional citations are not in fact a reliable way of tracking truthlikeness.

Your interaction with the value of that proposition in the bunker is thus not itself valuable in the way Bach's interaction with the beauty and richness of the Goldberg Variations is valuable. To be comparably valuable, your selection of that proposition would have to be appropriately informed by, and *guided* by, its value. You would have to be aware of the value and your awareness of the value would have to reliably guide your selection. You have to have some access to its value, and because that value is contingent and extrinsic to the object, you have to have some kind of access to those contingent facts.

Consider a famous example. Part of the value of Newton's alighting on his Law of Gravitation is surely the fact that he arrived at it by effectively *deriving* it from a body of empirical information, encapsulated in Kepler's Laws, together with some highly plausible theoretical assumptions. He did not pluck this star randomly out of the heavens. He selected it, logically guided by what was actually known about the movement of the planets. Of course, what was "known" about the planets was not quite true. The planets do not in fact carve out perfect ellipses for example. But Kepler's laws are pretty damn close to the truth, as is the Inverse Square Law which Newton derived from them (together with other quite truthlike theoretical assumptions). And that is what made his alighting on the Inverse Square Law such a great achievement. Others had speculated that the law of attraction might be an inverse square law. Newton, by contrast, derived it from the empirical phenomena by pure reason. He was tracking truthlikeness through deep acquaintance with that value as realized in Kepler's Laws.

This contrasts with the idea (suggested in Popper's early writings) that it is enough merely to toss out a bold conjecture of a truthlike hypothesis for your conjecturing to realize significant cognitive value. Conjecturing, however, can either be informed or idle. Alighting on a valuable proposition through some uninformed and fortuitous conjecturing does not seem to me to possess any particular cognitive value. It certainly does not possess the value of Newton's derivation of the Inverse Square Law.

Consider this counterargument. Suppose that someone tells the two of us, still trapped in that damned, air-conditioned, sealed-off concrete bunker, to guess what the weather outside is like. We offer our two propositions, mentioned above, as our guesses. Is your *guess* better than mine? If we say, yes, then it seems that merely selecting and grasping a good proposition is better than grasping and selecting a bad proposition, even if both are done at random.

Guess, like all propositional attitude terms, is, however, ambiguous. Sometimes it refers to the mental act of guessing, and sometimes to the propositional content of that guessing. In the content sense, your guess is better than mine, because the content of your guess is closer to the true than the content of my guess. But is your act of *guessing* your proposition any better than my act of guessing mine? Does the guessing inherit the value of the object of the guessing? In guessing something close to the truth, rather than far from the truth, was your mind *ipso facto* superior to mine, despite the fact your guess was not in any way informed by the value of the object guessed? It doesn't seem so to me. Merely selecting a valuable proposition is thus not in itself a valuable thing to do. You have to select it as valuable, because of its value, for the sake of its value. Merely taking some attitude or other, like guessing, to a valuable entity does not make your mental state any the more valuable. Your attitude has to be the appropriate attitude to an object with that kind of value and your taking the attitude has to be a response to the value.

If truthlikeness is valuable then embracing the more truthlike of two rivals for the truth would itself be valuable if the embracing were a response to the value of truthlikeness. There is a further question that needs to be answered. Is this kind of embrace of an extrinsically valuable proposition itself intrinsically or extrinsically valuable? That's certainly an interesting question, one worthy of further investigation, but one we can afford to leave open here.

2 If truthlikeness is valuable then the appropriate tracking of truthlike propositions is also valuable. So much, then, for the general problem of the value of truthlikeness. However, if tracking truthlikeness is a cognitive value, then one would also expect the details of the structure of truthlikeness to be a branch of value theory. In this second part of the paper I illustrate this possibility with a single example.

Let's begin with a sequence of propositions which have struck many (not all) as being illustrative of closeness relations among such. (David Miller is, of course, the most prominent hold-out here (Miller 1974). As we will see Miller's language-variance argument against these judgements and ones like them turns out to have interesting connections with a deep problem in the metaphysics of value.)

P_0 *hot* & *rainy* & *windy*
P_1 not-*hot* & *rainy* & *windy*
P_2 *hot* & not-*rainy* & *windy*
P_3 not-*hot* & not-*rainy* & *windy*
P_4 *hot* & not-*rainy* & not-*windy*
P_5 not-*hot* & not-*rainy* & not-*windy*

If P_5 is the truth then P_0, P_1, and P_3 approach the truth in that order, as do P_0, P_2, and P_4. The relative merit of other pairs in the sequence (P_1 and P_2 for example) presumably depends on the relative merit of true information about different weather factors. Similarly, if P_0 is the truth then P_4, P_3, and P_2 approach the truth in that order. To "capture" such relations amongst maximal conjunctions we can employ a simple additive measure, first suggested by Pavel Tichý, which can be generalized simply enough to other cases (see Tichý 1974). In the simplest case, the distance of a maximal conjunction from the true maximal conjunction is the total number of atomic propositions over which they disagree. This represents the qualitative ordering numerically in as much as it delivers all the intuitively clear judgements, but it goes beyond it by delivering a few more to boot. (It tells us, for example that P_1 and P_2 are equidistant from P_0 thus settling for us a judgement that we might well have hesitated to make, thereby effectively settling the issue of the relative merit of information about different weather factors.)

We can clearly enrich this simple atomic framework and generalize this simple additive measure of distance between maximal conjunctions in various ways. One would be to treat each weather factor (*heat, humidity, air movement*) as a weather determinable and realizations of the determinable as determinates of that factor. So the determinates of the determinable *heat* factor might be: *hot, warm, cold*; the determinates of the *humidity* factor might be: *rainy, drizzly, dry*; the determinates of the *air movement* factor might be: *windy, breezy, still*, and so on. Then there might be more complex sequences of closeness like the following:

Q_0 *hot* & *rainy* & *windy*
Q_1 *warm* & *drizzly* & *windy*
Q_2 *cold* & *rainy* & *breezy*
Q_3 *warm* & *dry* & *breezy*
Q_4 *cold* & *drizzly* & *still*

Such an ordering cold be captured by first assigning distances between determinates (e.g. *cold* is of distance 2 from *hot*, and *warm* is of distance 1; *dry* is of distance 2 from *rainy*, and *drizzly* is of distance 1; *still* is of distance 2 from *windy* and *breezy* of distance 1) and then adding the distances between the determinates of each factor to derive the overall distance between conjunctions.

Q_0 *hot* & *rainy* & *windy*   distance from Q_0: 0
Q_1 *warm* & *drizzly* & *windy*   distance from Q_0: 2
Q_2 *cold* & *rainy* & *breezy*   distance from Q_0: 3
Q_3 *warm* & *dry* & *breezy*   distance from Q_0: 4
Q_4 *cold* & *drizzly* & *still*   distance from Q_0: 5

Again, an additive measure of this kind goes beyond these particular qualitative judgements by implying additional qualitative judgements.

What about propositions that are not expressed by maximal conjunctions? We can capture sundry judgements by extending the simple additive measure on maximal conjunctions easily enough. Every proposition in this propositional framework is equivalent to a disjunction of maximal conjunctions. The distance of a *disjunction* of maximal conjunction from a maximal conjunction is presumably some function of the distance of the disjuncts. More generally, if we think of maximal conjunctions as proxies for *worlds* – maximal states of affairs, relative to some framework of basic states – then sets of worlds can stand as proxies for propositions. (Maybe sets of worlds just are propositions, but let's leave that issue to one side.) Then, if we have a suitable measure of distance between worlds, a measure of the distance of a proposition from a particular world is some function of distance of the worlds in it. Of course, there are various plausible functions, like the average of the distances of all worlds, the average of the minimum and the maximum world, or the sum of the distances of worlds, and so on. I will not go into these different proposals here.

Much of the early literature on truthlikeness consisted in testing general accounts of truthlikeness against low-level judgements ("intuitions") on orderings of particular cases involving simple examples like those presented above. What is striking about this, in retrospect, is that there was little to connect this kind of piece-meal testing with quite powerful results and techniques in value theory and the theory of measurement. But unless you think of the theory of truthlikeness as a sub-discipline of value theory then perhaps this is not surprising. I myself accidentally stumbled into this area years later when I began thinking seriously about some problems in applied ethics.

Ethical reasoning often employs so-called *bare difference* principles, also known as the *contrast* strategy. Suppose you suspect that some feature is valuable or disvaluable, in itself. Consider killing versus letting-die, for example. Many people think that killing someone is, in itself, much worse than letting someone die. This idea undergirds the commonly held view that active euthanasia is always and everywhere wrong, whereas passive euthanasia may be morally permissible. Enter James Rachels with his justly famous parable of Smith and Jones (see Rachels 1986, pp. 111–114). Each stands to gain a fortune if his nephew dies. Each plans to murder his nephew in the bath, and make it look like an accident. Smith carries out the plan. But as Jones enters the bathroom he sees his nephew slip and fall, hitting his head on the side of the bath. Jones's nephew begins to drown. Jones could prevent him from drowning but he stands by, ready to intervene and force his nephew's head under water should he recover, but finds that this is unnecessary. Smith killed his nephew. Jones merely let his nephew die. And yet the Jones episode seems no better than the Smith episode. This suggests, first, that it is not true that killing is worse than letting-die; and second, that killing is evaluatively equivalent to letting-die.

To derive these two different conclusions we need subtly different general principles (see Oddie 2001a and 2001b.). One is the principle (I call it the *bottom-up bare difference principle*) that if two situations A and B differ only over a single factor (of which F and G are determinates), such that A has F and B has G, and F is better (in itself) than G, then A is better than B. This principle enables us to use the Smith-Jones judgement to *refute* the claim that killing is worse in itself than letting-die. (If letting-die were better in itself than killing then the Jones episode would be better than the Smith episode, which it isn't.) The other is the principle (I call it the *top-down bare-difference principle*) that if two situations A and B differ only over a single factor F/G, such that A has F and B has G, and if A is better than (or equivalent to) B, then F is better than (or equivalent to) G. This enables to derive the evaluative equivalence of killing and letting-die from the Smith-Jones judgement.

What is really interesting here are the logical connections between these *bare-difference* principles (which have cropped repeatedly in the history of applied ethics), the *separability* principles that have been extensively investigated in economics, and the thesis of the *organic unity of value*, which enjoys a venerable tradition in value theory (see, for example Moore 1960). These connections I have explored in detail elsewhere, and it turns out that these connections have direct relevance to the measures of truthlikeness that have been at the center of the debate.

Let me describe just one such. A number of philosophers have assumed that the intuitive judgements of closeness of a maximal conjunction to the true maximal conjunction can be represented by an additive measure. What this presupposes is that each "bit" of true information is worth something, and that additional bits are worth more. That is to say, the value of the whole (maximal conjunction) is the sum of the value of its parts. This, of course, presupposes that there is no *interaction* between bits of true information as far overall value goes, that the value of truth does not conform to Moore's famous thesis of the organic unity of value.

Let's take one purported example of organic unity of value.

> Virtue is good. Furthermore, it is better for the virtuous to be happy than to be miserable. Since the addition of happiness to virtue makes a positive difference to value it seems to follow that happiness itself is valuable. But even though happiness is a valuable addition to virtue it is not clear that happiness always and everywhere adds value. Many (famously Kant) have denied that it is better for the vicious to be happy than for them to be miserable. Rewarding viciousness with happiness makes a bad situation worse. So even if happiness is valuable it does not invariably add value. (Oddie 2001, p. 313)

Of course, one could deny that happiness is good and then we do not have an instance of organic unity, characterized as "the value of the whole is greater or less than the sum of the value of its parts". But we do have a violation of the conjunction of the two bare-difference principles. We can preserve one of the principles only at the cost of sacrificing the other. However, we can show that on a plausible characterization of additivity of value factors, additivity entails both these bare-difference principles, and so we can apparently infer that this value ordering is not additive.

The conditions under which value orderings can be represented as additive have been quite thoroughly studied in economics (see Broome 1990 for an accessible introduction to some of these results). However those representations presuppose strong conditions of continuity which are absent in the discrete cases countenanced above. So the question arises, under what conditions are orderings in discrete spaces – involving a finite number of determinables each of which embraces a finite number of determinates – additive. In other words, our intuitive judgements on particular cases in a discrete space induces a possibly partial purely qualitative ordering on the elements of that space. Under what *qualitative* conditions will a partial ordering be susceptible to an additive representation such as the one suggested by Tichý to capture the intuitive judgements above? The known results developed in the economic literature on separability do not (as far as I could determine) directly address that.

It turns out that the answer can be derived from a powerful theorem of Dana Scott's in the theory of measurement (Scott 1964, pp. 233–247, theorem 1.3). The qualitative condition that is necessary and sufficient for an additive representation I call *the principle of recombinant values*. (For the derivation see Oddie 2001a.)

The basic idea is simple enough: if you have some bunch of complexes and you take them to pieces and reassemble them in some other way, then the overall value of the new set of complexes is equivalent to the overall value of the old set. This is an articulation, in a purely qualitative form, of the idea that the values of the individual components of a complex whole are totally independent of any surrounding factors. The value of a piece of true information does not depend, in itself, on whatever else happens to be true. The value of a fundamental bit of information, that it is hot for example, is not affected by whether it happens to be rainy or drizzly or dry, windy or breezy or still.

Of course, there might be good reasons to reject this strong independence of value of the basic bits of information, just as there might be apparently good reasons to reject the organic unity of value. If Kant is right about virtue and happiness then value is non-additive. Happiness adds value to the virtuous life, but subtracts value from the vicious life. And we can in fact reinterpret David

Miller's famous translation-variance against the judgements articulated in the P-example as a challenge to this strong independence. Miller asks us to consider two new weather states which are logically connected with *hot*, *rainy* and *windy*:

> *minnesotan* = *hot* if and only if *rainy*
> *arizonan* = *hot* if and if *windy*

Now let us recast our first sequence of worlds in terms of these weather states:

> T_0 *hot* & *minnesotan* & *arizonan*
> T_1 not-*hot* & not-*minnesotan* & not-*arizonan*
> T_5 not-*hot* & *minnesotan* & *arizonan*

If we stick with the judgements or relative merit with which we started (that P_5 (T_5), is further from the truth than P_1 (T_1)) then it is pretty clear that we cannot assign numerical values to *hot*, *minnesotan* and *arizonan* and make the truthlikeness of conjunctions depend on agreement between that conjunction and the true conjunction. If we frame the world in the *hot-minnesotan-arizonan* conceptual scheme but retain the intuitive judgements which we arrived at in the *hot-rainy-windy* scheme, then value is not additive, but rather exhibits organic unity.

Of course, one could also abandon the original judgements and endorse a new set of judgements in the *hot-minnesotan-arizonan* case: maintaining, for example, that if T_0 is the truth then T_5 is closer to the truth than T_1. It follows that if T_1 is P_1, T_5 is P_5 then we have a contradiction. (This was Miller's choice.) To avoid the contradiction we would have to deny the relevant identifications. (This was Tichý's choice.)

There is, however, a third possibility. Suppose that there is indeed a single consistent ordering of likeness to truth to which we have fallible access through our intuitive judgements. Then, given the ordering, whether or not value is additive will depend upon which factorization of the possibilities we utilize. If the original judgements (on the Ps) are correct then truthlikeness is additive, provided we use the *hot-rainy-windy* factorization. It will not be additive if we use the *hot-minnesotan-arizonan* factorization. Briefly: the question of whether or not value is additive depends, at least in part, on factorization. It is factorization relative. Indeed we can show that with a very liberal notion of factorization any ordering will be additive under some factorization. On more stringent and rather more plausible notions, some orderings will not be additive under any factorization. Such orderings will be absolutely organically unified.

Return to the case of virtue and happiness. Kant's judgements include the following (where > is *better than*)

virtuous & *happy* > *virtuous* & *miserable*
virtuous & *miserable* > *vicious* & *miserable*
vicious & *miserable* > *vicious* & *happy*

Clearly this is a non-additive ordering if we factor the space of possibilities using *virtue* and *happiness*. We cannot assign values to the parts – *virtue* and *happiness* – in such a way that the value of each whole is the sum of the values of its parts However, suppose that we frame a new state, one which I will call *just deserts* (or *deserving* for short). You are deserving, or you get your just deserts, just in case you are either virtuous and happy or else you are vicious and miserable. Briefly:

deserving = (*happy* if and only *virtuous*)

Now we can frame Kant's judgements as follows:

virtuous & *deserving* > *virtuous* & not-*deserving*
virtuous & not-*deserving* > *vicious* & *deserving*
vicious & *deserving* > *vicious* & not-*deserving*

Now we *can* represent the ordering in an additive way. (For example, assign 1 of value to virtue, −1 to vice; 0.5 to getting your just deserts, and −0.5 to not getting your just deserts.) What this suggests in this case is that even though virtue (having a good will) is a value, happiness is not a value. However, getting your just deserts is a value. Thus Kant's ordering enables us to identify the real values (virtue, desert) under the heuristic assumption that values must add up. In my (2001b) I suggest that even if we cannot prove additivity that we adopt additivity as a regulative ideal which will then enable us to identify what is in fact valuable in itself. Applying this to the case of truthlikeness, we could use the additivity of the value of truthlikeness as a heuristic principle to help us identify the valuable units of information.

3 These reflections by no means establish that all philosophy is a branch of axiology. But they do make it rather plausible that the problem of truthlikeness is, essentially, an inquiry into the nature and structure of value. And if that is true of this far-flung corner of the philosophical theoretician's domain, then that makes it a bit more plausible that it is also true of his entire kingdom.

University of Colorado at Boulder

References

Broome, John 1990: *Weighing Goods*, Blackwell, Oxford.
Miller, D. 1974: "Popper's Qualitative Theory of Verisimilitude", *British Journal for the Philosophy of Science* 25, 166–177.
Moore, G. E. 1960: *Principia Ethica*, Cambridge University Press, London.
Niiniluoto, I. *Is Science Progressive?*, D. Reidel, Dordrecht.
Oddie, Graham 2001a: "Recombinant Values", *Philosophical Studies* 106, 259–292.
Oddie, Graham 2001b: "Axiological Atomism", *Australasian Journal of Philosophy* 79, 313–332.
Rachels, James 1986: *The End of Life*, Oxford University Press, Oxford.
Scott, Dana 1964: "Measurement Structures and Linear Inequalities", *Journal of Mathematical Psychology* 1, 233–247.
Tichý, P. 1974: "On Popper's Definitions of Verisimilitude", *British Journal for the Philosophy of Science* 25, 155–160.

Jan Woleński

Induction and Metalogic

There are many modes of reasoning which maybe considered as kinds of inductive procedures. In order to simplify the following considerations I will restrict myself to so-called enumerative induction (in particular, I neglect statistical inference – perhaps the most important form of induction). Enumerative induction falls under the scheme:

(1) $$\frac{P(a_1), \ldots, P(a_n)}{\forall x P(x)}$$

The premises of this scheme, that is, sentences of the type $P(a)$, are singular statements with a as a proper name (an individual constant) or an other singular nominal expression (a definite description or an inexical expression, like "this", etc.) which refers to a concrete object as its denotation, for example 'London is a big city', 'the first European philosopher was a Greek' or 'this apple is sweet'. In order to skip trivial cases, let us assume that the domain of reference for the conclusion $\forall x P(x)$ is essentially larger than the set $\{a_1, \ldots, a_n\}$; in particular, the sentence $\forall x P(x)$ can be universally unrestricted, that is, its domain can consist of an infinite number of objects. Thus, using the traditional language, enumerative induction goes from the particular to the general or, otherwise speaking, it produces a general conclusion on the basis of its particular instances (it is only a temporary account, which will be modified later). This is important that the information covered by the conclusion exceed the content contained in the premises. Otherwise speaking, enumerative induction is creative and extend our knowledge. I regard (1) as displaying a form of empirical induction (hence, its domain will be symbolized by the letter **E**) and as giving rise to the famous epistemological question:

(*) How it is possible to justify empirical enumerative induction as a reasonable pattern of reasoning, that is, as providing support for its conclusion?

This question caused a controversy between inductionists (with Carnap as the main figure) and anti-inductionists (with Popper as the main proponent). The latter deny that science, at least mature science, employs induction at all and argue that it has no value in the justification of general statements, whereas the former maintain that we cannot avoid inductive procedures and try to defend its rationality with respect to empirical justification. I will not enter into this controversy except through parenthetical remarks. My general position is that (*) seems to be a sufficiently interesting philosophical question, independently of the possible application of induction in science or daily life.

It is instructive to compare enumerative induction with the form of inference known as mathematical induction, which also proceeds from the particular to the general in the domain **N** of natural numbers or, if also we add transfinite induction, in uncountable sets. Scheme (1) can be rewritten as:

(2) $$\frac{P(a_1), \ldots, P(a_n)}{P(a_{n+1})}$$

Assuming that for every $k < a + 1$, (2) holds, this scheme is equivalent to (1). Clearly, if we can pass from any k to its "successor" $k + 1$, we obtain the required general conclusion $\forall x P(x)$. The reverse implication is obvious. This situation almost exactly ("almost" will be explained in due course) recalls the equivalence of two forms of mathematical induction ($k, n \in \mathbf{N}$):

(3) $$\frac{P(1) \wedge \forall k\, (P(k) \Rightarrow P(k + 1))}{\forall n P(n)}$$

(4) $$\frac{\forall k < n\, (P(k) \Rightarrow P(k + 1))}{\forall n P(n)}$$

Yet (1) (or (2)) and (3) (or (4) constitutes a pattern of a deductive inference, but the former is of an inductive character; the difference between the shape of $- - - - -$ (inducibility) and ———— (deducibility) visualizes this fact. Clearly, empirical induction can lead to false conclusions from true premises (this means that this mode of reasoning is fallible), but mathematical induction always produces true conclusions from true premises (it is infallible).

The deductive character of mathematical induction has its source in the well-ordering of $\mathbf{N}$ (every subset of $\mathbf{N}$ contains the least element). Assuming the axiom of choice, one can prove that every set is well-ordered and this fact justifies the principle of transfinite induction, although it becomes a non-constructive mood of inference. On the other hand, the well-ordering of $\mathbf{N}$ is natural, so to speak, constructive and equivalent to (3) and the principle of the least number. In other words, the well-ordering of $\mathbf{N}$ is an essential property of this set. Nothing similar can be said about the empirical domain $\mathbf{E}$, because we do not know whether the empirical domain is ordered at all, but even if this were the case, there is the problem of proving that an ordering occurs. Just for this reason I wrote "successor" and "almost" in the previous explanations, because a_{n+1} is not the successor (in the sense of number theory) of a_n when empirical induction proceeds and thereby its similarity to mathematical induction is only partial.

This difficulty was famously indicated by Hume in his criticism of induction. In his own words, its summary is as follows (Hume 2000, pp. 28–31):

> *What is the nature of all our reasonings concerning matter of fact?* the proper answer seems to be that they are founded on the relation of cause and effect. When again it is asked, *What is the foundation of all our reasonings and conclusions concerning that relation?* it may be replied in one word, EXPERIENCE. But if we still carry on our sifting humour, and ask, *What is the foundation of all conclusions from experience?* this implies a new question, which may be of more difficult solution and explication. [...]. I shall pretend [...] to give a negative answer to the question here proposed. [...]. As to past *Experience*, it can be allowed to give *direct* and *certain* information of those precise objects only, and that precise time, which fell under its cognizance: But why this experience should be extended to future time, and to other objects [...] this is the main question on which I would insist. [...]. All reasonings may be divided into two kinds, namely, demonstrative reasoning, or that concerning relations of ideas [...] or that concerning the matter of fact and existence. That there are no demonstrative arguments in the case, seems evident; since it implies no contradiction, that, the course of nature may change. [...] Now whatever is intelligible, and can be distinctively conceived, implies no contradiction, and can never be proved false by any demonstrative or abstract reasoning *a priori*. [...] If we be,

therefore, engaged by arguments to put trust in past experience, and make it the standard of our future judgement, these arguments must be probable only, or such as regard matter of fact and real existence, according to the division above-mentioned. But there is no argument of this kind, must appear, if our explication of that species of reasoning be admitted as solid and satisfactory. We have said, that all arguments concerning existence are founded on the relation of cause and effect; that our knowledge of that relation is derived entirely from experience; and that all our experimental conclusions proceed upon the supposition, that the future will be conformable to the past. To endeavour, therefore, the proof of this last supposition by probable arguments, or arguments regarding existence, must be evidently going in a circle, and taking that for granted, which is the very point in question.

Hume pointed out that admitting an ordering by the principle of causality or uniformity of nature is inevitably circular. Hence, any justification of induction is fallacious, because it falls either under *regressum ad infinitum* or begs the question.

Hume's argument is sound to the effect that the ordering principle (the causality or uniformity of nature) cannot be taken for granted. On the other hand, his results shows only that induction is not deductive (see Stegmüller 1977). A closer analysis of Hume's criticism suggests that he restricted the scope of sound inferences to preserving truth in all cases when premises are true. Thus, for Hume, demonstrative (= deductive) reasoning is the only rational way of deriving conclusions from premises. Is this view compelling? I do not think so. Simple observation teaches us that (1) is better than the scheme

(5) $$\frac{P(a_1), ..., P(a_n)}{\forall x Q(x)}$$

because the former preserves truth more often than the latter. Using probability or the idea of verisimilitude (see Niiniluoto 1987 for perhaps the deepest analysis of this last concept) is one way of solving Hume's problem of induction. Its merit consists in a quantitative approach to induction by introducing the concept of the degree of confirmation (support, credibility, etc.) of an inductive conclusion by empirical data. I will not take this route. My proposal will go via some metalogical considerations.

There is an important difference between (1) and (5). In the first case, every premise logically follows from the conclusion, that is, for any i ($1 \leq i \geq n$), $P(a_n) \in Cn\{\forall x P(x)\}$, but this does not hold in the second case: we do not assume that the sentence $\forall x Q(x)$ entails any of $P(a_1)$, ..., $P(a_n)$. Intuitively, empirical support, if any, for a generalization obtained by enumerative induction depends on the logical relation holding between the premises and the conclusion of a given inference. This may accidentally occur in the case displayed by (5), but it is substantial in arguments falling under (1). The former scheme admits, for example, a passing from the premises 'London is a city', 'Paris is a city', 'Warsaw is a city' to the conclusion 'All students are interested in philosophy', but the proper scheme of enumerative induction qualifies this inference as having no value in supporting the conclusion. This difference between (1) and (5) suggests that perhaps induction is not so bad as anti-inductionists maintain. Now, if we consider (1) as a scheme of empirical justification, the succession in the pair $\{\{P(a_1), ..., P(a_n)\}, \forall x P(x)\}$ does not matter. We can start from the sentence $\forall x P(x)$ and look for its support by particular instances or begin with instances and generalize them to a universal statement. Hence, the words 'premises' and 'conclusion' are used in order to respect the tradition (see above) seeing enumerative induction as proceeding from the particular to the general. Speaking more precisely, this induction consists in providing empirical support for universal sentences by their particular instances. The logical relation displayed by Cn, and not the psychological direction of related inferences, constitutes the essence of the problem.

We can look at (1) in two ways. Firstly, this scheme might be taken as involving the assertion of sentences:

(6)
$$\frac{\mathbf{As}\ (P(a_1)), ..., \mathbf{As}\ (P(a_n))}{\mathbf{As}\ (\forall x P(x))}$$

According to (6), asserting the sentences $P(a_1)$, ..., $P(a_n)$ justifies the same attitude toward the statement $\forall x P(x)$. Nothing is said about truth here. This treatment would probably please philosophers who prefer anti-realism and wish to eliminate truth from methodology in favour of more pragmatic categories. The second account is displayed by

(7) $$\frac{\mathbf{Tr}\ (P(a_1)),\ ...,\ \mathbf{Tr}\ (P(a_n))}{\mathbf{Tr}\ (\forall x P(x))}$$

On this route, the truth of singular statements is somehow related to the truth of $\forall x P(x)$. I prefer this more semantic point of view, but I will show that both approaches satisfy the same metalogical construction.

In the analysis that follows I will use the Lindenbaum theorem on maximalization (the letters X, Y denote sets of sentences):

(L) If X is consistent, there exists Y such that $X \subseteq Y$ and Y is maximally consistent, that is, for any $A \notin Y$, $Y \cup \{A\}$ is inconsistent (equivalently: for any A, $A \in Y$ or $\neg A \in Y$).

The set Y is called the maximally extension of X. (L) can be rephrased as: every consistent set has its maximal consistent extension. Intuitively speaking, maximal consistency means that adding any sentence to a maximally consistent set produces a contradiction; hence, every A or its negation belongs to such a set. Of course, we assume that the language of X and Y is fixed. Any maximally consistent set is determined by the following procedure. The set of all the sentences of any language is infinite and countable. Hence, this set is enumerable and any sentence belonging to it can be numbered by a natural number ascribed to it. Let X be a consistent set of sentences. We form the infinite sequence of sets such that (a) $X = X_0$; (b) $X_{n+1} = X_n \cup \{A_{n+1}\}$, if $X_n \cup \{A_{n+1}\}$ is consistent; (c) $X_{n+1} = X_n$, if $X_n \cup \{A_{n+1}\}$ is inconsistent. One can prove that the sum of all sets formed in this way is maximally consistent. Informally speaking, if we have a consistent set X, we identify it with X_0. The next stage consists in taking the sentence A_1 and checking whether $X_0 \cup \{A_1\}$ is consistent or not. In the case of its consistency, we pass to $X_1 = X_0 \cup \{A_1\}$, but if X_1 is inconsistent, we stay with X_0; this procedure is "repeated" infinitely many times. It is important to note two things: (i) (L) is a purely syntactic theorem; (ii) (L) is not effective because it asserts the existence of an object without indicating the method of its construction. In fact, due to the second incompleteness theorem, there is no effective universal test of consistency. Hence, the phrases "we form" and "repeated" used in the description of the procedure should be taken *cum grano salis*.

Let T be an empirical theory based on axioms $H_1, ..., H_n$, assuming that each H_i is of the form $\forall x A(x)$. In principle, T can be practically equated with

$Cn\{H_1, \ldots, H_n\}$. The problem of the confirmation of T is reducible to giving inductive support for the sentences $H_1, \ldots, H_n$ conceived as universal empirical hypotheses. A set C of singular sentences is called a confirmation for T if and only if (a) C is finite; (b) C is consistent; (c) for any $A \in C$, $A \in CnT$; (d) no hypothesis of $H_1, \ldots, H_n$ belongs to C; (e) for any $A \in C$, A is true; (f) no counterexample for T is known, that is, there is no A such that $A \in CnT$ and A is false. The set of all confirmations of T in the above sense is denoted by the symbol **CONF**(T). Claim (a) asserts that only finite confirmations are effectively available, (b) – that confirmations are consistent (since they are finite, their consistency can be effectively checked); (c) – that elements of confirmations are entailed by a given theory; (d) – that axioms (universal hypotheses) are not consequences of confirmations; (e) – that confirmations are true (in fact, (e) entails (b)); (f) – that counterexamples are not known. I do not assume any special philosophy of science. In particular, singular sentences do not need to be protocols (in the sense of the Vienna Circle), epistemologically indubitable (in the sense of fundamentalism) or lacking theoretical import (in Popper's sense). The only limitation (except those listed in (a) – (d)) consists in assuming that all elements of confirmations are singular, that is, that they refer to concrete objects or experiences (reference to previous theories is not excluded). Condition (f) does not entail that a theory T must be rejected after finding a counterexample. It can be modified, suspended or accepted temporarily only (as, for example, Rutherford's planetary model of the atom after discovering the first quantum effects and before the quantum theory). The only insight suggested by (f) consists in changing the former evaluation of T as a satisfactory theory.

Let T be a theory and C its confirmation. The problem of giving an empirical support for T by enumerative induction (and perhaps by other methods as well) may be formulated as follows: is $C \cup \{H_1, \ldots, H_n\}$ a subset of a maximally consistent set? Since C is true, it is also consistent (if only consistency is involved, there are many such extensions). By (L) it has the maximally consistent extension, which covers all truths and only truths. We need to find a procedure for arguing that T belongs to the maximally consistent extension of C. Denote this extension by **MC**. If $H_1, \ldots, H_n$ is in **MC**, the same concerns all the logical consequences of T. However, **MC** also contains true sentences that are irrelevant for T, that is, independent of it. The following procedure can be suggested. We take C as C_0. Since all sentences are enumerated and their indexing by numbers is arbitrary, we take an arbitrary singular sentence A as A_1 and investigate whether $A_1 \in CnT$ or not. If it belongs, we obtain a new confirmation $C_1 = C \cup \{A_1\}$. By definition, $C_1 \in$ **CONF**(T) and $C \subseteq C_1 \subseteq$ **MC**; otherwise, that is, if $A_1 \notin CnT$, we stay with C and apply the procedure with respect to A_2. This is a general way of obtaining a sequence of confirmations $C \subseteq C_1 \subseteq C_2 \subseteq \ldots \subseteq C_i \subseteq \ldots$; of course, each $C_i \subseteq$ **MC**. Since we cannot assume that T is decidable, we have no algorithm to check for an arbitrary A whether $A \in CnT$ or $A \notin CnT$ (I owe this point to Elżbieta Kałuszyńska). However, we can proceed by various semantic arguments, for example, by building possible countermodels. This last remark shows the advantage of the semantic approach. Moreover, assuming that all theories are first-order, we have a theoretical guarantee that every truth is provable. On the other hand, nothing essentially new arises when we base the construction on (6). The requirement that confirmations should be true is replaced by the contraint (e'): elements of confirmations are accepted. Although acceptance does not imply consistency, the construction of **CONF**(T) = $C \subseteq C_1 \subseteq C_2 \subseteq \ldots \subseteq C_i \subseteq \ldots$ is basically the same, perhaps with some additional difficulties stemming from the plurality of maximally consistent extensions and concerning the checking of consistency and entailmenthood.

Theorem (L) assures that **MC** with respect to a given C exists. Hence, the question whether $H_1, \ldots, H_n$ belong to **MC** is legitimate and, at least theoretically, has a "yes" or "no" answer. The sum $C \cup C_1 \cup C_2 \cup \ldots \cup C_i \cup \ldots$ does not exhaust **MC**, because it omits irrelevant elements for T. On the other hand, the construction of **CONF**(T) selects the cases of valuable induction from the whole set of the consistent extensions of C. Clearly, any effective process of extending C provides only a finite number of elements of **CONF**(T); nothing more can be achieved for logical reasons and provided that confirmations are realized in the limited time. In fact, since **CONF**(T) is an infinite family of sets, there is always an infinite number of possible confirmations between a given C_0 and T. Speaking more technically, the cardinal degree of completeness of any reachable C_i, that is, the power of the set of its consistent extensions is at least countably infinite. Another way of expressing this situation consists in saying that T is the set-theoretical limit of its confirmations. This means that almost all – that is, except the finite number – confirmations are still to be constructed. It is important for the problem of justification that T be true. Full justification cannot be achieved, because it requires the effective construction of an infinite set of confirmations. This apparently seems to confirm Hume's skepticism. In a suitable translation, it says that passing from C_i to C_{i+1} does not provide new reasons for accepting T as true, but invoking that it does consists in a persuasion only.

Nevertheless, I do not consider Hume's skepticism to be justified. Popper, in the spirit of Hume, argued that the initial probability of any universal hypothesis H is equal to zero or close to it, independently of arbitrary empirical data e. However, this leads to a strange consequence, namely that $p(\neg H, e) \approx 1$, that is, the existence of a counterexample for T is at least almost certain. I see no reason for assuming that this is the case. The importance of (L) for (*) consists in the fact that for any C there is always a more extensive confirmation and ultimately its **MC** including T (there is no reason for assuming *a priori* that it is not included in **MC**). This conclusion holds independently of whether reality is ordered and how or whether future experience is similar to past empirical data and exhibits the weak points of Hume's arguments against induction. Let me recall the former observation that Hume only proved that induction is not deduction. The proper perspective for looking at the power of enumerative induction requires one to be conscious of what can be achieved in the process of giving support, namely that, as I already pointed out, we can only provide a finite initial segment of **CONF**(T). There is another way of arguing that $T \subseteq \mathbf{MC}$ that does not require reviewing new confirmations. Assume that at a given moment t only k confirmations are effectively available. If we provided $i = k$ confirmations, empirical data, all possible at t and supporting T, are given. We can say on the basis of (f) that we excluded all possible counterexamples against T. Since usually $i < k$, the construction of C_i means the elimination of a certain counterexample; similarly for any passing from C_i to C_{i+1}. Since the set **Tr** of truths is compact, that is, it possesses the property "all its elements are true" if and only every finite subset of **Tr** has this property, extending C's is a method of arguing that T is true. Thus, we propose

(**)   If C_i, $C_{i+1} \in \mathbf{CONF}(T)$ and $C_i \subset C_{i+1}$, then C_{i+1} is a better confirmation of T, that is, C_{i+1} provides better reasons for accepting T as true than C_i,

as an answer to the question (*). It must be very strongly stressed that this proposal considers confirmation qualitatively, not quantitatively.

Jagiellonian University

References

Hume, D. 2001: *An Enquiry Concerning Human Understanding*, ed. by T. L. Beauchamp, Clarendon Press, Oxford (originally published in 1748).

Niiniluoto, I. 1987: *Truthlikeness*, Kluwer Academic Publishers, Dordrecht.

Stegmüller, W. 1977: "The Problem of Induction: Hume's Challenge and the Contemporary Answers", in W. Stegmüller, *Collected Papers on Epistemology, Philosophy of Science and History of Philosophy*, D. Reidel, Dordrecht, pp. 68-136.

III
Epistemology, Culture, and Religion

Juha Manninen

Beginning the Logical Construction of Cognition*

1. Rudolf Carnap had some problems with the title of his famous book, *Der logische Aufbau der Welt* (1928). He had written a first, unpublished, sketch planning such a project, entitled "Vom Chaos zur Wirklichkeit", already in the summer of 1922. That early construction began with lived experiences, with reality as a middle-level, proceeding then to psychology and ultimately to the formal structures of physics, purified completely from qualia.

In a publication from the year 1924, Carnap developed the idea distinguishing two different levels: the primary world of experience and the secondary world of experience.[1] The first was the "immediately given" in its original order, no longer chaotic, and the second contained the further synthetized everyday world with its objects together with the conventional choises of mathematical physics. Both the ordinary "world of things" and the developed "world of physics" were also constructions, not givens. Carnap continued:

> Welches ist nun die wirkliche Welt, die primäre oder die sekundäre? Nach der übereinstimmenden Auffassung der idealistischen und der realistischen Philosophie, sowie der in der physikalischen Forschung und im gewöhnlichen Leben üblichen Ansicht führt die Konstruktion der sekundären Welt zum Aufbau der 'Wirklichkeit'. Die positivistische Philosophie dagegen erkennt nur dem Primären Wirklichkeitswert zu, die sekundäre Welt ist nur

---

* I would like to thank Georg Gimpl, Alaric Hall, Thomas Mormann and Panu Raatikainen for their comments and suggestions and the Archives for Scientific Philosophy (University of Pittsburgh) and the Vienna Circle Archives (Rijksarchief in Noord-Holland, Haarlem) for their permission to quote unpublished archive sources.

[1] R. Carnap, "Dreidimensionalität des Raumes und Kausalität", *Annalen der Philosophie und philosophische Kritik* 1924: 4, pp. 106ff.

eine willkürliche, aus Gründen der Ökonomie ausgeführte Umgestaltung jener. Wir überlassen diese im eigentlichen Sinne transzendentale Frage der Metaphysik; unsere immanente Erörterung hat es nur mit der Beschaffenheit der Erfahrung selbst zu tun, insbesondere mit der Unterscheidung ihrer Formfaktoren in notwendige und wahlfreie, die wir primäre und sekundäre nennen, und mit den Beziehungen beider Arten.

The cognitively primary level was not without form, although it was subjective and qualitative. Still, only the second one was subject to lawful and deterministic connections. Order in the primary world made possible relational primitive concepts and a constitutional system for physics using explicit definitions.

Carnap continued to call his project *Prolegomena zu einer Konstitutionstheorie der Wirklichkeit*. The large manuscript for a *Konstitutionstheorie* opened his career in Vienna in 1926. It was discussed extensively in the Vienna Circle, doubtless benefiting from new insights. Unfortunately, this manuscript has not yet been found,[2] but the distinction between the cognitively primary and secondary systems and the transition from the former subjective to the latter objective one must have been a part of it, because it remained as Carnap's main agenda even in the next stage of the development of his views.

As it proved impossible to find a publisher for the book, Carnap wrote a shortened version in Davos during the winter 1927–28, simultaneously adapting the text to the state of the discussions in Vienna. The official head of the Circle, Moritz Schlick, certainly was no friend of *a priori* philosophy. A book entitled *Konstitutionstheorie* sounded too much like neo-Kantianism. Already in 1926, Schlick had suggested the title *Der logische Aufbau der Welt* for Carnap's book.[3] "Constitution theory" could be mentioned in the subtitle. Carnap

---

[2] R. Carnap to R. Rand, November 5, 1935: "Beim Durchsehen meiner Bücher, sehe ich, dass Sie von mir noch einen von Reininger, Külpe I und ein MS 'Konstitutionstheorie' haben. Das MS können Sie für dauernd behalten...". Archives for Scientific Philosophy of the Twentieth Century, University of Pittsburgh, Special Collections (=ASP), RR 3-9-14. The manuscript has not been found among Rose Rand's papers.

[3] M. Schlick to R. Carnap, March 14, 1926: "Der Titel Ihrer Arbeit, auf den bekanntlich in mancher Hinsicht viel ankommt, scheint mir nicht sehr praktisch gewählt zu sein, da auch ein chemisches oder medizinisches Werk 'Konstitutionstheorie' heißen könnte. Ein Name, der über den philosophischen Charakter der Schrift

agreed. He adopted the subtitle *Versuch einer Konstitutionstheorie der Begriffe* to express his intentions.

However, when finalizing the book in Davos, Carnap grew dissatisfied with this solution. What he was doing was, in his opinion, actually *Der logische Aufbau der Erkenntnis*, and he was already planning a sequel with an alternative, equally possible physical or "materialistic" basis. Schlick's old suggestion would be an excellent title for the sequel, not for the present one with a beginning in the solipsistic lived experiences.[4] Schlick did not see any reason for a change. The second book could be *Der logische Aufbau der Natur*.[5] Carnap followed the advice. When the book finally appeared in August 1928, the publisher had dropped the subtitle, without asking a permission from Carnap.[6]

---

keinen Zweifel lässt, wäre gewiss praktisch. Wie wäre es mit 'Der logische Aufbau der Welt'? Dass es sich um eine Konstitutionstheorie der Erkenntnisgegenstände handelt, könnte dann der Untertitel sagen." – ASP RC 029-32-17.

[4] R. Carnap to M. Schlick, December 23, 1927, 'Frage über die Wahl des Buchtitels': "Das Konst[itutions]system des Buches hat eine eigenpsychische ('solipsistische') Basis. Reihenfolge der Gebiete: Eigenpsych[isches], Phys[isches], Fremdpsych[isches], Geistiges. Diese Reihenfolge ist bedingt durch die Absicht des Systems, die erkenntnismässige Ordnung der Begriffe darzustellen.

An einer Stelle des Buches deute ich kurz an, dass ein anderes Konst[itutions]-system möglich ist; mit physischer ('materialistischer') Basis. Reihenfolge: Physisches, Psychisches (ohne Unterscheidung des Ich), Geistiges. Die Leistung dieses Systems ist eine andere: es dient nicht der Erkenntnistheorie, sondern der Realwissenschaft. Es hat als Basisgebiet dasjenige Gebiet, das als einziges eine durchgehende eindeutige Gesetzmäßigkeit seiner Vorgänge besitzt. Die psych[ischen] u[nd] geist[igen] Gegenst[ände] werden hier aus den phys[ischen] konstruiert; dadurch werden sie mit eingeordnet in das *eine* gesetzmäßige Gesamtgeschehen.

Welches der beiden Systeme verdient mehr den Namen eines 'Aufbaus der Wirklichkeit'? [...]

Ich möchte den *Buchtitel nun schon mit Rücksicht auf diesen späteren Plan wählen*. Vielleicht jetzt 'Erkenntnislogik'; das Spätere: 'Wirklichkeitslogik'? Dazu der frühere Untertitel? Oder: 'Der logische Aufbau der Erkenntnis', später: 'Der log[ische] Aufbau der Welt'? Ist 'Erkenntnislogik' oder 'Logik der Erk[enntnis]' zu blass?" – The Vienna Circle Archieves in the Rijksarchief in Noord-Holland, Haarlem (=VCA).

[5] M. Schlick to R. Carnap, January 4, 1928. – ASP RC 029-30-36.

[6] R. Carnap to M. Schlick, August 6, 1928: "Den Untertitel des Buches [...] hat Benary weggelassen; ich weiß nicht, ob mit Absicht oder aus Versehen. Merkwürdigerweise hab ichs bei der Korrektur des Titelblattes gar nicht bemerkt, sondern erst nachher, als es zu spät war. Vielleicht schadet es aber nicht viel." – VCA.

Carnap opened the book by explaining that a cognitive constitution system construes the objects from level to level from the basic objects of the system, the experienced "primary world" as he used to call them. Again, in the sense of Carnap's earlier "secondary world", now secondary "system", the result of these transformations would be the following:

> Obwohl der subjektive Ausgangspunkt aller Erkenntnis in den Erlebnisinhalten und ihren Verflechtungen liegt, ist es doch möglich, wie der Aufbau des Konstitutionssystems zeigen soll, zu einer intersubjektiven, *objektiven* Welt zu gelangen, die begrifflich erfassbar ist und zwar als eine identische für alle Subjekte.[7]

2. Carnap's *Aufbau* is often seen as a classic or even a paradigm of analytic philosophy. It has been less clear, however, what the aim of the book in fact was. Carnap himself was soon dissatisfied with his major work. He was a professor in Prague when Herbert Feigl sent to him one of his papers for comments, from America, in 1934. Carnap observed dryly: "Die Terminologie und Gesichtspunkte meines *Aufbaus* und der früheren Wiener Diskussionen mag ich schon gar nicht mehr hören. Du weißt: einen überwundenen eigenen Standpunkt verabscheut man mehr als den des Gegners."[8]

But he added, curiously – at a time when he was already orienting his thinking towards that in the USA: "Dabei ist der Standpunkt eigentlich gar nicht mal wirklich 'überwunden' oder widerlegt; nur die Aspekte sind heute ganz andere. In vielem ist es wohl auch beabsichtigte Anpassung an amerikan[ische] philos[ophische] Fragestellung."

W. V. O. Quine was one of the philosophers who visited Carnap in Prague. For a long time, the dominant interpretation of the *Aufbau* was connected with his name. According to this interpretation, Carnap was the staunchest of all empiricists, unique for his attempt to make true Bertrand Russell's program of logical constructions, beginning with the "given", but then, arriving at physics, failing in his method.

---

[7] R. Carnap, *Der logische Aufbau der Welt*, Weltkreis-Verlag, Berlin-Schlachtensee 1928, § 2.

[8] R. Carnap to H. Feigl, April 29, 1934. – ASP HF 03-69-03.

Recently, a neo-Kantian reading of Carnap has been gaining momentum, developed especially by Michael Friedman.[9] In this reading, the structure guaranteeing objectivity was seen as the most important ingredient in the *Aufbau*, despite the fact that the *Aufbau* attempted – albeit incoherently – to be an empiricist epistemology. Even this interpretation sees the book as a failure, but not in the same way as Quine: Carnap's purely structural definite descriptions did not succeed in eliminating the incoherent components.

Of course, Carnap's *Aufbau* was not a neo-Kantian book, but the advantages of the neo-Kantian line of interpretation have been presented very clearly in modern literature. Accepting much of the structuralism, but emphasizing Carnap's opposition to all traditional philosophy and interpreting the *Aufbau* as still, in essence, as an empirical enterprice Christopher Pincock has presented a "reserved" interpretation, actually a new naturalistic reading of Carnap's controversial book.[10] Pincock draws his evidence mostly from Carnap's work on a physical construction system. The idea of such an alternative system was presented already in the *Aufbau*. Pincock has resisted the temptation to a purely structural reading. Carnap's own move to that effect was optional, not a part of the main argumentation of the book. The primitive relations of the *Aufbau* are given in a sense that is assumed as basic and undefined. Accordingly, the theory of the construction systems was planned in connection with and as a part of contemporary science.[11]

---

[9] M. Friedman, *Reconsidering Logical Positivism*, Cambridge University Press, Cambridge 1999.

[10] C. Pincock, "A Reserved Reading of Carnap's *Aufbau*", *Pacific Philosophical Quarterly*, 2005, pp. 518–543.

[11] The number of different interpretations of the *Aufbau* is definitively bigger, cf. C. U. Moulines, "Making Sense of Carnap's Aufbau", *Erkenntnis* 35, 1991, 263–286. After that, the debate has very much intensified, enriched with a number of new readings, see e.g. the presentation of the significance of geometrical conventionalism for Carnap in T. Mormann, "Geometrical Leitmotifs in Carnap's Early Philosophy", *Cambridge Companion to Carnap*, ed. by R. Creath and M. Friedman, Cambridge University Press, Cambridge 2007, and for a re-evaluation of quite other aspects, T. Mormann, "Werte bei Carnap", *Zeitschrift für philosophische Forschung*, 2006: 2. One could say that seeing the *Aufbau* as a theory of many construction systems is a feature that unites recent interpretations.

3. Aside from Moritz Schlick's support for Carnap's book, the earliest known reactions to it came from Otto Neurath, Eino Kaila and Ludwig Wittgenstein. For Neurath, according to his enthusiastic review in the social democratic *Der Kampf*, the good news was exactly Carnap's structuralism:

> Er unternimmt es, die Sinneseindrücke auf Grund bestimmter Anordnungen zu kennzeichnen, von Anordnungen, in denen 'rot', 'hart', 'laut', 'Cis' usw. *nicht* auftritt, sondern nur das, was an diesen Tatbeständen mathematisch-logisch erfassbar ist – *und das reicht aus*! [...] Von Einfühlung jeglicher Art, von der persönlichen Einstellung als Ausgangspunkt, kehrt sich Carnap bewusst ab. Er kennt nur Einsicht, die von Menschen jeglicher Art erfasst werden kann! Anordnung ist das Allgemeinste, das Universellste, das wir an den Dingen erfahren![12]

However, Carnap's claims about his "ideal language" went too far. It would not be applicable to the social sciences. In their case, it was impossible to avoid taking into account "wie man nun die Erkenntnis fördern könne, solange man 'saubere' und 'unsaubere' Denkweisen in bunter Mischung benützen müsse, was übrigens vielleicht immer nötig sein wird!"[13] The last mentioned was a reference to Neurath's view of the "aggregations", *Ballungen*, of the historically developed natural languages, actually very much the same as Pierre Duhem's *bon sens* as a starting point for science. This was a topic that Neurath would never give up and Carnap never understand.

But Neurath was not worried. The foremost of his heroes had turned an idealist upside down. Maybe he would succeed in doing the same to Carnap.

4. Eino Kaila's correspondence with the new group of philosophers dates back to 1923, when he wrote to Hans Reichenbach, the author of *Relativitätstheorie und Erkenntnis Apriori* (1920), deeply impressed by the book. In the following years, he corresponded even with Schlick and finally with Carnap.[14]

---

[12] O. Neurath, *Gesammelte philosophische und methodologische Schriften*, Vol. 1, ed. by R. Haller and H. Rutte, Verlag Hölder-Pichler-Tempsky, Vienna 1981, p. 296.

[13] Ibid.

[14] For an early study of Kaila's connections with the Vienna Circle, see I. Niiniluoto, "Eino Kaila und der Wiener Kreis", in *Weder – Noch. Tangenten zu den finnisch-österreichischen Kulturbeziehungen*, ed. by G. Gimpl, Deutsche Bibliothek, Helsinki 1986,

Immediately after the *Aufbau* and its companion *Scheinprobleme in der Philosophie* appeared in August 1928, Carnap sent copies of them to Kaila. Again, Kaila was impressed. He sent Carnap a long letter. Only Carnap's friendly answer from 18.9.1928 is preserved. Carnap was delighted to hear that their basic attitudes were very near each other, especially in seeing in the question of realism as nothing but a pseudoquestion.

In addition to being a philosopher, Kaila was a competent experimental and theoretical psychologist. Concerning Carnap's discussion of the visual space, Kaila had given references to counterexamples that did not fit to Carnap's construction. Carnap admitted that in case of the state of affairs mentioned by Kaila another way of making the construction should be chosen. Or the constitution could be roughly the same, only with a proviso added.

Kaila had also enquired about the truth value as the meaning of a sentence. Carnap admitted that even others had been critical of this Fregean approach in the *Aufbau*. Now he had adopted a new concept, analyticity, to do the same work for the constitutional system:

...eine Aussage heiße *analytisch*, wenn sie bei bloßer Kenntnis der erforderlichen Definitionen der in ihr vorkommenden nichtlogischen Zeichen bewiesen werden kann (der Beweis geschieht, indem schrittweise für jedes vorkommende nichtl[ogische] Z[eichen] sein Definiens eingesetzt wird, 'konstitutionale Rückübersetzung', wodurch man schließlich zu einer Tautologie gelangt). Eine Aussage heiße *synthetisch*, wenn zum Beweis außer der Def[inition] auch nichtlogische Axiome erforderlich sind (die konst[itutionale] Rückübers[etzung] führt dann nicht zu einer Taut[ologie], sondern zu solchen empirischen Axiomen). (Die logischen Definitionen und Axiome werden in beiden Fällen, wie ja überhaupt für jede Deduktion, vorausgesetzt.) Ist 'p ≡ q' (p äquiv[alent mit] q, d.h. von gleichem Wahrheitswert) eine analyt[ische] Aussage, so sollen p und q '*analytisch äquivalent*'

---

pp. 223-241. See also I. Niiniluoto, "Eino Kaila and Scientific Realism", in *Eino Kaila and Logical Empiricism*, ed. by I. Niiniluoto, M. Sintonen and G. H. von Wright, Acta Philosophica Fennica 52, Philosophical Society of Finland, Helsinki 1992 and I. Niiniluoto, "Descriptive and Inductive Simplicity", in *Logic, Language, and the Structure of Scientific Theories: Proceedings of the Carnap-Reichenbach Centennial, University of Konstanz, 21-24 May 1991*, ed. by W. Salmon and G. Wolters, University of Pittsburgh Press / Universitätsverlag Konstanz, Pittsburgh, Pa. 1994, pp. 147-170.

heißen. Und nun sagen wir von zwei anal[ytisch] äquivalenten Aussagen (anstatt früher von zwei äquivalenten), dass sie die *gleiche Bedeutung* haben. Und entsprechend sage ich jetzt auch bei Gegenstandszeichen, dass sie die gl[eiche] Bed[eutung] haben, wenn sie anal[ytisch] identisch sind; und bei Aussagefunktionen, wenn sie analytisch generell äquivalent sind. Die drei Fälle können wird jetzt zusammenfassen [...].[15]

Further, Carnap explained his topological transformations, concluding:

Die Redeweise von der 'Umordnung' besagt somit nur, dass in den betreffenden Bereichen (z.B. im Fremdpsychischen) nichts elementar Neues vorkommen kann, und dass die dort zu betrachtenden Relationen andere sind als die früher konstituierten. Neue Eigenschaften oder Beziehungen werden aber damit den Elementen nicht beigelegt, sondern nur solche Beziehungen neu hervorgehoben, die durch die früheren Beziehungen schon bestimmt sind.[16]

It was Carnap's hope that this would help Kaila to formulate his counter-argument against the empirical concept of reality.

A month later, Kaila expressed his doubts about the *Aufbau* in an article entitled "Die Logisierung der Philosophie und die Überwindung des Gegensatzes zwischen Realismus und Phänomenalismus". He suggested to Schlick that the paper could be printed as a booklet, perhaps together with Carnap's replies.[17] Carnap was somewhat surprised that Kaila was prepared to change friendly correspondence to a public controversy.[18] Kaila's paper remained unpublished and it is not preserved. Anyhow, Carnap and, officially, Schlick invited Kaila to speak in Vienna about the constitution theory.

---

[15] R. Carnap to E. Kaila, September 18, 1928. – G. H. von Wright's collection, Helsinki University Library (=GHvW).

[16] Ibid.

[17] E. Kaila to M. Schlick, September 28, 1928: "In der Grundeinstellung bin ich mit Dr. Carnap einverstanden, insbesondere hat er mich davon überzeugt, dass der Realismus-Phänomenalismus-Streit in seiner traditionellen Form in der Tat gegenstandslos gewesen ist. Ich muss also meinen eigenen Standpunkt recht wesentlich modifizieren. Andererseits sind jedoch auch Punkte vorhanden, in bezug auf die ich vorläufig nicht Herrn Dr. Carnap zustimmen konnte." – VCA.

[18] R. Carnap to M. Schlick, October 27, 1928. – ASP RC 029-30-23.

Late in the spring of 1929, Kaila's selected topic in the Circle was probability, undeniably a weak point in the *Aufbau*. Kaila's intention was to present his own research, especially as he had a few years earlier published a book devoted to the topic: *Die Prinzipien der Wahrscheinlichkeitslogik* (1926). In that book Kaila had characterized his position as "logical empiricism" in contrast to earlier positivisms and psychologisms, apparently the first occurrence of the term.[19] His conceptual toolbox included Schlick's critical realism and also much of Ernst Cassirer's structural analysis of science, including his understanding of "verificationism" as the necessity of predictions for the testing of hypotheses. Kaila's reasons for a number of these principles were pragmatic:

> Um handeln zu können, um die exakte Wissenschaft auf die Empirie anwenden zu können, muss vorausgesetzt werden, dass von der Vergangenheit und Zukunft, von einer gesetzmäßigen Außenwelt Wahrscheinlichkeitserkenntnis erlangt werden kann. Es wird aber im Anschluss an Hume zugegeben, dass dies keine *logische* Begründung der in Frage stehenden Voraussetzungen ist.

Schlick was forced to leave for his trip to America, but in a long letter Herbert Feigl explained to him what had happened in Vienna:

> In Wien gab es noch einige hochinteressante Diskussionen, zunächst mit Kaila – und später, als er abgereist war, auch in unserem Kreis. Kaila, der in der ersten Woche unseren Standpunkt beinahe schon zugegeben hatte, wurde später extrem rückfällig – und es war zum Schluss fast ein wenig traurig, als er wieder zu seiner Wahrscheinlichkeitslogik Zuflucht nahm, um mit ihrer Hilfe die Realismusthese als *sinnerfüllt* zu erweisen. Immerhin gab er zu, er müsse nun ein ganzes Jahr darüber nachdenken. […] Wir haben enorm viel diskutiert und das Problem nach allen Seiten gewendet, haben ihn auf die Zulässigkeit der Metaphysik als Poesie hingewiesen, – er war

---

[19] E. Kaila, *Die Prinzipien der Wahrscheinlichkeitslogik*, Annales Universitatis Fennicae Aboensis, Ser. B, Tom. IV, N:o 1, Turku 1926, p. 35: "Dieser *logische* Empirismus [unter]scheidet sich aber scharf von dem klassischen *psychologistischen* Empirismus." For a study of psychologism, see M. Kusch, *Psychologism. A Case Study in the Sociology of Philosophical Knowledge*, Routledge, London / New York 1995.

nicht zu trösten.[20] [...] Angeregt durch die Fragen und Einwände von Kaila hatte Carnap eine logische Analyse der induktiven Sätze gegeben: Der Satz: 'morgen wird die Sonne aufgehen' ist nach Carnap überhaupt *sinnlos*, – *sinnerfüllt* hingegen ist allein der Satz 'morgen wird die Sonne *wahrscheinlich* aufgehen' und zwar liegt der *Sinn* dieses Satzes allein in den vollständigen Induktionen über *vergangene* Sonnenaufgänge und ihren relativen Häufigkeiten – und was man sonst hier noch an (*vergangenen*) astronomischen Fakten als Induktionsgrundlage hinzunehmen will. Dass sich der Satz auf die schlechthin unbekannte *Zukunft* bezieht, hat keinen *theoretischen* Sinn, sondern hier liegen ganz wie beim Fremdpsychischen *nur Begleitvorstellungen*, Einstellungen, Handlungsweisen (belief, expectation) vor. – Analog beziehen sich Vergangenheitsaussagen eigentlich auch nicht auf die 'wirklich' *vergangene* Zeit, sondern ihr theoretischer Sinn liegt gleichfalls in den *gegenwärtigen* Erinnerungen. Die Frage nach einer *absoluten* Erinnerungstreue ist selbstverständlich metaphysisch, es gibt nur die Korrektur durch spätere Erlebnisse, aber das ist ein ganz harmloses empirisches Vorgehen"[21]. [...] Es handelt sich hier also um die Durchführung der *Instantan*-solipsistischen Basis, eine Sache, die wie Sie wohl sehen werden, nicht ohne Schwierigkeiten und Bedenken ist. Nach Carnaps Sprachanalyse ist eben der Sinn jedes Satzes (wenn er überhaupt einen haben soll) in dem bisherigen Weltprotokoll (=Relationsliste der Elementarerlebnisse) gegeben. Aus diesem Schatz spinnen wir unsere ganze Weisheit heraus. Aber nicht nur unser *Wissen* um *wahre* Sätze, sondern auch den *Sinn möglicher* Sätze. Kaila hingegen konnte sich angesichts der Tatsache der ständig wachsenden Protokolle nicht mit Carnap's These der Abgeschlossenheit dieses Protokolls zufrieden geben, er betrachtet es eben nur als einen Ausschnitt aus einem viel umfassenderen, *möglichen* Weltprotokoll. Carnap gibt das ständige Wachstum der Liste zu, formuliert dies aber so: Jedes neue Elementarerlebnis ist eine nichtlogische Konstante, eine neue Vokabel unserer Sprache. Solange wir diese Vokabel nicht haben, können wir nicht mit ihr sprechen, weshalb also Zukunftsaussagen sinnlos sein müssen.[22]

---

[20] H. Feigl to M. Schlick, July 21, 1929. – VCA.
[21] Ibid.
[22] Ibid.

Feigl was not a completely impartial observer. His sympathies were on the side of Schlick's earlier critical realism which Schlick had now given up, mainly because of Carnap. But it must be agreed that a "solipsism" of the present moment was a more accurate description for Carnap's starting point than the "phenomenalism" known from Carnap's intellectual autobiography. Assuredly, Carnap would have added the qualification "methodological" even in 1929.

In addition to Kaila, Feigl said, there was also another person arguing against Carnap: "Dagegen wendet sich nun besonders scharf auch Waismann, der im Anschluss an Wittgenstein in den Zukunftsaussagen Sätze mit Variablen sieht, die er als vollkommen sinnerfüllt betrachtet."[23]

5. Already well before his time in Vienna, Kaila had heard that Wittgenstein was dissatisfied with the starting point of Carnap's *Aufbau*. In a letter to Schlick he had written:

> Auf die faszinierende Gestalt Wittgensteins bin auch ich seit einigen Jahren aufmerksam geworden; leider ist seine Schreibweise für mich allzu orakelhaft; ich erwarte deshalb mit Spannung die Waismannsche Darstellung und Erläuterung der Gedanken von Wittgenstein. [...] Ganz besonders interessiert mich der von Ihnen angedeutete, von Wittgenstein *gegen die Konstitutionstheorie* erhobene Einwand, dass es nämlich *unmöglich sei, 'von den Momentan-Gesamt-Erlebnissen als Basis auszugehen'*.[24] (Italics added.)

Kaila's quotation from Schlick shows that Wittgenstein was not so completely unreceptive to the discussions of Schlick's Circle as the latter day perspectives seem to make him. Of course, there is no need to suppose that Wittgenstein was well informed about the discussions concerning the constitution theory in the Circle. In addition to Wittgenstein's *Tractatus*, Carnap's *Konstitutionstheorie* (his lost *Habilitationsschrift*) had been the main subject of discussions in the Circle. Schlick and Waismann and even Feigl had a lively communication with Wittgenstein. "Konstitutionstheorie" could mean the views expressed in the manuscript, but also those of the *Aufbau*. Carnap's *Aufbau* had appeared already several months before Wittgenstein left for

---

[23] Ibid.
[24] E. Kaila to M. Schlick, January 13, 1929. – VCA.

Cambridge in January 1929. The "given" was defined there as the total experience during an instant. Wittgenstein's reference to Carnap is quite clear, at least if we have any reason to believe Schlick who was an authority to both of them.

There is another piece of evidence that attests to Wittgenstein's dissatisfaction with the *Aufbau*. It dates from the very same weeks as Schlick's and Kaila's letters about this matter. Carnap wrote in his notes that there had been a discussion in which Waismann had desperately tried to formulate Wittgenstein's opinion on the book, Wittgenstein's criticism against Carnap's method of quasi-analysis in the study of the given.[25]

Wittgenstein did not tolerate Carnap and he had closed Carnap out of his communication. But as can be seen from his letter to Schlick from February 18, 1929, his contact to Schlick's Circle was still excellent otherwise: "Ich habe mich entschlossen, ein paar Termine hier in Cambridge zu bleiben und den Gesichtsraum und andere Dinge zu bearbeiten [...] Bitte grüßen Sie die Tafelrunde [= Schlick's Circle, J.M.] und Herrn Waismann ganz besonders; ich hoffe und freue mich, Sie alle in einem Monat wiederzusehen."[26]

The phenomenology of the visual space was something of a special interest to Schlick and his Circle. Carnap had presented a highly technical account of it and its relation to the cognitive system as a whole. What was Wittgenstein's argument against Carnap? Schlick and others agreed on the significance of the specious present, especially as regarding verification. Could Wittgenstein, the natural born genius, present something which made Carnap's system unacceptable?

---

[25] "(Nachdem ein [fruchtloses] Gespräch vorhergegangen, in dem Waismann sich vergeblich bemühte, Wittgensteins Meinung deutlich zu formulieren.)" And further: "*Über Extensionen*. Veranlasst durch Wittgensteins Kritik der Quasianalyse (über die wir gestern gesprochen haben) wollen wir einige logische Fragen klären." R. Carnap, "Gespräch mit Waismann", ASP RC 102-76-03. Later on, in a meeting of the Circle in 1931, Waismann presented criticisms of Carnap's approach but still added: "Die Bestrebungen, ein Konstitutionssystem aufzustellen sind auch die Wittgensteins. Die ganzen Differenzen beziehen sich auf die Art der Durchführung der Idee, nicht auf die Idee selbst." See the notes by R. Rand in F. Stadler, *Studien zum Wiener Kreis. Ursprung, Entwicklung und Wirkung des Logischen Empirismus im Kontext*, Suhrkamp, Frankfurt am Main 1997, p. 299.

[26] F. Waismann, *Wittgenstein und der Wiener Kreis*, ed. by B. F. McGuinness, Basil Blackwell, Oxford 1967, p. 17.

The main bulk of Wittgenstein's notebooks during his first year in Cambridge was concerned with the philosophy of mathematics. At the beginning of the year, and also when he grew tired of his mathematical explorations, Wittgenstein continued his remarks on cognition. During his Christmas holiday in Vienna they gained predominance. In the weeks and months before his return to Vienna, it is easy to see these manuscripts as preparations for his discussions with Schlick and Waismann. Most of the discussions were apparently spontaneous, but Wittgentein never forgot his special relation to Schlick. He was definitively a member of Schlick's network, although he would certainly not sit together with Carnap in any kind of meeting.

The notebooks show that in the beginning Wittgenstein did not have any clear argument against Carnap. The name of Carnap is not mentioned at all, but it is well-known that Wittgenstein was not fond of giving names and exact references. Carnap's formalisms were skipped, but some of the basic epistemological distinctions survived. Wittgenstein's discussions with Carnap had come to an abrupt end, but Carnap's impact upon the Circle – on Schlick, Waismann, Feigl and others – was momentous. Questions common to the majority of the Circle could not be avoided. Of course, Wittgenstein was questioning them in his own way, but still using the terminology proposed by Carnap.

Consider, for example, his discussion on pages 85-88 of his first Cambridge notebook:

> Gibt es überhaupt Zeit im ersten System? Kann man von einem Ereignis oder vielmehr von einer Tatsache im System der Data sagen 'es war'?
>
> Wenn ich die Tatsachen des ersten Systems mit Bildern auf der Leinwand und die Tatsachen im zweiten System mit den Bildern auf dem Filmstreifen vergleiche, so gibt es auf dem Filmstreifen ein gegenwärtiges Bild, vergangene und zukünftige Bilder; auf der Leinwand aber ist nur die Gegenwart.
>
> Das eine Chrarakteristische an diesem Gleichnis ist, dass ich darin die Zukunft als präformiert ansehe.
>
> Es hat einen Sinn zu sagen, die zukünftigen Ereignisse seien präformiert wenn es im *Wesen* der Zeit liegt, nicht abzureißen. Denn dann kann man sagen: 'Etwas wird geschehen, ich weiß nur nicht, was.': Und in der Welt der Physik kann man das sagen.

> Wie aber in der Welt der Data? Reißt diese Welt nicht wirklich ab?
> Kann man von einem Datum sagen, es sei früher als ein anderes?
> Ich habe eben *gegenwärtige* Sinnesbilder und *gegenwärtige* Erinnerungsbilder. Kann ich nicht sagen, dass ich aus diesen nur im 2. System die Zeit konstruiere?
> Man kann aber auch sagen, dass ich aus diesen gegenwärtigen Data ein zeitliches 2. System konstruieren *kann*, sagt etwas über das 1. System aus und was es aussagt drücke ich in den Worten aus: Das erste System ist zeitlich geordnet.[27]

During the year 1929 Wittgenstein would repeatedly speak of the first and second "systems" or "worlds" in his epistemological remarks, following the Carnapian terminology. The first one was roughly speaking the phenomenological one and the second the physical, not unlike the Carnapian construction, albeit more straightforward.[28] The problem to be solved was the relation of these two different systems. Both Wittgenstein and Carnap were concentrating on the primary significance of the specious present.

There was even in Cambridge a very important young person who was using the distinction between "the primary system" and "the secondary system". F. P. Ramsey had been active in different ways in making Wittgenstein's ideas known in his own country and also making possible the philosopher's new career in Cambridge. Ramsey was responsible for advising Wittgenstein academically. In 1929, Ramsey wrote an essay on "Theories". He was using the very same dichotomy between the two systems, well aware of its Carnapian origin. The primary system was expressed in an observational language and the secondary in theoretical. In Carnap's sense, Ramsey attempted to formulate a theoretical system using only explicit definitions. But his conclusion was that then the system would lose all its vitality, remaining static

---

[27] L. Wittgenstein, *Wittgenstein's Nachlass*, The Bergen Electronic Edition, Oxford CD-ROM 2000, Item 105.

[28] No special importance should be attached here to the difference of the words "phenomenal" and "phenomenological". L. Boltzmann had used the designation "general phenomenology" of the view that seeks to represent phenomena without going beyond experience. Wittgenstein had read his Boltzmann. See e.g. L. Boltzmann, *Theoretical Physics and Philosophical Problems*, ed. by B. McGuinness, Reidel, Dordrecht and Boston 1974, p. 95.

and uninteresting.[29] It is probable that Ramsey and Wittgenstein discussed the two systems with each other. However, Wittgenstein did not develop his ideas concerning the duality using Ramsey's logic, and, in any case, he was acquainted with the Carnapian distinction already before he came to Cambridge.

Wittgenstein believed that there was a special phenomenological language, in addition to the physical one. In his very first philosophical notes in Cambridge, after some delightful comments about Ramsey, Wittgenstein tried to explain the relation of the two languages as follows:

> Es scheint viel dafür zu sprechen, dass die Abbildung des Gesichtsraumes durch die Physik wirklich die einfachste ist. D.h. dass die Physik die wahre Phänomenologie wäre.
>
> Aber dagegen lässt sich etwas einwenden: Die Physik strebt nämlich Wahrheit, d.h. richtige Voraussagungen der Ereignisse, an, während *das* die Phänomenologie nicht tut, sie strebt *Sinn* nicht *Wahrheit* an.
>
> Aber man kann sagen: Die Physik hat eine Sprache und in dieser Sprache sagt sie Sätze. Diese Sätze können wahr oder falsch sein. Diese Sätze bilden die Physik und ihre Grammatik die Phänomenologie (oder wie man es nennen will).
>
> Die Sache schaut aber in Wirlichkeit schwieriger aus durch den Gebrauch der mathematischen Terminologie. Wenn z.B. die Wissenschaft zweifelt, ob die beobachteten Erscheinungen durch die Elektronen- oder durch die Quantentheorie richtig zu beschreiben sind, so scheint es auf den ersten Blick, als handelte es sich um eine Entscheidung in der Grammatik.
>
> Es gibt eine bestimmte Mannigfaltigkeit des Sinnes und eine *andere* Mannigfaltigkeit der *Gesetze*.
>
> Die Physik unterscheidet sich von der Phänomenologie dadurch, dass sie Gesetze feststellen will. Die Phänomenologie stellt nur die Möglichkeiten fest.
>
> Dann wäre also die Phänomenologie die Grammatik der Beschreibung derjenigen Tatsachen, auf denen die Physik ihre Theorien aufbaut.[30]

---

[29] N.-E. Sahlin, *The Philosophy of F. P. Ramsey*, Cambridge University Press, Cambridge 1990, p. 133.

[30] Wittgenstein 2000, Item 105, pp. 4–5. (15.2.1929)

Further quotations from the spring of 1929 show how and why Wittgenstein was experimenting with the phenomenological language:

Die phänomenologische Sprache beschreibt genau das gleiche wie die gewöhnliche, physikalische. Sie muß sich nur auf das beschränken, was verifizierbar ist.

Ist das überhaupt möglich?

Vergessen wir nicht, dass die physikalische Sprache auch wieder nur die primäre Welt beschreibt und nicht etwa eine hypothetische Welt.[31] [...]

Die Sprache selbst gehört zum zweiten System. Wenn ich eine Sprache beschreibe, beschreibe ich wesentlich etwas Physikalisches. Wie kann aber eine physikalische Sprache das Phänomen beschreiben?

Ist es nicht so: das Phänomen (specious present) enthält die Zeit, ist aber nicht in der Zeit?[32] [...]

Die Verifikation der Sprache – also der Akt, durch den sie ihren Sinn erhält – geht allerdings in der Gegenwart vor sich.

Aus dem vorigen geht hervor – was übrigens selbstverständlich ist – dass die phänomenologische Sprache das Selbe darstellt wie unsere gewöhnliche physikalische Ausdrucksweise und nur den Vorteil hat, dass man mit ihr manches kürzer und mit geringerer Gefahr des Missverständnisses ausdrücken kann.[33]

Much later, probably already in the summer, Wittgenstein came exactly to the contrary conclusion: "Die Beschreibung der Phänomene mittels der Hypothese der Körperwelt ist unumgänglich durch ihre Einfachheit verglichen mit der unfassbar komplizierten phänomenologischen Beschreibung."[34]

Wittgenstein kept repeating his central metaphor intended to clarify the relation of the two languages: "Wir befinden uns mit unserer Sprache sozusagen nicht im Bereich des projizierten Bildes sondern im Bereich des Films."[35] He was doubtful: "Und doch kann es eine phänomenologische Sprache geben. (Wo muss diese Halt machen?) [...] Oder ist es so: Unsere gewöhnliche Sprache

---

[31] Ibid., p. 108.
[32] Ibid., p. 114.
[33] Ibid., pp. 120, 122.
[34] Wittgenstein 2000, Item 106, p. 102.
[35] Wittgenstein 2000, Item 107, p. 2.

ist auch phänomenologisch, nur erlaubt sie es begreiflicherweise nicht, die Sinnesgebiete deren gesamte Mannigfaltigkeit die ihre ist, zu trennen."[36]

When the year went on, Wittgenstein's verificationism hardened: "Die Verifikation ist nicht *ein* Anzeichen der Wahrheit, sondern *der* Sinn des Satzes. (Einstein: Wie eine Größe gemessen wird, das ist sie.)"[37]

This meant that there was no longer any need for a phenomenology that would give the sense of the sentence, separate from its truth. In October 1929, Wittgenstein drew the conclusions:

Die Annahme dass eine phänomenologische Sprache möglich wäre und die eigentlich erst das sagen würde, was wir in der Philosophie ausdrücken wollen ist – glaube ich – absurd. Wir müssen mit unserer gewöhnlichen Sprache auskommen und sie nur richtig verstehen. [...] Ich meine: was ich Zeichen nenne, muss das sein was man in der Grammatik Zeichen nennt, etwas auf dem Film nicht auf der Leinwand. [...] Jeder Satz ist ein leeres Spiel von Strichen oder Lauten ohne die Beziehung zur Wirklichkeit und seine einzige Beziehung zur Wirklichkeit ist die Art seiner Verifikation.[38]

It was wrong to claim that only the specious present was real. Such an approach led to a solipsism of the present which was a misunderstanding of the work of our language:

Wenn man sagt, die *gegenwärtige* Erfahrung nur hat Realität, so muss hier schon das Wort 'gegenwärtig' überflüssig sein [...] Die Gegenwart, von der wir hier reden ist nicht das Bild des Filmstreifens, das gerade jetzt im Objektiv der Laterne steht, im Gegensatz zu den Bildern vor und nach diesem, die noch nicht oder schon früher dort waren, sondern das Bild auf der Leinwand das mit Unrecht gegenwärtig genannt wurde, weil gegenwärtig hier nicht zum Unterschied von vergangen und zukünftig gebraucht wird. Es ist also ein bedeutungsloses Beiwort.[39]

Wittgenstein was now back in Vienna, prepared to explain his argument, and his argument had taken a definite shape. Our ordinary language did not

---

[36] Ibid., p. 3.
[37] Ibid., p. 143.
[38] Ibid., p. 176. (22.10.1929)
[39] Wittgenstein 2000, Item 108, pp. 2–4.

contain any distinction between "the primary" and "the secondary".[40] There was no "primary language" in contrast to our ordinary physical language, only certain forms of expression which easily led to distortions. "Die Sätze unserer Grammatik haben immer die Art physikalischer Sätze und nicht die 'primären' und vom Unmittelbaren handelnden Sätze."[41]

Merrill B. Hintikka and Jaakko Hintikka have described a turn in Wittgenstein's thought towards the end of the year. They even observe the similarity of Wittgenstein's terminology to that of Carnap and Ramsey,[42] but they do not seriously consider the possibility that Wittgenstein actually was developing an argument specifically against Carnap. Could it be possible that Wittgenstein was simultaneously struggling against his own alter ego, the author of the *Tractatus*, as Hintikka and Hintikka suggest, dismissing Carnap? It is certainly true that Wittgenstein was making such revisions in many respects. As there is no space to go to the interpretations of the *Tractatus*, I must simply use Occam's Razor to distance the *Tractatus* from this game. However, as concerns the history of the Vienna Circle the turn towards physicalism is of special importance. What was the role of Wittgenstein in it, if any?[43]

On his Christmas holiday in Vienna 1929, Wittgenstein explained extensively his new views to Schlick and Waismann: "Ich glaube, dass wir im Wesen nur eine Sprache haben und das ist die gewöhnliche Sprache. Wir brauchen nicht erst eine neue Sprache zu erfinden oder eine Symbolik zu konstruieren, sondern die Umgangssprache *ist* bereits *die* Sprache, vorausgesetzt, dass wir sie von den Unklarheiten, die in ihr stecken, befreien."[44]

Wittgenstein did not deny the existence of the phenomenological or its importance for verification. But if his standpoint proved to be true, it would in any case be bad news for Carnap's constructive project.

---

[40] Ibid., p. 46.

[41] Ibid., p. 60. (4.1.1930)

[42] M. B. Hintikka and J. Hintikka, *Investigating Wittgenstein*, Basil Blackwell, Oxford 1986, pp. 141–142.

[43] For a recent review of the continuing debate, see David Stern, "Wittgenstein versus Carnap on Physicalism: A Reassessment", in *Cambridge Companion to Logical Empiricism*, ed. By T. Uebel and A. Richardson, Cambridge University Press, Cambridge 2007.

[44] Ibid., p. 45.

6. Early in the year 1930, Neurath was organizing Marx evenings in Vienna for his philosophical friends, including Carnap. Neurath had been planning a book for the scientific series edited by Schlick and Philipp Frank. It would be entitled *Der wissenschaftliche Gehalt der Geschichte und Nationalökonomie* and it would contain something that nobody had done earlier, a synthesis of the exact thought of unified science with Marxist ideas. Neurath had no academic position, but now the book was in the process of writing. Actually, the book proved to be more militant than exact. Schlick did not accept it for publication.[45]

On 16.7.1930, Schlick wrote to Frank a letter and a report about the manuscript, horrified. There is no need to go into details, but the report makes it plain that even others were aware of the product of the self-styled materialist:

> ...der erste Abschnitt scheint mir für das ganze zum großen Teil entbehrlich, vor allem aber ist er für den uneingeweihten Leser lange nicht verständlich genug. Auf diese letzteren Punkte haben besonders auch Carnap und Waismann hingewiesen. Neurath sagte mir, die beiden seien mit dem Buch im wesentlichen einverstanden. Mir gegenüber hat aber Carnap sich weniger positiv, Waismann deutlich negativ ausgedrückt, dass hier ein Missverständnis vorliegen muss...[46]

The book was predominantly about history and social science on a materialistic base, following Carnap's hints towards the possibility of a materialistic system but not the technicalities of his approach. The first chapters contained Neurath's views about philosophy. Neurath was especially angry about the prospect of leaving these parts out. By following Schlick's advice, he could easily have written a book on sociology, but other parts of the book had been more difficult.[47] The conflict was not a minor matter, soon to be forgotten. Not even Neurath's old friend Frank could help.

Neurath was soon writing a new book about empirical sociology. It appeared the next year in the series where he had originally planned to publish

---

[45] See J. Manninen, "Wie entstand der Physikalismus?", in *Nachrichten. Forschungsstelle und Dokumentationszentrum für Österreichische Philosophie* 10/2002, pp. 22–52, also in www.austrian-philosophy.at, and J. Manninen, "Towards a Physicalistic Attitude", in *The Vienna Circle and Logical Empiricism. Re-evaluation and Future Perspectives*, ed. by F. Stadler, Kluwer, Dordrecht / Boston / London 2003, pp. 133–150.

[46] M. Schlick to Ph. Frank, July 16, 1930. – ASP RC 029-30-08.

[47] O. Neurath to R. Carnap, July 22, 1930. – ASP RC 029-19-12.

the rejected manuscript. Already on 16.8.1930 Neurath wrote to Carnap: "Ich habe inzwischen das Buch in gänzlich anderer Fassung für die Sammlung fertig."[48] This was hard to believe, but Carnap could tell Schlick the following in a letter from 28.8.1930:

> Gestern war ich mit Feigl bei N[eurath]. Wir haben ihm beruhigend zugeredet. Er war nämlich immer noch erregt und glaubte, dass nicht nur dem MS, sondern ihm persönlich Ungerechtigkeit widerfahren sei. Und nun hatte er tatsächlich schon ein beinahe fertiges neues MS! Er hat uns verschiedene Kapitel daraus vorgelesen. Und wir waren erstaunt zu sehen, dass er hier wirklich in ganz anderem Tone geschrieben hat, sachlich argumentierend, und ernsthaft in der Formulierung. Der Inhalt ist jetzt ein ziemlich anderer, mit gleichen Grundgedanken. Es sind hier keine ausführlichen erkenntnistheor[etischen] Erörterungen mehr, sondern Darstellung der Sache selbst...[49]

What was the epistemology that caused the worries in the earlier manuscript? How did Neurath understand the materialistic base of unified science? It did not have much to do with material things as such. According to Neurath, all science was concerned with spatio-temporal processes and their relations.

From Carnap's point of view, both cognitive and physical construction systems were legitimate. *Aufbau*'s cognitive system had as its basic empirical relation recollected similarity applied to one individual's elementary experiences. The physical construction system should use as its building blocks two basic relations on worldpoints, coincidence and local time order.

In contrast to Carnap's autopsychological and physical starting points, Neurath described science as starting with everyday language:

> Der Aufbau der Einheitswissenschaft geht ungefähr so vor sich: Wir gehen aus von den Aussagen des Alltags, sprechen wir in aller Harmlosigkeit – es ist die Harmlosigkeit von Menschen, die aus der Überlieferung von Jahrtausenden gelernt haben – vom 'quadratischen Querschnitt eines blauen Tischfußes'. So können wir nachträglich daraus Verschiedenes abspalten,

---

[48] ASP RC 029-14-06.
[49] R. Carnap to M. Schlick, August 28, 1930 – VCA.

bis wir zu einer streng wissenschaftlichen Formulierung gelangen. Rückblickend erscheint die Ausgangsformulierung als eine 'Ballung', in der neben dem 'quadratisch' (einer aus vielen Beobachtungen abgeleiteten geometrischen Konstruktion), noch das 'blau' auftritt; dies werden wir späterhin zu spezifischen 'Daten' rechnen, die innerhalb der wissenschaftlichen Aussagen durch mathematische Größen, z.B. Schwingungen, ersetzt werden.[50] [...] Was wir sehen, hören usw. wird allmählich seiner Besonderheit entkleidet und *'vereinheitlicht'*. Die 'Ballung', die als 'quadratischer Querschnitt eines blaues Tischfußes' auftritt, weicht allmählich einer *'Vereinheitlichung'*, in der alles das zusammengefasst ist, *was beliebigen Sinnesgebieten gemeinsam* ist und durch Zeichen ausdrückbar ist. Der eine kann mit Hilfe mehrfacher Augenbeobachtungen zu einem quadratischen Gebilde gelangen, der andere mit Hilfe von Druckerfahrungen, der dritte, indem er bestimmte Hörversuche anstellt. Alle sprechen unter gewissen Kautelen von *'demselben* viereckigen Gebilde', indem sie ein-eindeutig einander zuordenbare Aussagen einander mitteilen.[51]

Neurath left no room for alternative construction systems. Everything would be contained within the one domain, described with a quotation from Carnap: "Das materialistische Konstruktionssystem hat den Vorzug, dass es dasjenige Gebiet (nämlich das physische) als Basisgebiet hat, das als *einziges* eine *eindeutige Gesetzmäßigkeit* seiner Vorgänge besitzt."[52]

One can summarize Neurath's view with three points.

First, the language of the unified science should strive for universality: "Die Einheitswissenschaft kennt nur *eine* Art von Aussagen, die sie miteinander kombiniert, sie kennt nur ein Begriffssystem. Ihr ist ein 'monistischer' Zug eigen, der sich in allem geltend macht."[53] But in order not to do harm to the scientific enterprise, this should not be pressed too hard. A relevant self-consciousness among scientists was enough. Qualifications to the lawfulness were needed, but all scientific formulations had to do with spatio-temporal processes.

---

[50] O. Neurath, Der wissenschaftliche Gehalt der Geschichte und der Nationalökonomie. – VCA K.2, p. 10.
[51] Ibid., p. 12.
[52] Ibid., p. 10.
[53] Ibid., p. 23.

Second, the language of science was intersensual. Neurath explained this point extensively and enthusiastically:

> In der wissenschaftlichen Optik kommt nichts vor, was ein Blinder nicht verstehen und weiterführen könnte. Er kann Konsequenzen neu finden, wenn ihm die Grundlagen bekannt sind. Der Blinde kennt 'neben' und 'zwischen', 'größer' und 'kleiner', 'soviel', kurzum das, was zum Aufbau der physikalischen Aussagen nötig ist.
>
> Der Blinde kann aber nicht nur theoretisch die Optik erfassen, er kann grundsätzlich die Erfahrungen sammeln, welche der Sehende macht. Wenn der Sehende einen Interferenzversuch unternimmt, der 'hell' und 'dunkel' in gewissen Perioden liefert, kann der Blinde etwa durch Einschaltung von Selenzellen in ein Telephonsystem die Periodizität von 'laut' und 'still' erhalten. Wir sind alle elektrizitätblind und verfügen über eine vollkommene elektrische Theorie. [...] Jedenfalls ist die Physik nicht an spezifische Daten gebunden, die Optik hat nichts mit dem Auge, die Akustik nichts mit Hören zu tun, sondern mit bestimmten Schwingungsarten. Mach, dem die hier vertretene Anschauung so viel verdankt, hat dieser materialistischen Basis zu wenig Rechnung getragen, als er die einzelnen physikalischen Gebiete bestimmten Sinnesorganen zuwies.[54]

In Neurath's case, this point had also a personal dimension. Neurath's wife was a blind logician who participated in the meetings of Schlick's Circle.

Third, science was intersubjective: "Die Einheitswissenschaft kennt 'Wirklichkeit' nur sofern sie in sich widerspruchslose kontrollierbare Aussagen machen kann. Darüber hinaus hat das Wort 'Wirklichkeit' in der Wissenschaft keinen angebbaren Sinn. [...] Was als Aussage über irgend etwas auftritt, unterliegt grundsätzlich der Kontrolle aller."[55]

Neurath did not use the terms "universal", "intersensual" and "intersubjective". But they certainly catch what Neurath meant in the quoted passages. Later on, together with references only to the spatio-temporal processes, they were used to define what is meant by "physicalism".

---

[54] Ibid., p. 11.
[55] Ibid., p. 11, 14.

7. Carnap was very helpful when Neurath had his conflict with Schlick. He even promised that in an extreme case he could rewrite the manuscript so that it would be acceptable to the series of the Circle. In August, he would have time for that.[56] Neurath was confident that he could do the job himself,[57] as he did.

In the Pittsburgh archives for scientific philosophy there is an early manuscript written by Carnap and entitled "Die physikalische Sprache als Universalsprache der Wissenschaft", catalogue number ASP RC 110-03-22. In 1932, Carnap published in the journal *Erkenntnis* a paper under the same title. The manuscript and the publication are substantially different. Later on, the published version was translated into English as a small book entitled *The Unity of Science* (1934). The manuscript is a first draft of the published version.

Somehow this first draft has so far escaped the attention of researchers. The final draft is actually a new paper, completely rewritten. When Carnap sent this new formulation of his ideas to Neurath, he related: "Als ich diesen vor 1 ½ Jahren geschriebenen Aufsatz jetzt für den Druck fertig machen wollte, stellte sich nämlich heraus, dass ich ihn von A bis Z neu schreiben musste."[58]

Meanwhile, Neurath's book *Empirische Soziologie* had appeared and Neurath had also written a paper "Soziologie im Physikalismus", to be published in *Erkenntnis*.[59] Towards the end of the year 1931 the Austrian social democratic journal printed his article "Weltanschauung und Marxismus", summarizing his views. Carnap commented the last mentioned: "Deinen Beitrag im *Kampf* habe ich mit Vergnügen und voller Zustimmung gelesen."[60]

There is no need for the mention of the time span of "1½ years" to be exact. It would mean June 1930. But Carnap afterwards added in shorthand some explanations in the margin of the original draft. He remarked that the draft was written on 3.6.1930. He mentioned further that the paper was an

---

[56] R. Carnap to O. Neurath, July 25, 1930. – ASP RC 029-14-11.
[57] O. Neurath to R. Carnap, July 28, 1930. – ASP RC 029-14-10.
[58] R. Carnap to O. Neurath, January 16, 1932. – ASP RC 029-12-70.
[59] Already in his letter to Carnap on August 16, 1930, Neurath expressed his hope to write to *Erkenntnis* a companion paper to his book. Together with Reichenbach, Carnap was editing this journal of the Circle. – ASP RC 029-14-06. During the spring of 1931, Neurath began to propagate his ideas in the Circle under a new title: "physicalism".
[60] As in note 58.

improved version of the lectures that he had held in the Verein Ernst Mach on 20.5.1930 and in Karl Bühler's colloquium on 28.5.1930. It is plausible that the original draft took shape soon after the lectures – in any case, during the summer of 1930. Carnap informed Schlick in September: "Ich habe meinen Kön[igsberger] Vortrag jetzt ausgearbeitet. Ein Aufsatz 'Die physikalische Sprache als Universalsprache der Wissenschaft' ist fertig; ein zweiter, der sich an den ersten anschließt, eine ausführliche Ausarbeitung dessen, was ich bei Bühler vorgetragen habe ('Die Psychologie in physikalischer Sprache'), ist beinahe fertig."[61]

In the background, there was Neurath's work with his own book. Already in a letter to Carnap in March 1930, Neurath was expecting problems with Schlick and even censorship: "Hätte er nur einmal einer Marx-Besprechung beigewohnt, hätte ich ihm doch mit grosser Freude mein Buch vorgetragen."[62] Then, in April, a draft of the book seems to have been in the final stage of preparation: "Ph. Frank schrieb mir inzwischen [...] von Schlick's Wunsch [...] mein Manuskript vor Abgehen nach Berlin zu lesen."[63] In fact, Schlick did read the manuscript only much later. But Carnap was well aware of a possible forthcoming conflict.

It was in these circumstances that he gave the lectures and wrote the original draft. The paper illuminates a transformation of Carnap's thought – and obviously also what he would have wanted to do with Neurath's manuscript, if he had been asked. Neurath, again, apparently knew that there was a significant disagreement between his own and Carnap's views.

The first draft was not an attempt to build a physical construction system. The language of empirical science was Carnap's challenge. Carnap was defending the unity of science: "...alle Sachverhalte sind von *einer* Art, nach *einer* Methode erkennbar, in *einer* Sprache ausdrückbar."[64] But he still maintained a two-track approach. There were two universal languages, the phenomenal and the physical, each translatable into the terms of each other.

---

[61] R. Carnap to M. Schlick, September 1930. – ASP RC 029-30-02.
[62] O. Neurath to R. Carnap, March 25, 1930. – ASP RC 020-14-19.
[63] O. Neurath to R. Carnap, April 12, 1930. – ASP RC 029-14-18.
[64] R. Carnap, Die physikalische Sprache als Universalsprache der Wissenschaft. [1930] – ASP RC 110-03-22, p. 1.

It was possible to understand the sense of a sentence if and only if one knew how it was verified. Verification, again, presupposed the phenomenal language: "Um einen Satz *p* zu verifizieren, muss ich *p* mit dem Sachverhalt vergleichen, den *p* besagt. Ein Sachverhalt liegt unmittelbar nur im Gegebenen vor. Daher muss ich *p* zum Zweck der Verifikation in einen Satz *p′* meiner phänomenalen Sprache übersetzen; *p′* spricht dann von meinen Erlebnisinhalten."[65]

Only such a translation made possible the comparison of a sentence with a state of affairs: "Da ein Erlebnissachverhalt nur dem erlebenden Subjekt selbst gegeben ist, so geschieht Verifikation stets monologisch."[66]

Carnap's perspective was still the autopsychological one of the *Aufbau*. The phenomenal language spoke of the "given", i.e. the immediate lived experiences of one individual. However, the nature of the given was an open question. Mach had thought that the given consisted of the simplest sense data. This was no longer acceptable, as Carnap had observed already in the *Aufbau*. But the psychological base appeared now to have been complicated:

> Das Problem einer genaueren Bestimmung der Art des "Gegebenen" ist gegenwärtig noch nicht gelöst. [...] Die neuere Psychologie ist bei aller Verschiedenheit ihrer Richtungen doch übereinstimmend zu der Auffassung gekommen, dass wir zu den Einzelempfindungen erst durch abstraktive Zerlegung gelangen, dass sie nicht das unmittelbar Gegebene sind. Als solches sind vielmehr umfassendere Komplexe aufzufassen; dabei ist noch nicht klargestellt, ob die Sehgestalten oder das gesamte Sehfeld als Einheit oder das Gesamterlebnis eines Augenblicks als Einheit, noch unzerlegt in Sinnesgebiete, das Gegebene bilden.[67]

Despite its necessity for verification, the phenomenal language was always the language of a single subject, "sie ist eine monologische, solipsistische Sprache, keine intersubjektive Sprache".[68] Or in another, less technical for-

---

[65] Ibid., p. 9.
[66] Ibid.
[67] Ibid., pp. 6–7.
[68] Ibid., p. 10.

mulation, "sie dient zum Sprechen von mir zu mir, kann nicht zur Verständigung von Subjekt zu Subjekt verwendet werden".[69]

Positivistic and realistic ontologies were to be rejected as metaphysics, but Carnap was prepared to call his own epistemology "methodological positivism" or "methodological solipsism", because of its thesis that "...die phänomenale Sprache eine Universalsprache ist, d.h. dass jeder Satz irgend eines Wissenschaftsgebietes sich schrittweise in diese Sprache übersetzen lässt; und dass diese Übersetzung den Weg zur Verifikation weist, und damit die erkenntnistheoretische Fundierung gibt".[70] In a footnote, Carnap gave a reference to the *Aufbau*.

In contrast, the simplest physical sentences concerned spatial and terporal determinations of properties. In prescientific language, there were still qualitative terms, but in science these were turned over into quantitative ones, expressible by numbers and connected by laws. Much like Neurath, Carnap thought that science was free of qualities and consequently not bound to any specific sensory areas. He discussed this topic at length. Then he remarked that there was a happy empirical fact about the world-order, the following:

> ...dass es möglich ist, physikalische Zustandsgrößen derart anzusetzen, dass die erlebten Qualitäten in eindeutiger funktionaler Abhängigkeit von ihnen stehen, und zwar sämtliche erlebten Sinnesqualitäten, sofern die physikalischen Zustandsgrößen vollständig gegeben sind. In unserem Beispiel [given in Carnap's earlier discussion, J.M.]: man kann die Skalen von Tast-Spektroskop, Hör-Spektroskop und Photo-Spektroskop (mit bloßen Helligkeitsunterschieden) so aufeinander abstimmen, dass diese Apparate in jedem Einzelfall dasselbe Ergebnis liefern. Diesen Tatbestand, dass dieselben physikalischen Bestimmungen für jedes Sinnesgebiet gültig sind, drücken wir dadurch aus, dass wir sagen: *Die physikalischen Bestimmungen gelten intersensual.*[71]

It was still another happy regularity of nature that the physical magnitudes were also independent of the persons who were doing the research:

---

[69] Ibid., p. 5.
[70] Ibid., p. 11.
[71] Ibid., p. 16.

Beginning the Logical Construction of Cognition 279

> Wenn zwei Subjekte verschiedener Meinung sind, in bezug auf die Länge eines Stabes, die Temperatur eines Körpers, die Frequenz einer Schwingung, so wird ein solcher Meinungsunterschied in der Physik niemals als unbehebbare subjektive Differenz hingenommen, sondern stets versucht, durch ein gemeinsames Experiment zu einer Einigung zu kommen. Die Physiker sind der Ansicht, dass grundsätzlich stets eine solche Einigung möglich sein muss, und dass, wo sie praktisch nicht gelingt, nur technische Schwierigkeiten (Unvollkommenheit der technischen Hilfsmittel, Mangel an Zeit u.dgl.) im Wege stehen. Diese Ansicht hat sich bisher in allen Fällen, die man mit hinreichender Gründlichkeit überprüfen konnte, bestätigt. Wir drücken diesen Tatbestand dadurch aus, dass wir sagen: *Die physikalischen Bestimmungen gelten intersubjektiv.*[72] [...] Wie aber steht es mit der qualitativen Vorstufe der quantitativen physikalischen Sprache, also mit jener Sprache, die man im täglichen Leben zur intersubjektiven Verständigung anzuwenden pflegt? Wir werden sehen, dass derartige qualitative Bestimmungen zwar intersubjektiv sinnvoll sind, aber beim Übergang von einem Subjekt zu einem andern nicht stets gültig bleiben, also keinen Anspruch auf intersubjektive Geltung erheben können.[73]

Everyday communication was used in intersubjective communication, but it was not intersubjectively valid – rather it was, at best, only approximately so. Unlike Neurath, Carnap did not see any wisdom incorporated in everyday language. Science was a system of intersubjectively valid true sentences and the only intersubjective universal language was the physical one. Anything that could be expressed at all could be expressed in it. In the rest of his paper, Carnap attempted to demonstrate this.

When all states of affairs were expressed in the physical language, only objects of one kind were left: the spatio-temporal, physical processes. Without rejecting the universal but subjective phenomenal languages, Carnap had a firm standpoint concerning the language and object domain of the empirical sciences: "Indem wir die Wissenschaft auf der Basis der physikalischen Sprache aufbauen, gelangen wir also zur intersubjektiven Einheitswissenschaft mit

---

[72] Ibid., p. 17.
[73] Ibid., p. 18.

einem Objektbereich von durchgängiger Gesetzmäßigkeit. Daher ist diese Basis die geeignetste für die Realwissenschaft in sachlicher Einstellung."[74]

One might ask whether logic was not, for Carnap, a universal language, the most universal of all. But it was something different. With a reference to Frege, Russell and Whitehead, and Wittgenstein Carnap explained:

> Diese Analyse hat gezeigt, dass es nicht neben oder über den Fachwissenschaften eine Philosophie als eigenes System philosophischer Sätze geben kann. Vielmehr ist die Philosophie eine Tätigkeit, nämlich die der Klärung der Begriffe und Sätze der Wissenschaft. [...] Die Sätze der *Logik und Mathematik* sind Tautologien, analytische Sätze, die allein auf Grund ihrer Form gültig sind. Sie teilen keinen Sachverhalt mit, sind ohne Aussagegehalt... [...] Die gehaltvollen Sätze, also die Sätze, die einen Sachverhalt zum Ausdruck bringen, gehören zum Bereich der Realwissenschaft.[75]

Logic was presupposed in both of the two universal languages and also in the partial languages of the special sciences. It was an activity of clarifying all languages. But it expressed no states of affairs.

8. Towards the end of the year, Kaila published in Finland an enlarged and polished version of his critical study of Carnap's *Aufbau*, entitled *Der logistische Neupositivismus*. It was discussed twice in Reichenbach's seminar in Berlin where Carl G. Hempel wrote a criticism of one aspect of its logic: logical relations do not have direction.[76] In December, Rose Rand presented Kaila's arguments to

---

[74] Ibid., p. 30.

[75] Ibid., pp. 2–3.

[76] This critique was quoted by Carnap in Vienna. However, Carnap thought, correctly, that Hempel probably did not have as critical an attitude towards Kaila's logic as he had. Kaila's attempt had been to describe the lived experience of time, using some observations of Edmund Husserl: "Schon durch die Einsinnigkeit der Zeit ist demnach auch die Zukunft da. In seiner bildreichen Sprache der 'strengen Phänomenologie' drückt Husserl dies so aus: 'Jede Wahrnehmung hat ihren retentionalen und protentionalen Hof.' [...]

Oben sahen wir, dass wir die 'Richtung' einer Relation nur durch den Hinweis auf die Erlebniszeit, in der das Denken der Relation stattfindet, 'erklären' können. Der

the Vienna Circle.[77] There followed a discussion in which Carnap, Hans Hahn and Kurt Gödel participated.[78] The very next day, Carnap wrote a letter to Kaila.

Kaila had immediately understood that Carnap's book was in a completely different category from most of what was called philosophy. It deserved great respect. Still it was, in Kaila's opinion, a failure. But it failed in a way that was analogous to Kant's *Kritik der reinen Vernunft*. Kant had given an absolute significance to Newton's physics and Carnap was doing the same with certain principles which were actually anchored in contemporary exact sciences.

The question of realism could not be rejected as metaphysical. If Carnap's results could be accepted "so bedeuten sie in der Tat das Ende aller Philosophie. Und noch mehr: wenn sie richtig sind, so sind sie geeignet, auch die empirische Forschungsarbeit ihres *élans* zu berauben; denn die 'realistische Sprache' der Wissenschaft ist in Wirklichkeit weit mehr als eine blosse Sprechweise: sie ist der Ausdruck der lebendigen *Seele* der Wissenschaft."[79]

Kaila commented on a number of issues, including the external world, other minds, and the nature of experienced time. Two of his criticisms had special weight, one of them logical and the other psychological, but both of them essential for Carnap's constructive project.

Kern der Erlebniszeit ist die Einsinnigkeit des Zeitflusses. Diese Einsinnigkeit wiederum enthält, dass eine 'Zukunft' da ist.
Hieraus folgt, dass die Frage, inwiefern wissenschaftliche Aussagen, in denen die 'Zukunft' ihrem gemeinen Sinne nach vorkommt, legitim sind, zu bejahen sind." – E. Kaila, *Der logistische Neupositivismus. Eine Kritische Studie*, Annales Universitatis Aboensis, Ser. B, Tom XIII, Turku 1930, pp. 50–51. An English translation of Kaila's study is available in E. Kaila, *Reality and Experience. Four Philosophical Essays*, ed. by R.S. Cohen with an introduction by G.H. von Wright, Vienna Circle Collection 12, Reidel, Dordrecht / Boston / London 1979.

Hempel wrote to Kaila a six page techical letter on January 3, 1931. Now he expressed: "Wir müssen im Erkenntnisprozess das Auftreten etwa des Erlebnisses 'Quecksilberspiegel oberhalb Nullstrich' von dem des Erlebnisses 'Nullstrich oberhalb Quecksilberspiegel' unterscheiden können. Dieser Unterschied lässt sich aber mit der extensionalen Methode der Quasianalyse nicht fassen." And after further comments: "Allem Anschein nach liegt hier also ein Versagen der Quasianalyse vor." – GHvW.

[77] Her referate is preserved, together with Waismann's notes, see ASP RR 9-14. Carnap seems to have thought that it is better that he himself gives the reply to Kaila.

[78] Stadler, *Studien zum Wiener Kreis*, pp. 176–178.

[79] Kaila, *Neupositivismus*, p. 11.

The idea of analytical equivalence could indeed be found in modern science. However, Carnap had put it to a use that was analogous to Kant's error. Carnap's given meant the complete cross-sections of experience, the epistemic primacy of the total impressions as elementary experiences. The various sense modalities had to be abstracted from these basic elements using the method of quasi-analysis. But as the construction proceeded, no new elementary features could be added to what could already be found in the given. Such new elements would have contradicted the principle of analytical equivalence.

An equivalence of meaning was built into Carnap's principle:

> Falls also behauptet wird, zwei analytisch äquivalente Begriffe könnten doch einen verschiedenen 'Inhalt', einen verschiedenen 'Sinn' haben, oder, es werde doch unter den Begriffen Verschiedenes 'vorgestellt', so muss gesagt werden, dass ein derartiger Unterschied wissenschaftlich bedeutungslos ist, weil er ja – voraussetzungsgemäß – *nicht darstellbar* ist. [...] ...so sind z.B. die Meinungsverschiedenheiten zwischen den (erkenntnistheoretischen) Realisten und Idealisten, zwischen den Solipsisten und nicht-Solipsisten usw. überhaupt nicht formulierbar; sie betreffen nicht darstellbare Bedeutungen, sondern nicht-darstellbare ('subjektive') 'Inhalte'.[80]

Among the further consequences were the following:

> ...dass alle derartigen Sprechweisen ihrer logischen Bedeutung nach letzten Endes stets nur Aussagen über die Beziehungen der Grundelemente sein müssen. Die sogenannten wahrgenommenen Dinge und Vorgänge sind durchaus nicht 'gegeben' im Sinne der Konstitutionstheorie, sie sind aus den Grundelementen gewonnene verwickelte logische Konstruktionen. Alle 'Ergänzungen' des 'Wahrgenommenen' sind deshalb auch keine Überschreitungen des Gegebenen im strengen Sinn, sondern logische Konstruktionen, die aus dem Gegebenen genau so gewonnen werden wie die sogenannten 'wahrgenommenen' Dinge und Vorgänge.[81] [...] dass es keine (prinzipiell) *nicht* entscheidbare Fragen geben *kann*. Denn jede Aussage, die nicht bedeutungsleer ist, muss sich in eine bedeutungsgleiche Aussage über die Grundelemente übersetzen lassen; dazu braucht man ja nur die im Laufe

---

[80] Ibid., p. 18.
[81] Ibid., p. 31.

der Konstitution eingeführten Begriffe durch die definierenden Aussagefunktionen zu ersetzen, bis nur solche Aussagen vorliegen, deren Zutreffen oder nicht Zutreffen aus dem 'Protokoll' über das Gegebene unmittelbar zu ersehen ist...[82] [...] Alle Zukunftsaussagen müssen, falls sie überhapt begründet sind, ihre Gründe selbstverständlich in den *vergangenen* Erfahrungen haben. Jede Voraussage etwa der Wissenschaft ist deshalb gewissen Aussagen über die Vergangenheit eindeutig zugeordnet. Diese Zuordnung ist eine Äquivalenz, und zwar, weil der Begriff der 'Zukunft' im Protokoll über das Gegebene nicht vorkommt, sondern durch jene Zuordnung erst definiert wird, eine *analytische* Äquivalenz. Daraus folgt, dass alle Zukunftsaussagen ihrer logischen Bedeutung nach Vergangenheitsaussagen sind.[83]

Carnap was not astounded by anything that Kaila said commenting the *Aufbau*. As concerns the analyticity of his approach, Carnap commented serenely:

Den ersten Teil 'Zur Darstellung' finde ich ausgezeichnet. Es interessierte mich besonders, wie von Ihrem Gesichtspunkt aus das Prinzip der analytischen Äquivalenz zum wichtigsten Prinzip der Theorie wird, während ich ihm gar kein besonderes Gewicht beigelegt hatte. Es schien mir so einleuchtend, dass ich es beinah für trivial hielt. Ich bin auch jetzt noch so fest von seiner unbedingten Geltung überzeugt, dass es mir schwer wird, auch bloß zu denken, wie man eine andere Meinung haben könnte. Auch Ihre Hervorhebung der übrigen erkenntnislogischen Prinzipien (Extensionalitätsthese, Forderung der strukturellen Kennzeichnung, Entscheidbarkeitsthese), scheint mir sehr bemerkenswert.[84]

The next year Carnap published in *Erkenntnis* a review of Kaila's book. Now he wrote, still more sharply: "Das Wesentliche ist jedoch, dass auch, wenn alle inhaltlichen Bemerkungen Kailas zu Recht bestünden, dadurch die Grundgedanken der Konstitutionstheorie, insbesondere das Prinzip der analytischen

---

[82] Ibid., pp. 32–33.
[83] Ibid., p. 35.
[84] R. Carnap to E. Kaila, December 12, 1930. – GHvW.

Äquivalenz, nicht berührt werden würden."[85] Carnap could not guess that later in his life he would hear much more about his concept of analyticity.

After a few critical remarks and a long quotation of Hempel's argument against Kaila, Carnap's letter began commenting on other aspects, for instance the respective limits of the construed and the phenomenal world:

> Über diese Frage habe ich, wie Sie sich entsinnen werden, vor längerer Zeit mit Ihnen und mit Herrn Waismann diskutiert. Meine damals geäußerte Auffassung ist aber nicht dogmatisch gemeint. Ich sehe die Frage durchaus noch als offen an. Aber die Unbegrenztheit der phänomenalen Welt kann in dem Sinne, in dem sie meiner Meinung nach allein in Betracht zu ziehen ist, nicht eine unendliche Anzahl der Elemente bedeuten.[86]

There was a reservation, perhaps because Carnap had just returned to Vienna from Warsaw with a handful of new logical ideas: "Wenn hier von einer Unbegrenztheit gesprochen werden darf (was mir, wie gesagt, noch nicht ganz klar ist), so nur in dem Sinne, dass die Syntax der phänomenalen Sprache die Bildung unbegrenzt vieler Formen zulässt, deren System die phänomenale Welt bildet."[87]

Another main target of Kaila's critique was Carnap's psychology. The *Aufbau* was built on certain presuppositions which Carnap described as results of contemporary psychology.

Was there any empirical reason to begin the reconstruction of cognition with the autopsychological? The psychology of children's development indicated rather that the other persons were primary in relation to the own self.[88] And how could the "I" ever be captured with Carnap's logic of classes and relations?

Most importantly of all, the cross-sections of total experiences at a certain point of time, Carnap's "givens", were not at all what Gestalt psychology or any other form of modern psychology meant with the structure of experience. Kaila documented this at length. He summarized:

---

[85] R. Carnap, "Review of E. Kaila, Der logistische Neupositivismus", *Erkenntnis*, 1932, p. 77.
[86] See note 83.
[87] Ibid.
[88] Kaila, *Neupositivismus*, pp. 38–39.

Aus der extensionalen quasianalytischen Methode folgt, dass jede interne Mannigfaltigkeit der Grundelemente in der Konstitutionstheorie unberücksichtigt bleiben muss. Carnap jedoch glaubt darüber hinaus, dass seine Annahme, die 'Querschnitte des Erlebnisstroms' seien schlechthin einfache Quales, mit den Anschauungen der modernen Psychologie, insbesondere der Gestalttheorie, übereinstimme. Er scheint anzunehmen, dass die Zurückweisung der Lehre, die Erlebnisse seien aus irgendwelchen 'psychischen Elementen' zusammengesetzt, gleichbedeutend sei mit der Aussage, dass sie – von allen 'begrifflichen Verarbeitungen' abgesehen – ungegliederte Totaleindrücke (ohne innere Mannigfaltigkeit) seien.

Hier liegt ein fundamentales Missverständnis vor.

Die Zurückweisung der 'atomistischen' Psychologie, der 'Mosaik-These', ist ja nur die negative Seite der Gestalttheorie. Viel wichtiger ist natürlich die positive These, dass die Erlebnisse typisch eben die Form von 'Gestalten' haben, d.h. eine (physiologisch direkt bedingte) originäre innere Mannigfaltigkeit aufweisen. [...] Gerade dies ist ja das Wichtige an der Gestalttheorie, dass sie über das vage Gerede der früheren 'nicht-atomistischen' Psychologie ('einheitliche Totalimpressionen' u.dgl.) hinausgegangen ist und eine originäre, bestimmte Gesetze befolgende *Strukturiertheit* auch der ersten sinnlichen Phänomene lehrt; nur dürfen diese Strukturen eben nicht als 'additive' Konglomerate von Stücken, sondern als gegliederte Ganze, als 'Gestalten', in denen 'jedes Glied seine Eigenart nur durch und mit den anderen besitzt', aufgefasst werden.[89]

Kaila's conclusion was that Carnap begins his quasi-analysis with the false givens, "too soon", as he said. It is probable that Kaila had presented this point already in his original paper criticising Carnap or when he met Carnap in Vienna. He had in fact contributed papers to the Berlin Gestalt psychologists' journal and was able to see that Carnap's "given" had nothing to do with contemporary empirical research.

The first draft of the physicalism paper had shown Carnap hesitating: "...dabei ist noch nicht klargestellt, ob die Sehgestalten oder das gesamte

---

[89] Ibid., pp. 73–74.

Sehfeld als Einheit oder das Gesamterlebnis eines Augenblicks als Einheit, noch unzerlegt in Sinnesgebiete, das Gegebene bilden."[90]

In his review of Kaila's book, Carnap admitted:

> Wir wissen heute noch nicht, welche Form und welchen Inhalt die Atomsätze haben, ob sie eine primäre, interne Mannigfaltigkeit der Erlebnisse beschreiben (wie Kaila meint), oder die Erlebnisse als unzelegbare Einheiten nehmen (wie in meinem Versuch angenommen wurde). Hiervon hängt auch ab, ob und auf welcher Stufe Quasianalyse anzuwenden ist. [...] Die Entscheidung dieser Fragen [...] wird vielleicht in manchem der Auffassung von Kaila recht geben und nicht den Annahmen, die ich in meinem Aufbauversuch zum Zweck der Illustration gemacht habe. In solchen Punkten wird dann das Konstitutionssystem entsprechend abgeändert werden müssen.[91]

But this did not mean that there was anything wrong with the method of quasi-analysis or with the logical method in general.

Kaila's idea of realism was too vague to be discussed. One of Carnap's further moves also deserves mention. Why should one think that the order of perceptual qualities is a key to the order of quantities and corresponds to it? For Carnap this was a happy coincidence. In the review, however, he admitted that it was an open and debatable question.

9. Neurath returned to the discussions of the Circle in the spring of 1931. He was defending and developing the views of the opening chapters of his unpublished manuscript, now under the title of "physicalism".

During one of the lively discussions, on 26.2.1931, Carnap concluded: "Bei der Erörterung der Frage, ob das urprünglich Gegebene schon strukturiert ist, zeigt sich die Problematik dieser ganzen Diskussion."[92] Neurath was critical of this whole approach: "Eine Bezugnahme auf das Gegebene ist in jedem Sinne

---

[90] See above.
[91] Review of Kaila, p. 77.
[92] Stadler, *Studien zum Wiener Kreis*, p. 289.

überflüssig. Es gibt nur Aussagen und zwar Protokollaussagen und physikalische Aussagen. In diesen Aussagen kommt das Gegebene nicht vor."[93]

Before the next meeting, on 4.3., Neurath, Hahn and Carnap had a preparatory discussion about physicalism. Neurath explained again that the starting point was the historical, natural language. Beginning with its *Ballungen*, its exact and unexact concepts, taking from them certain elements, our language was unified step by step. Neurath was lecturing, presenting a number of theses:

> Der Unterschied zwischen solipsistischer und intersubjektiver Sprache fällt weg. [...] Die physikalische Sprache ermöglicht die weitgehendste Verständigung der Menschen untereinander und des Menschen mit sich selbst. Sie ist die Universalsprache unter allen Umständen. [...] Es gibt keine Erlebnisaussagen, sondern nur Feldaussagen. In den Feldaussagen wird immer die Relation der Gegenstände zum Individuum mitberücksichtigt. [...] Eine Person, ein Ich ist ohne Beziehung zu Gegenständen sinnlos. [...] Die phänomenale Sprache ist eine unvollkommene physikalische Sprache. Man könnte die phän[omenale] Sprache als phys[ikalische] Spr[ache] erlernen. – Jede neue Protokollaussage muss mit der Gesamtheit der bisherigen konfrontiert werden. Keine Protokollaussage ist physikfrei; eine unphysikalische Sprache gibt es nicht. [...] Es gibt keine Erläuterungen, auch nicht solipsistisch. Wenn man mit der einen Hälfte der Sprache über die andere Hälfte spricht, ist alles in Ordnung; aber man kann nicht über die ganze Sprache von außerhalb sprechen. [...] Die Einheitssprache ist mit dem Physikalismus identisch. Die Einheitssprache ist eine verbesserte Ballung.[94]

On the day after that Carnap presented his physicalism to the Circle, very much in the same way as he had written in the previous summer, but not quite following all of Neurath's advice. He made a significant move:

> Ich nenne jetzt die Ausgangsbasis des Konstitutionssystems nicht mehr das Eigenpsychische und verwende das Wort 'Eigenpsychisches' nur für konstruierte Systeme. Das Psychische in der Psychologie ist also nur eine abgekürzte Redeweise für physikalische Vorgänge. [...] Das Wesentliche der

---

[93] Ibid., p. 290.
[94] Besprechung über Physikalismus am 4.3.1931. – VCA 192.

physikalischen Sprache ist, dass wir von dem einzelnen Sinnesbereich unabhängig werden.⁹⁵

In the next meeting, 12.3., Carnap was explaining the necessary modifications to the system of his *Aufbau*. As concerned the logic, extensionality was enough. He took a further step in this way: "Der Terminus 'Sinn' in der alten Bedeutung fällt weg. Zwischen dem eigentlichen logischen Gehalt und dem rein psychologischen Vorstellungsgehalt liegt nichts dazwischen."⁹⁶

There was actually a great number of open questions. Carnap no longer believed that one basic relation would be enough, if the beginning was to be made with elementary experiences.⁹⁷ There were alternative ways to start the construction, but somehow solipsism and perhaps even the Wittgensteinian prelinguistic elucidations would be necessary.⁹⁸

These discussions were the end of *Aufbau*. During the spring, everybody in the Circle had been talking about syntax. This was the way out that Carnap found. The famous passage in his 'Intellectual Autobiography' is the following:

> ...the whole theory of language structure and its possible applications in philosophy came to me like a vision during a sleepless night in January 1931, when I was ill. On the following day, still in bed with fever, I wrote down my ideas on forty-four pages under the title 'Attempt at a metalogic'. These shorthand notes were the first version of my book *Logical Syntax of Language*.⁹⁹

Looking at Carnap's participation in the discussions of the Circle during the next few months it is hard to believe that the possible applications became clear to him in an immediate flash. Some of them can indeed be found in his farewell to the *Aufbau* system. In June 1931, Carnap was lecturing to the Circle about the details of his new metalogic, touching slightly also the applications.

Meanwhile, Neurath was anxious to publish the programmatic ideas about physicalism which were left out of the book on sociology. He did this in a

---

⁹⁵ Stadler, *Studien zum Wiener Kreis*, pp. 293, 296.
⁹⁶ Ibid., p. 298.
⁹⁷ Ibid.
⁹⁸ Ibid., pp. 299–301.
⁹⁹ R. Carnap, "Intellectual Autobiography", in *The Philosophy of Rudolf Carnap*, ed. by P. A. Schilpp, Open Court, La Salle, Ill. 1963, p. 53.

paper "Soziologie im Physikalismus" which he submitted to *Erkenntnis* in May 1931.[100] Carnap, on the other hand, was obsessed with his metamathematical approach to logic which he first called "metalogic". The trip to Warsaw had enriched him with Alfred Tarski's syntactical outlook.

He had not forgotten his paper about the physical language. He discussed the topic in the summer with Feigl and towards the end of the year with Hempel. Then, he decided to write the paper anew. In January 1932, he sent the new version with the old title to Neurath who was for a time in Moscow organizing work on visual statistics. Carnap expressed his delight that

> ...die These jetzt schärfer formuliert und genauer begründet ist. Ich glaube, dass die Übereinstimmung zwischen unseren Auffassungen dadurch noch enger geworden ist. Deine energischen und temperamentvollen Äußerungen haben meinen Überlegungen einen sehr wesentlichen Anstoß gegeben, wofür ich Dir sehr dankbar bin. Die Ausarbeitung der Metalogik im Sommer hat mir auch bei der Klärung dieser Fragen sehr geholfen. Ich glaube, dass jetzt durch die Gegenüberstellung der 'formalen' und 'inhaltlichen' Redeweise vielen Lesern die Sache klarer werden wird. [...] Es ist mir jetzt ganz klar, dass Du mit Deiner Warnung vor den bei uns üblichen 'Erläuterungen', Recht gehabt hast.[101]

The Wittgensteinian "elucidations" to *the* language would no longer be needed, as Neurath has always said. Carnap added that his paper would soon appear in the same issue of *Erkenntnis* as Neuraths "Soziologie im Physikalismus".

After that letter Carnap was bombarded with telegraphs and letters from Moscow. Neurath was worried that now everybody would think that his paper was only a weak imitation of Carnap's excellent analysis, although it was written independently and much earlier than Carnap's new version. Neurath knew Carnap's original draft of the paper. Now the paper incorporated a radical modification, although there was much that was common to both of the papers.[102] Carnap added to his paper a long footnote where he gave a great part

---

[100] R. Carnap to O. Neurath, March 2, 1932. – ASP RC 029-12-60.
[101] R. Carnap to O. Neurath, January 16, 1932 – ASP RC 029-12-70.
[102] O. Neurath to R. Carnap, January 25, 1932: "Wenn man die beiden Aufsätze liest, wird nicht klar, dass der Aufsatz über 'Universalsprache' Gedanken des anderen

of the priority to Neurath. He had been a lonely researcher "on the icy slopes of logic", as he said in a letter, and he did not want to lose his friendship with Neurath. In fact, he expressed the unity of his and Neurath's views more strongly than was the case.

When the paper appeared, Carnap sent an offprint to Wittgenstein. The result was a bombardment from Cambridge. Wittgenstein sent letters to Schlick and Carnap. His worries were very much the same as Neurath's. He had browsed through Carnap's paper and thought that his new, unpublished ideas of recent years had been stolen. As the first among these he mentioned in one of the letters Carnap's physicalism. Waismann had kept the Circle informed about Wittgenstein, under Schlick's supervision.

The truth was different.

Carnap did not understand what was going on. Wittgenstein's attack came to him as a complete surprise. His own comments are to be found in a letter which he wrote to Schlick in September 1932: "Ich finde im Tractatus keine deutliche Äußerung zum Physikalismus. Ja ich weiß bis heute noch nicht, wie W[ittgenstein]s Auffassung hier ist, da Waismann ja nicht deutlich Stellung genommen hat, sondern nur sagte, dass er (und W[ittgenstein]) gewisse Bedenken haben, ohne sie jedoch formulieren zu können."[103]

This was obviously a reference to the discussion early in 1929 when Waismann tried to explain Wittgenstein's argument against the *Aufbau*. It is no wonder that Waismann's attempt was unsuccessful. Wittgenstein himself was not yet in clear waters about this.

During the spring of 1931, two camps could be discerned in the discussions of the Circle. Schlick and especially Waismann were representing Wittgenstein's opinions – or rather they were seen as doing so. Neurath was pretty much alone with his physicalism. Only Carnap tried, half-heartedly, to give him support. In May, Waismann's 'Theses' on Wittgenstein were discussed intensively in the Circle, up to thesis 13. After that, Waismann had to give the floor to Carnap's metalogic.[104]

---

weiterführt. Vor allem die Behauptung, dass die Protokollaussagen ein *Teil* der physik[alischen] sind." – ASP RC 029-12-69.

[103] R. Carnap to M. Schlick, September 28, 1932. – VCA.

[104] See R. Rand's notes of the whole period, published in Stadler, *Studien zum Wiener Kreis*, pp. 275–334.

Carnap had a copy of the theses. One can find in them, for instance, the following: "Was die Elementarsätze beschreiben, sind die Phänomene (die Erlebnisse). [...] Der Sachverhalt – das Phänomen..."[105] In fact, this had still been Carnap's view about the phenomenal universal languages in the summer of 1930. In the new draft of his paper on physicalism, the relations between the physical and phenomenal languages were changed. In a letter to Neurath in March 1932, Carnap explained:

> Neuraths Thesen bestimmten vor allem die Richtung und den Zielpunkt von Carnap's Überlegungen; diese Thesen bestärkten ihn dann in der Überzeugung von der Richtigkeit des Ergebnisses seiner Überlegungen, als dieses Ergebnis mit Neuraths Standpunkt zusammentraf. Kurz: Neuraths Behauptungen haben sich für Carnaps Untersuchungen als fruchtbare Anregungen erwiesen und sind durch sie in wesentlichen Punkten bestätigt worden.[106] [...] Ohne die Metalogik wäre ich z.B. gar nicht dazu gekommen, die scharfe Trennung zwischen formaler und inhaltlicher Redeweise zu machen und die letztere abzulehnen. Diese Trennung kommt zwar in ihrem Ergebnis in die Nähe Deiner These, dass man nicht über 'Erlebnisse' sprechen darf, aber sie ist bei Dir nicht zu finden [...] Erst auf Grund dieser Trennung und der Verwerfung der inhaltlichen Redeweise ist aber die Überwindung des Dualismus der beiden Sprachen möglich.[107]

Carnap's and Wittgenstein's developments towards physicalisms of some kind were independent of each other.[108] In addition, there was no identity of views. Wittgenstein's end point was the physicalism of ordinary language, Carnap's that of physics. When Wittgenstein attained more information about

---

[105] F. Waismann, Thesen aus dem Zirkel, 1930, pp. 20, 22. – ASP RC 102-76-02. Also in Waismann, *Wittgenstein und der Wiener Kreis*, pp. 249, 251.

[106] R. Carnap to O. Neurath, March 2, 1932. – ASP RC 029-12-61.

[107] Ibid.

[108] Carnap's use of the concept of "hypothesis" in the final publication is a more complicated case. It would require a separate treatment. It was among Wittgenstein's accusations, but even it can hardly be called a plagiarism as it was in general use at least since Poincare.

Carnap than his first browsing of Carnap's paper, he rejected all metalogic vehemently.[109]

Carnap kept repeating the psychological alternatives to the "given" of the *Aufbau*, pointed out by Kaila. It was an open question until Carnap's Paris lecture in 1935. One cannot say that Carnap ever solved it, but in that lecture he excluded psychology completely from the logic of science.[110]

The contingent fortunate conditions for intersensuality and intersubjectivity remained in Carnap's final publication – the very same conditions as can be found first in Neurath's rejected manuscript and then in the original draft of Carnap's paper: "...these facts, though of empirical nature, are of far wider range than single empirical facts or even specific natural laws."[111]

---

[109] The use of the term "metalogic" that is to be found in Wittgenstein's notebooks was mainly directed against his own earlier views. It was the idea that there is something to be found behind the signs, say, a phenomenological language. See e.g. the chapter on "Metalogic and the Domain of Logic: the Shift to Ordinary Language", in S. S. Hilmy's *The Later Wittgenstein. The Emergence of a New Philosophical Method*, Basil Blackwell, Oxford 1987. In addition to the Bergen electronic edition of Wittgenstein's *Nachlass* and the complete publication of Wittgenstein's correspondence in 2002, also as a CD-ROM, further information can be found in the book *The Voices of Wittgenstein. The Vienna Circle*, ed. by G. Baker, Routledge, London 2003. Unfortunately, Baker knew no manuscript sources outside of Oxford.

[110] See e.g. R. Creath, "Carnap's Program and Quine's Question", in *Carnap Brought Home. The View from Jena*, ed. by S. Awodey and C. Klein, Open Court, Chicago and La Salle, Ill. 2004, and M. Friedman, "Hempel and the Vienna Circle", in *Logical Empiricism in North America*, ed. by G. L. Hardcastle and A. W. Richardson, Minnesota Studies in the Philosophy of Science XVIII, University of Minnesota Press, Minneapolis 2000. For further information about Carnap's attempts to solve the problems pointed out by Kaila, see A. W. Carus, *Carnap and Twentieth-Century Thought*, Cambridge University Press 2007 (forthcoming), Chapters 7 and 8.

[111] R. Carnap, *The Unity of Science*, transl. with an introduction by M. Black, Kegan Paul, London 1934, p. 65. From the point of view of his neo-Kantian interpretation, Richardson finds here nothing but a reason for scorn: "But this is perilously close to nonsense. Carnap is now saying that two empirical facts ground the possibility of objective science." A. W. Richardson, *Carnap's Construction of the World. The Aufbau and the Emergence of Logical Empiricism*, Cambridge University Press, Cambridge 1998, p. 203.

The neo-Kantian interpretations of this phase of Carnap's development cannot be accurate. This explains why the sequel to *Aufbau* was never written. Carnap wrote in his introduction to the English edition of *The Unity* as follows: "...we give no answer to philosophical questions, and instead *reject all philosophical questions*, whether of

We have now arrived at the point after which the *Physikalismusstreit* became completely public in the pages of *Erkenntnis* and elsewhere. There are excellent and detailed studies of this controversy,[112] but so far its prehistory has remained a matter of guess-work.

In the published version of his "Die physikalische Sprache als Universalsprache der Wissenschaft", Carnap adopted a predominantly one language solution. The protocols of scientists were a part of the physical language. But there was still something left of his earlier full-blown two-track solution. He commented on the protocol language, that: "Sie wird auch als 'Erlebnissprache' oder 'phänomenale Sprache' bezeichnet; weniger bedenklich ist die Bezeichnung 'erste Sprache'."[113]

This gave Neurath the opportunity to strike back, now in public controversy.

Some years later Wittgenstein was arguing against what he called "the private language". Who was it who had believed in monological languages?

Metaphysics, Ethics or Epistemology" (p. 29). In addition to the sentences of logic and mathematics there was only empirical science, strict *Realwissenschaft*, and, of course, the activity of logical analysis, but no traditionally conceived philosophical sentences. However, in the opening essay of the new journal *Philosophy of Science* Carnap first explained the background of this "antiphilosophy", in fact a resolute reading of Wittgenstein's *Tractatus*: "Wie widerlegt nun Wittgenstein jenen Einwand, dass auch seine eigenen Sätze sinnlos seien? Überhaupt nicht; er stimmt ihm zu!" Then Carnap went on: "Wir wollen im Folgenden versuchen, an Stelle dieser radikal negativen Antwort eine *positive* Antwort auf die Frage nach dem Charakter der Sätze der Wissenschaftslogik und damit der Philosophie geben." Quoted according to the German original, in R. Carnap, *Scheinprobleme der Philosophie und andere metaphysikkritische Schriften*, ed. by T. Mormann, Felix Meiner Verlag, Hamburg 2004, pp. 113–114.

[112] T. E. Uebel, *Overcoming Logical Positivism from Within. The Emergence of Neurath's Naturalism in the Vienna Circle's Protocol Sentence Debate*, Rodopi, Amsterdam 1992.

[113] See *Erkenntnis* 1931, p. 438. The issue appeared actually in 1932.

Maybe Descartes and the Cartesians, as the followers of Gilbert Ryle suggested? I leave it for the reader to find the solution to this puzzle.[114]

Helsinki Collegium for Advanced Studies

---

[114] Actually, Gordon Baker suggested that it is fruitless to seek Wittgenstein's opponent among Descartes, Locke, etc.: "Why not rest content with his treating concrete problems apparent among some of his contemporaries (Russell, Schlick, Carnap) or even participants in his discussions at Cambridge?" G. Baker, *Wittgenstein's Method. Neglected Aspects*, ed. by K. Morris, Blackwell, Oxford 2004, p. 138. Baker does not give any references, but, aside with Carnap, one could look at B. Russell, *The Philosophy of Logical Atomism and Other Essays 1914-19*, ed. by J. G. Slater, George Allen & Unwin, London 1986, p. 176: "A logically perfect language, if it could be constructed [...] would be very largely private to one speaker." For the correct answer to the puzzle, see also David Pears, "Private Language", in *The Philosophy of Jaakko Hintikka*, The Library of Living Philosophers, Vol. XXX, ed. by Randall E. Auxier and Lewis Edwin Hahn, Chicago and La Salle, Ill. 2006, p. 393.

Uskali Mäki

Putnam's Realisms:
A View from the Social Sciences

I

For the last three decades, the discussion on Hilary Putnam's provocative suggestions around the issue of realism has raged widely. Putnam's various formulations of, and arguments for, what he called internal realism in contrast to what he called metaphysical realism have been scrutinised from a variety of perspectives. One angle of attack has been missing, though: the view from the social sciences and the ontology of society. This perspective, I believe, will provide further confirmation to the observation that Putnam's two concepts of realism are all too aggregative in that they conflate elements that had better be kept distinct, at least for many important purposes (e.g. Niiniluoto 1996).

The present essay can be read as an argument for a topic-specific examination of realism. This means that it challenges the overall validity of Richard Boyd's arguments against what he calls "realism about", that is, against "a certain fragmented conception of scientific realism, one according to which realism is deeply topic specific" (Boyd 1990, p. 175). Boyd takes the 'x' in "realism about x" to designate two kinds of things: entities postulated in scientific theories ("realism about the ether" and "realism about higher taxa") and scientific disciplines ("realism about physics" and "realism about biology") (ibid., pp. 175, 190). He also suggests that "it seems possible cogently to accept realism about the natural sciences while denying it about at least some of the social sciences" (ibid., p. 191). In Boyd's view, the possibility of realism about social sciences is undermined by their weak instrumental success; it is the instrumental reliability of method that he takes as the basis for a realist account of science. I do not underwrite this connection between realism and instrumental success, thus I am not compelled to reject realism about social sciences on these grounds; and, as we shall see, social sciences have other peculiarities as well (see Mäki 1996). I think we should, at least to some degree, disaggregate and relativize the issue of realism. I do have sympathy with Boyd's strategic arguments against "realisms about", but yet I believe realists will be better off by adopting a more concrete and localized approach to the issue (Mäki 2005). Social objects and social sciences deserve a separate treatment.

There are two strategies that can be followed in examining a realism about x (type of entity, theory, discipline). One is to first fix the meaning of 'realism' and then specify the extension of 'x'. This is Boyd's strategy. The other is to fix some part of the extension of 'x' and then adjust the meaning of 'realism' so as to accommodate whatever peculiarities the x in question may have (and finally to check whether the adjusted meaning of 'realism' meets the criteria of a minimal notion of realism; it would be here that we encounter a relatively fixed idea of some minimal realism, or a set of realist intuitions). The latter is my approach here.

I begin with the interconnected ideas of "realism about social objects" and "realism about social sciences". I am not only interested in what comes after 'about'; I am also interested in what goes before it, namely 'realism' itself. Here is one premise of the discussion: the issue of proper forms of realism for dealing with specific problems is an issue to be settled at least partly *a posteriori*. Here is the problem for which a solution will be sought: what constraints, if any, does what we know about the social sciences and society (as depicted by the social sciences and our commonsense views) impose upon the forms of realism that can justifiably be adopted about them?

This paper suggests to offer only some partial insight into this issue. It seeks to do so indirectly by discussing Putnam's characterizations of internal and metaphysical realism. The focus will be on two aspects of Putnam's realisms: [1] the role and kinds of independence and dependence (in relation to the human mind and related things); and [2] the possibility of error. It is shown that the nature of social objects has implications concerning the appropriate constraints on [1] and [2], and thereby on the kinds of realism that are available to social scientists.

II

For later commentary, we need a list of some of Putnam's characterizations of aspects of the two realisms. Consider first "metaphysical realism".

Metaphysical Realism, aspect 1. "... the world consists of some fixed totality of mind-independent objects." (1981, p. 49)

Metaphysical Realism, aspect 2. "... the world is, after all, being claimed to contain Self-Identifying Objects, for this is just what it means to say that the *world*, and not thinkers, sorts things into kinds." (1981, p. 53)

Metaphysical Realism, aspect 3. "There is exactly one true and complete description of 'the way the world is'." (1981, p. 49)

Metaphysical Realism, aspect 4. "Truth involves some sort of correspondence relation between words or thought-signs and external things and sets of things." (1981, p. 49)

Metaphysical Realism, aspect 5. "... *truth* is supposed to be *radically non-epistemic* – we might be 'brains in a vat' and the theory that is 'ideal' from the point of view of operational utility, inner beauty and elegance, 'plausibility', simplicity, 'conservatism', etc., *might be false*." (1978, p. 125)

Of these aspects, 1, 2, and 5 will be the most relevant to our endeavour.[1] That is, in Metaphysical Realism, we will focus on the ideas of mind-independent and self-identifying objects and the idea that even an ideal theory may be false. Let us then turn to a few of Putnam's characterizations of "internal realism", year model 1981:

Internal Realism, aspect 1. "'Objects' do not exist independently of conceptual schemes." (1981, p. 52)

Internal Realism, aspect 2. "*We* cut up the world into objects when we introduce one or another scheme of description." (1981, p. 52)

---

[1] There is no reason to divide aspect 3 into two further aspects – one concerning truth, the other concerning completeness – because the idea of only one true description seems to be intelligible at most if it is also required to be a complete description. Yet, this idea is intelligible only in a weak way; it is not obvious that any philosopher ever has held it. As soon as aspect 1 is further decomposed into the mind-independence, totality, and fixed list of objects (as it should), similar queries can be raised; for example, why should the list of objects be fixed? (See, e.g. Hacking 1983, p. 94; however, see Smart 1995, p. 311.)

Internal Realism, aspect 3. "'Truth' ... is some sort of (idealized) rational acceptability – some sort of ideal coherence of our beliefs with each other and with our experiences as those experiences are themselves represented in our belief system – and not correspondence with mind-independent or discourse-independent 'states of affairs'." (1981, pp. 49–50)

The import of these three aspects of Internal Realism is intuitively sufficiently clear, but let it be separately emphasized that, according to aspect 3, because the truth of a theory amounts to its idealized rational acceptability, based on the perception that it meets a set of theoretical and operational constraints, no theory that is ideal in this sense can fail to be true. This denies aspect 5 of Metaphysical Realism.

III

Consider then the nature of social objects. Let us focus on mind-independence in aspect 1 of Metaphysical Realism. We are immediately able to question this idea: social objects are typically depicted as being mind-dependent rather than mind-independent by social scientists (and philosophers writing on social ontology). I think it is fair to say that this peculiarity is regularly neglected in the writing on realism in general; there tends to be an implicit premise of physicalism involved in the general characterizations of realism in terms of mind-independence. An example is Rozema (1992, p. 293) who characterizes Putnam's Internal Realism as "essentially a post-Kantian form of idealism" which claims that

[us1] "it is the cognitive, linguistic, and physical behaviour of *us* that determines the fundamental structure and existence of *all* 'things'".

This perspective is too aggregative, exemplified here by expressions such as "us" and "all". [us1] presents itself as a global thesis about all there is. Now if we replace "all" by "social" we end up with a claim that is not a post-Kantian form of idealism and is most likely true in light of what the social sciences say about the nature of the social world. Indeed, from a realist point of view, it seems unproblematic to say that

[us2] It is the cognitive, linguistic, and physical behaviour of *us* that determines the fundamental structure and existence of *social* things.

I say [us2] is unproblematic from a realist point of view because it seems to describe an evident feature of social reality. On the other hand, it is a highly problematic statement for a realist, because it is also an invitation to reconsider the very notion of realism in this context.

There are many approaches to defining social objects of various kinds that conform to [us2]. Among the definientia of the concept of a social object we find mutual expectations, consensual beliefs, common knowledge, shared meanings, collective acceptance, plural subjects, collectively intentional design, unintended consequences of intentional actions, and other similar notions (see, e.g., Lewis 1969; Collin 1997; Gilbert 1989; Tuomela 1995; Searle 1995). Whatever the details of any particular account, it shares with others the idea that the existence and properties of social objects are dependent on human minds. Take away the minds of people, and you will have taken away social reality. In other words, [us2].

This means that aspect 1 of Metaphysical Realism is incorrect and [us1] as a reading of the respective element of Internal Realism is correct about social objects. Such objects do not exist mind-independently, regardless of whether physical objects do.

IV

Let us then look at social objects from the point of view of the possibility of being mistaken about their existence and characteristics. Putnam's Metaphysical Realist thinks that even an ideal theory may be false. This means that a non-ideal theory may also be false – even more likely so. The Metaphysical Realist must think that whatever the warrants of a theory, it is always possible that the world is not the way it is described to be by the theory. Now consider an argument that implies that this Metaphysical Realist picture does not fit with a certain class of social objects. The argument is David-Hillel Ruben's (1989, pp. 70–74), the conclusion concerning Metaphysical Realism is mine.

Ruben discusses social objects defined in terms of their ascribed purposes or functions, such as money (or prison or purchase). Functions can be divided into those that objects actually have and those that they are thought or intended to have. Social objects can be divided into types and tokens (such as money as a generally accepted medium of exchange and a particular bank note in my pocket now). Consider first social token objects. The question is whether we can be mistaken in classifying certain particular items as money. And it seems that we indeed can be so mistaken. Even though I and all others involved may believe that a certain item is a token of money, an instance of the social kind of money, and proceed using it as if it were, it still may fail to be a money token – because it is a counterfeit, for example. We may fail to correctly classify the item, even if we all believe to be correct about it. Our singular classificatory beliefs don't make the item a money token. There is a sense in which money tokens would seem to be among the objects that aspect 2 of Metaphysical Realism talks about.

The same conclusion does not hold for social types. General classificatory beliefs, if held by everybody, cannot be mistaken in the same way as singular beliefs can. It is not the case that if everyone in a society shares a particular general classificatory belief, then this belief may be false. As Ruben puts it, "there is no distinction at this point to be drawn between a general illusion about the social world and the reality of such a world" (Ruben 1989, p. 72). If everyone in a society believes that there is the institution of money in this society and acts accordingly – that is, according to the rules and conventions defining money – then it has to be the case that the institution of money obtains.[2] Even if general error were possible concerning the physical world, this does not seem possible in the case of (many) social types. To put it in other words: collectively held beliefs often constitute social objects; claim [us2] is true about these objects. Metaphysical Realism does not fit.

These reflections have further implications for the issue of truth. Consider the following formulation of aspect 4 of Putnam's Metaphysical Realism: "There is supposed to be something that the world is like independent of how we conceptualize it. To express a truth, then, should be to give an accurate account of what that unconceptualized world is like. Giving such an account requires a correspondence between an utterance and the unconceptualized world." (Heller 1988, p. 114) If this is how a Metaphysical Realist should think about truth, then it seems that Metaphysical Realism is not an option when it comes to the truth about a certain class of social objects. As we have seen, these objects are not independent of how we conceptualize them.

---

[2] Further qualifications are needed on this statement, but they will be given elsewhere. Without such qualifications, the idea would be disputable.

It should be noted that while aspect 3 of Internal Realism defines truth in terms of idealized rational acceptability, I have not invoked this notion in the above argument. I have suggested that at least some social objects exist in the sense required by aspect 1 of Internal Realism. However, this is not yet to say that the beliefs constitutive of those objects meet the condition of ideal rational acceptability. If it is the case that "the only criterion for what is a fact is what it is [ideally] rational to accept" (1981, p. x), then I am afraid we have to conclude that this criterion is typically not met by the relevant beliefs under consideration. It is often sufficient for the existence and properties of a social object that there is collective acceptance, without the constituent beliefs satisfying such ideal rationality requirements. Thus, it would appear that there are social objects about which aspect 1 (and perhaps aspect 2) of Internal Realism is correct, while aspect 3 is not. Interestingly, this appears to make the conclusion even more radical than if aspect 3 were also satisfied. Mere consensual acceptance may be sufficient to constitute the "reality" of at least some social objects.

V

There is an important issue concerning claim [us2] that must not be neglected. Consider again [us2], "it is the cognitive, linguistic, and physical behaviour of *us* that determines the fundamental structure and existence of social things". The issue I have in mind is related to the extension of 'us' here. There are two relevant options available, namely people in roles other than social scientists and people in their roles as social scientists. If 'us' is taken to refer to human beings *qua* "members" of society, then the version of (internal) realism that the argument outlined in Section IV above supports is not a form of *scientific* realism. It is rather an idea about the general mind-dependent character of social objects. In this case we would be entitled to accept claim [us3]:

[us3]   It is the cognitive, linguistic, and physical behaviour of people other than social scientists *qua* social scientists that determines the fundamental structure and existence of social things.

If, on the other hand, 'us' is taken to refer to social scientists, then what would be at stake would be a form of *scientific realism*. The issue would be about claim [us4]:

[us4] It is the cognitive, linguistic, and physical behaviour of social scientists *qua* social scientists that determines the fundamental structure and existence of social things.

In this case, the argument of Section IV would be insufficient to resolve the issue; the issue would have many other facets that would require a separate treatment. This is because the existence of the kinds of social objects that we talked about above does not seem to be essentially dependent on "the cognitive, linguistic, and physical behaviour" of social scientists *qua* social scientists. This requires a qualification that will be discussed next.

VI

Another important ambiguity about [us4] concerns the notion of *determination*. Let us distinguish between two types of determination or "construction", and let us call them conceptual and causal construction or determination. According to the first idea, social reality is conceptually constructed by theoretical representations of it. According to the second one, social reality is causally constructed by theories about it. (See Mäki 2002; 2005.)

The argument from *conceptual construction* is a version of the idealist doctrine according to which the world is essentially dependent on our thoughts of it, e.g., the world is a creation of our thinking. The version that we are discussing makes the world dependent on our ways of conceptually representing it. There cannot be any gap between the world and our representations, since we make worlds by means of conceptual representations. Accordingly, conceptually representing amounts to world-making rather than world-uncovering or world-discovering.

The argument from *causal construction* is a popular claim and has sometimes been used against realism about social sciences. The implementation of a plan based on a theory is an exemplification of this idea: the world is shaped after a blueprint suggested by the theory. The phenomena of so-called self-fulfilling and self-defeating prophesies are also often referred to in this context. In general, the idea is that of the causal materialization of thoughts (plans, expectations, hopes, fears) that are inspired or shaped by social scientific theories.

The first clarificatory point to make is this. It is certainly not the case that *theories* causally produce their objects in society, even though the idea is often formulated in this fashion. Unfortunately, the very phrasing of one of the ideas is misleading here: it is as if the prophesies themselves fulfil or defeat themselves. At most, if the argument is to make sense, it has to be phrased so as to state that *people* having the contents of those theories as the contents of their beliefs may act so as to causally help produce social realities and changes in them in ways that have consequences for the truth values of those theories.

This suggests a distinction between two statements:

(A) People inspired by theory T or the prediction it entails engage in action that produces or reproduces a social fact described by T – and they act in that way *because* they are inspired by T or the prediction. T is made true – or the object of T is made real – by people acting on T.

(B) The act of representing the social world by T creates a social fact – not only does it (re)conceptualize our view of the world, but it also (re)constitutes the structure of the world. T is made true – or the object of T is made real – by people inventing or holding T.

By incorporating this ambiguity about 'determination', we get two specified versions of [us4]:

[us5] It is the cognitive, linguistic, and physical behaviour of social scientists *qua* social scientists that *causally determines* the fundamental structure and existence of social things.

[us6] It is the cognitive, linguistic, and physical behaviour of social scientists *qua* social scientists that *conceptually determines* the fundamental structure and existence of social things.

It seems that [us5][3] is consistent with much of Metaphysical Realism, albeit not with mind-independence, viz. its aspect 1. In particular, claim [us5] is consistent with the denial of aspects 1 and 2 of Internal Realism, provided that the relevant specifications have been made: it is possible, without contradicting oneself, to accept [us5] and reject both "Social 'objects' do not exist independently of social scientific conceptual schemes" (aspect 1) and "Social scientists cut up the world into objects when they introduce one or another scheme of description" (aspect 2).

---

[3] It is clear that [us5] would require further qualifications. For example, it is usually not the cognitive, linguistic, and physical behaviour of *social scientists alone* that might have the causal potency in question.

On the other hand, claim [us6] about conceptual determination is clearly in line with aspects 1 and 2 of Internal Realism and is not open for an externalist interpretation, thus it is incompatible with Metaphysical Realism. Now [us6] is beset with two interesting problems. First, because [us6] talks about social sciences and the role of scientific theories in constructing social reality, it reintroduces the issue of ideality of rational acceptance. Recall this idea was part of Putnam's formulation of Internal Realism. The problem is that it seems that in the social sciences, meeting the condition of ideal conditions or even telling what it would mean to meet it, is harder than in much of the natural sciences. This may be taken as an additional problem to the Internal Realist.

Second, a further complication arises as we add to the picture another characterization of Metaphysical Realism: its aspect 5 says, "the world is, after all, being claimed to contain Self-Identifying Objects, for this is just what it means to say that the *world*, and not thinkers, sorts things into kinds" (1981, p. 53). Elsewhere Putnam characterizes Metaphysical Realism in terms of a pre-categorized, ready-made world (Putnam 1982). Now we have seen that social reality is largely pre-categorized, pre-conceptualized – the subject matter of the social sciences has its own version of what social reality is like: the world and thinkers are not distinguishable as it is the latter that help constitute the former. The social world has its own self-conceptualization and at least many of its objects are self-identifying. Interestingly, then, it appears that a tenet of Metaphysical Realism – described metaphorically as Putnam does – is almost trivially met in the social sciences!

Now we should further disentangle two versions of the idea that social objects are self-identifying. Both can be put in terms that Putnam himself uses (but does not utilize to separate the two versions I am suggesting). One idea is that the social world is pre-categorized. The other is that it is ready-made. To hold that the social world is *pre-categorized* is neutral with respect to the question of whether it should be *re*-categorized by social scientists in terms of theories. To say that society is *ready-made* may be taken to imply that no (or at least no radical) re-conceptualization is called for, rather it is the task of social scientific theorizing to mirror the ready-made categorization out there in society.

Once again, the prevalent views held in the social sciences and their philosophy appear to be mixtures of Internal Realism and Metaphysical Realism. Many of those who are identified as holding a realist philosophy of social science combine the ideas of pre-categorization and re-categorization: the social world is pre-conceptualized by social actors, but because this "commonsense" view is not flawless (it might be radically misguided in some cases), it is the task of social science theory to offer a less flawed picture of social reality. On the other hand, people like Alfred Schutz and Peter Winch, customarily identified with phenomenology and hermeneutics rather than scientific realism, view the social world as ready-made and impose it as the duty of social theory to conform to its inviolable pre-conceptualization (Schutz 1945; Winch 1958). It is the latter positions that, somewhat surprisingly perhaps, turn out to be closer to the composite view Putnam calls Metaphysical Realism.

VII

Summing up, some sort of internal realism seems appropriate for characterizing at least some classes of social objects: *realism about social objects* should be of a sort that allows their existence to be dependent on "the cognitive, linguistic, and physical behaviour of" people other than social scientists *qua* social scientists. On the other hand, this concession does not yet settle the other issue of the appropriate sort of (scientific) *realism about social scientific theories*. The above reflections appear to suggest that there is a fairly clear difference between the issue of realism about social objects and the issue of realism about social scientific theories. This is, indeed, how I am inclined to think. The reasons for thinking so have to do with some peculiarities that characterize social sciences: First, the social sciences deal with a mind-dependent and pre-conceptualized subject matter. Second, much of social science modifies those prior conceptualizations and revises commonsense beliefs, but rarely postulates objects of an entirely new kind relative to the commonsense objects (Mäki 1996).

The simple upshot of the observations of this essay is that the nature of social reality and social sciences as we know them appear to fit with Putnam's characterizations of both metaphysical and internal realism, albeit with different aspects of each. An obvious conclusion is that Putnam's two realisms are overly burdened artificial conglomerates and that their components have to be disentangled and treated separately. Moreover, to develop an adequate realism about social objects and social sciences is a challenge that has to be met in a genuinely "naturalist" spirit as a local and *a posteriori* project.

Academy of Finland

References

Boyd, Richard 1990: "Realism, Conventionality, and 'Realism About'", in G. Boolos (ed.), *Meaning and Method: Essays in Honour of Hilary Putnam*, Cambridge University Press, Cambridge, pp. 171-195.
Collin, Finn 1997: *Social Reality*, Routledge, London.
Gilbert, Margaret 1989: *On Social Facts*, Routledge, London.
Hacking, Ian 1983: *Representing and Intervening*, Cambridge University Press, Cambridge.
Heller, Mark 1988: "Putnam, Reference, and Realism", *Midwest Studies in Philosophy* XII, 113-127.
Lewis, David 1969: *Convention*, Blackwell, Oxford.
Mäki, Uskali 1996: "Scientific Realism and Some Peculiarities of Economics", in R. S. Cohen et al. (eds.), *Realism and Anti-Realism in the Philosophy of Science*, Boston Studies in the Philosophy of Science, Vol. 169, Kluwer, Dordrecht, pp. 425-445.
Mäki, Uskali 2002: "Some Non-reasons for Non-realism About Economics", in U. Mäki (ed.), *Fact and Fiction in Economics. Realism, Models, and Social Construction*, Cambridge University Press, Cambridge, pp. 90-104.
Mäki, Uskali 2005: "Reglobalizing Realism by Going Local, or (How) Should Our Formulations of Scientific Realism be Informed about the Sciences", *Erkenntnis* 63, 231-251.
Niiniluoto, Ilkka 1996: "Queries about Internal Realism", in in R. S. Cohen et al. (eds.), *Realism and Anti-Realism in the Philosophy of Science*, Boston Studies in the Philosophy of Science, Vol. 169, Kluwer, Dordrecht, pp. 45-54.
Putnam, Hilary 1978: *Meaning and the Moral Sciences*, Routledge and Kegan Paul, London.
Putnam, Hilary 1981: *Reason, Truth, and History*, Cambridge University Press, Cambridge.
Putnam, Hilary 1982: "Why There Isn't a Ready-Made World", *Synthese* 51 (2), 141-167.
Putnam, Hilary 1987: *Realism with a Human Face*, Open Court, LaSalle.
Rozema, David 1992: "Conceptual Scheming", *Philosophical Investigations* 15, 293-312.
Ruben, David 1989: "Realism in the Social Sciences", in Hilary Lawson and Lisa Appignanesi (eds.), *Dismantling Truth*, Weidenfeld and Nicolson, London, pp. 58-75.
Smart, J. J. C. 1995: "A Form of Metaphysical Realism", *Philosophical Quarterly* 45, 301-315.
Searle, John 1995: *The Construction of Social Facts*, Free Press, New York.
Tuomela, Raimo 1995: *The Importance of Us*, Stanford University Press, Stanford.

Sami Pihlström

Values as World 3 Entities

1. Introduction

The deep and pervasive issue of realism, with which Ilkka Niiniluoto has struggled for decades,[1] is a general problem in metaphysics, semantics, and

---

[1] Having started my studies in theoretical philosophy at the University of Helsinki in Fall 1989, I attended Professor Niiniluoto's class on ontology in Spring 1990. This was the first time I came into contact with him – and also, in any significant detail, with Karl Popper's ontology of the three worlds, which he defended in the lectures (partly on the basis of his then recent Finnish collection of essays: see Niiniluoto 1990). It might be worth recalling that I first personally met him only after the spring term was over, in August 1990, when the Philosophical Society of Finland celebrated the centennary of the birth of its former President, the late Professor Eino Kaila. I remember introducing myself to Ilkka Niiniluoto at the Hietaniemi Cemetary, where a delegation of the members of the Society, led by Niiniluoto, visited Kaila's grave. As a young student, I was working for the summer in the newspaper *Helsingin Sanomat*, and I briefly interviewed Niiniluoto on this occasion on Kaila's importance in the history of Finnish philosophy. (The interview was published in *Helsingin Sanomat* on August 10, 1990.) Later, I became better acquainted with Niiniluoto and, when preparing my Master's Thesis on Hilary Putnam's internal realism, started to study his writings on realism in close detail. I had the pleasure of working as Niiniluoto's research assistant in Summer 1993, and when I defended my Ph.D. thesis on the problem of realism in the pragmatist tradition (see Pihlström 1996) in May, 1996, I was happy to have Niiniluoto both as my supervisor and as the Custos of the disputation. Since then, I have had the privilege of cooperating with him in several activities at the Department of Philosophy and at the Philosophical Society, which he is still chairing. He also kindly asked me to comment in detail on the manuscript of his *Critical Scientific Realism* (Niiniluoto 1999), which I was pleased and honored to do. I am sure I am not the only one who missed his tolerant personality when he, in 2003, left the Department after having been elected as the *rector magnificus* of our university. It is largely thanks to Niiniluoto that I am able to see my own philosophical work, since the early 1990s, as focusing on one single question, the problem of realism – albeit in a great variety of different manifestations in different areas of philosophy, approached by means of somewhat different philosophical methodologies (e.g., pragmatism and transcen-

epistemology but has interesting and well-known applications in the philosophy of science, the philosophy of religion, metaethics and value theory, and elsewhere. This paper will focus on a particular application of the realism issue, namely, on what may be called the problem of *axiological realism*. This is the problem of whether there are such things as values, and if so, what kind of things they are: are they mind-independent in some sense, are there truths and falsehoods about them,[2] do they exist as "abstract objects", and so on? In particular, I will discuss the issue of *moral realism*, but in principle my considerations can be extended to non-moral values, such as aesthetic or epistemic values.

Instead of going through the vast literature on moral realism and its alternatives,[3] I will take up one interesting theory defending the reality of values, the theory of values as "World 3" entities, based upon Sir Karl Popper's work. This is a theory Ilkka Niiniluoto has vigorously defended over the years (cf. Niiniluoto 1981; 1990; 1994; 1999), although his explicit pronouncements in the area of metaethics have remained (especially in his writings in English) relatively scarce. If it turned out that values can be plausibly seen as "emergent" World 3 objects in Popper's and Niiniluoto's sense, then we might have secured a version of axiological (moral) realism.

When discussing the reality of values from this perspective, I am primarily interested in the basic ontological or metaphysical question itself: are there

---

dental argumentation). It has been and still is a great honor to work with a philosopher like Niiniluoto, to learn from him, to disagree with him, and have someone like him commenting on one's own ideas.

[2] Putative moral truths or falsehoods include, for instance, such statements as "Murder is wrong", "Taxation is criminal", or "Eating whale meat is intolerable" – that is, statements involving either "thin" or "thick" ethical concepts, or both. In truths or falsehoods like this, values such as life, justice, or the avoidance of suffering are implicated, and moral realists see such statements as being about, or referring to, such values (whether or not they succeed in this).

[3] See several important papers in, e.g., Sayre-McCord (1988) and Kotkavirta & Quante (2004). The former is a collection of (now) classical articles, whereas the latter includes more recent reflections with both historical and systematic comparisons between different kinds of moral realism, drawing from sources as different as Aristotle, the pragmatists, Ludwig Wittgenstein, John McDowell, and Charles Taylor. One of the few recent substantial defenses of a non-naturalist form of moral realism in the tradition of G. E. Moore is Shafer-Landau (2003).

values, and what kind of entities are they? Although I am opposed to too easy dichotomies between metaphysical and methodological or epistemological issues (see Pihlström 2004), I am tempted to see my own recent work on moral realism (Pihlström 2005) as focusing on the *methodological* issue of what it means for us to commit ourselves ethically – what it means for us to understand ethics as irreducible to anything non-ethical, or autonomous and "overriding", extending to each and every (even the most "purely factual") corner of our human form(s) of life – whereas the present essay more explicitly takes up the *metaphysical* question of what values are and in what sense they exist or are real. Partly defending the Popper-Niiniluoto view of values as World 3 cultural formations, I will, without discussing these views in any detail here, arrive at an alternative to (i) value nihilism or antirealism, such as J. L. Mackie's (1977) "error theory", (ii) merely "procedural" (as distinguished from "substantive") moral realisms *à la* Christine Korsgaard (1996), and (iii) Hilary Putnam's recent (2004) suggestion of defending "ethics without ontology", among others – although otherwise I do find Putnam's criticism of the fact/value dichotomy extremely helpful in developing a sensible form of moral realism.[4] I will, in brief, articulate and to some extent defend a pragmatic, yet ontological, realism about values. As I end up with an entanglement of metaphysical, epistemological, and (meta)ethical considerations during this project, my results will in the end be sharply critical of Niiniluoto's views, too.

My inquiry will proceed as follows. Sections 2 and 3 will outline the theory of the three worlds and the resulting ontology of cultural entities, including values. Section 4 will focus more closely on the issue of moral realism and consider Niiniluoto's moderate moral relativism. Section 5 will draw a parallel between Niiniluoto's commitments in the ontology of mathematical entities and in the philosophy of value, suggesting that a relativist rejection of realism is no more plausible in ethics than it is in the philosophy of mathematics. In section 6, I will briefly compare these issues with some of Putnam's arguments. Section 7 further argues that moral realism (which Niiniluoto rejects) is in fact

---

[4] In Pihlström (2005), I offer a pragmatic reconstrual of moral realism partly based on Putnam's arguments but rejecting his non-ontological approach. See that book also for the recent literature on these topics, with which I am unable to deal with here. If moral realism is articulated in a pragmatist manner, it will inevitably differ significantly from, say, Shafer-Landau's (2003) version of realism.

a more fundamental commitment than his principal philosophical thesis, critical scientific realism, because scientific realism presupposes a realist understanding of value – though *both* forms of realism, in my view, can and should be pragmatically articulated. The paper closes with the concluding section 8, which pulls some of these threads together.

2. Popper's World 3

Many (perhaps most) of us, excluding only the most rigorous reductionists and eliminationists, think that there are, in addition to physical and mental objects, states, and processes, also cultural (or socio-cultural) entities, such as scientific theories, social institutions, linguistic meanings, works of art, and moral and aesthetic values. Human persons might also be included in this list. There are different opinions about what kind of cultural entities there actually are, but many philosophers feel that a comprehensive ontological theory should somehow account for their existence. Let us, therefore, take a look at a famous – or, for some, notorious – theory that promises to do so.

In a series of writings from the late 1960s to the early 1990s, Karl Popper argued that reality consists of three relatively autonomous domains or "worlds":[5] World 1 ("nature", *phusis*) contains physical objects, states, and

---

[5] Popper first formulated the theory in his address "Epistemology without a Knowing Subject" in 1967 (see Popper 1972, ch. 3). For further developments of the theory, see Popper (1974), pp. 1049ff., (1978), (1980), (1984), chs. 1 and 12, and (1992), pp. 180–196. In particular, Popper's contribution to Popper & Eccles (1977) is an important source (see pp. 1–224). See also Niiniluoto (1999), pp. 23ff. Popper's three worlds ontology is, of course, connected with his general theory of science (cf. Sahavirta 2006), with his view that the growth of knowledge is an objective affair, not a matter of individual scientists' beliefs, i.e., something that takes place through conjectures and refutations at the level of the objective contents of theories (in World 3). This is not the right place to discuss Popper's rejection of "subjective" theories of knowledge or general epistemological and methodological problems. One hardly needs to accept Popper's falsificationist criterion of demarcation in the philosophy of science in order to defend the reality of what he places in his Worlds 1, 2, and 3. While I shall mainly discuss World 3 as an ontological concept, it should be kept in mind that it does not seem to be clear even for Popper himself whether the theory is primarily ontological or epistemological. Sahavirta's recent comprehensive study (ibid.) illustrates this fact.

events; World 2 ("mind", *psykhe*) contains subjective conscious states, experiences, and processes; and World 3 ("culture") is the realm of human-made "*objective contents of thought*" (Popper 1972, p. 106; emphasis in the original). The cultural entities belonging to World 3 include such human-made constructs as artifacts, tools, numbers, myths, artworks, institutions, and – for Popper himself most importantly – problems, problem situations, critical arguments, and theories.[6] Thus, the "contents" of thoughts, statements, propositions, books, and libraries are typical World 3 entities. World 3 is *dependent on* Worlds 1 and 2: it would not exist, had these ontologically more fundamental worlds not existed. Yet, in some sense, World 3 is *irreducible to* the more basic layers of reality. It has been and continually is created by human beings, but as Popper puts it, it "transcends" its makers. Many of its elements are unintended results of human actions; thus, it is ontologically "autonomous" (ibid., p. 161), or, in Niiniluoto's preferred phrase, "relatively autonomous".[7] These charac-

---

[6] It is worth noting that the concept of World 3 enables Niiniluoto (1999, p. 30) to defend an inclusive ontology of *properties*: the properties of physical things in World 1 are *tropes* (property-individuals, abstract particulars) – indeed, Niiniluoto favors a bundle theory according to which physical objects are bundles of tropes – but properties (including uninstantiated ones) in a more traditional sense exist in World 3, as abstractions from World 1 property-instances, hence as *universals* (though, in this theory – unlike in typical Aristotelian realism about universals – properties as universals, unlike tropes, are not mind-independent).

[7] The relative autonomy of World 3 can be compared to what John McDowell (1996) calls our "second nature": according to McDowell's influential neo-Aristotelian proposal, conceptual and normative capacities operating within what Wilfrid Sellars labeled the "space of reasons" are *natural* for humans (parts of "human nature") but require *education* (or *Bildung*, as the German tradition puts it) in order to arise. The rational, conceptual, and normative order captured in the notion of the space of reasons is not Platonic but requires human subjectivity (ibid., p. 92; see Thornton 2004, p. 45). As Thornton (ibid., p. 63) puts it, McDowell seeks "a middle ground between the radical independence and the complete dependence of the moral world on moral subjects" – just as Popper's concept of World 3 seems to do. Values, for McDowell, are "secondary qualities", that is, "not *brutely* out there", yet "out there independently of individual judgements" (ibid., p. 79), again precisely like World 3 entities are. Values and moral reasons are, in words echoing Popper, "semi-autonomous" (ibid., p. 228). However, the contrast between primary and secondary qualities invoked here is not sharp, because both are "constitutively apt for conceptualization within the rational space of reasons" (ibid., p. 82). The McDowellian theory is by no means reductive, as the evaluation of virtues is argued to already presuppose a moral standpoint; moral

terizations can be captured by saying that World 3 *emerges* out of Worlds 1 and 2 (in the same sense in which World 2 emerges out of World 1).[8]

A crucial feature of Popper's theory of the three worlds is that World 3 causally influences Worlds 1 and 2 (and that, similarly, there is a causal influence from World 2 to World 1). This *downward causation* is a most important characteristic of emergentism in general, and there is a hot debate going on in the philosophy of mind and general metaphysics whether such a concept is acceptable for a naturalist or materialist.[9] Popper's position differs in this regard from that of most recent contributors, because he explicitly rejects the principle of the causal closure of the physical world (or World 1) – a principle that most contemporary naturalists or physicalists (e.g., Kim 1998) believe to be necessary

---

upbringing (as a "second nature" process) leads to a practical rationality which enables us to sensitively apply concepts to particular situations (ibid., pp. 83, 92–93). Moreover, there is nothing "queer" about values, if understood in this sense, because they are rooted in natural human reactions and dependent on our sensibilities (see de Gaynesford 2004, pp. 166–167, 174–176); yet, the space of reasons is autonomous and irreducible to anything non-normative (ibid., pp. 63, 148–150). For more reflections on McDowell's relevance to the current debate over moral realism, see Pihlström (2005), ch. 2. Incidentally, we may wonder why there is so little explicit discussion of pragmatism to be found in McDowell's or his commentators' (like Thornton's and de Gaynesford's) works, given their emphasis on pragmatist themes such as second nature, habituation, the autonomy of normativity, etc. For a number of discussions of the ways in which normativity becomes a problem for reductive scientific naturalists, see the essays in de Caro & Macarthur (2004); Kelly's (2004) contribution to that volume is particularly relevant to the issue of ethical naturalism.

[8] This paper will leave the concept of emergence unproblematized. I am aware of the fact that Popper's treatment of emergence is far from satisfactory. His loose characterizations are not very helpful in the current rather technical discussion of emergence. As I have argued elsewhere, the problems of emergence and realism ought to be connected (see El-Hani & Pihlström 2002; Pihlström 2002a). As a working definition of emergence, for the purposes of the present discussion, we may adopt the relatively simple one favored by Niiniluoto (e.g., 1990, p. 38): P is an emergent property of an entity (or a system) x, if and only if the proposition $P(x)$ cannot be deduced from the (complete) true theories about the parts (or constituents) of x. A lot remains open here, especially the concept of truth, at least if one refuses to join Niiniluoto in defending a Tarskian-styled correspondence theory.

[9] See Kim (1998), Stephan (1999), El-Hani & Pihlström (2002), and Pihlström (2002a).

for a properly scientific worldview.[10] Contrary to what the physicalists suppose, World 1, according to Popper, is not causally closed, but World 2 and World 3 entities, especially scientific problems and theories, can play a causal role in affecting the behavior of physical things. There is no direct interaction between Worlds 1 and 3, however; their causal relations must always be mediated through World 2. (Popper 1992, pp. 184-185; Popper & Eccles 1977, pp. 19ff., 36-41, 57, 207, 537ff.)

Clearly, Popper is a *realist* about all the three worlds. Insofar as (a capacity for) causal efficacy is required for an entity to be real, this requirement must be extended to World 3 entities as well. Furthermore, it should be noted that realism about the three worlds is compatible with scientific realism: World 1 may contain unobservable entities and processes postulated in scientific theories (as Popper and Niiniluoto claim it does), i.e., entities to which the theoretical terms of those theories refer; somewhat more controversially, even World 3 may likewise be partly populated by theoretical entities postulated by the human sciences (e.g., "implied readers" and "implied authors" postulated in literary theory, or social group pressures postulated in sociology).[11]

Unlike eliminative or reductive materialists, Popperian *emergent materialists* – such as Niiniluoto – argue that while mental phenomena are ontologically and causally dependent on physical phenomena (they need a physical basis) and while cultural phenomena are similarly dependent on both physical and mental phenomena (requiring a basis there), these "higher-level" phenomena are neither conceptually nor ontologically reducible to the "lower-level" ones. World 1 "supports" Worlds 2 and 3, but the latter two are genuinely something "more" than their supporting basis – something quite real and (in the case of World 3) objective. Some of the cultural entities to be found in World 3 are abstract (e.g., numbers), while others are embodied (e.g., works of art); in both cases, World 3 entities must be suitably manifested in or supported by material

---

[10] The rather unorthodox flavor of Popper's position is manifested in the fact that another major philosopher attacking the causal closure principle is William Hasker (1999), who defends a form of "emergent dualism" and theism, thus explicitly detaching his emergentism from materialism (which, however, remains the standard framework for emergence theories).

[11] Cf. Pihlström (1996), ch. 5. Mathematical entities, such as numbers, would provide a more obvious example, if we want to say that mathematicians "postulate" such entities.

things and processes. This holds even for those World 3 entities which are not (directly) embodied, e.g., such abstract entities as numbers: there must be a relevant "supporting" material practice in Worlds 1 and 2 for such entities to come into existence and to go on existing.

Niiniluoto (1999, p. 24) formulates his main argument for the reality of World 3 as follows: "[O]ur natural and scientific languages ontologically commit us to such things as numbers, concepts, novels, symphonies, and societies, and attempts to reduce them to physical or subjective mental entities have been utterly implausible."[12] For example, the uniqueness of such World 3 entities as works of art requires a realist interpretation: a symphony is not a spatially located object, though it may be destroyed in the course of time; a particular symphony may retain its identity as a cultural entity despite – or, rather, because of – its multiple instantiations (performances, note scripts, etc.) in Worlds 1 and 2 (ibid.). There can be better and worse performances of the *same* symphony (see also Popper & Eccles 1977, pp. 449–450), which seems to require that the work itself exists as a cultural entity in World 3, ontologically independently of its performances (or other materializations and documentations) – though, presumably, *not* independently of the habit or tradition of performing it, or at least of maintaining it as materially documented in musical libraries and archives.[13] To take another example, one may argue that a social institution, as a cultural entity, survives the replacement of all its members by new ones. Thus, the University of Helsinki is the same entity as the Royal Academy first founded in Turku (Åbo) in 1640, although no people or university buildings from that period exist any longer. At least, according to Niiniluoto (1999), the reductive materialists who claim that such entities as

---

[12] In fashionable contemporary terms, this might perhaps be rephrased as a truthmaker argument (see Armstrong 2004): truths about the kinds of things Niiniluoto mentions, if such truths there are, require as their truthmakers irreducible World 3 entities, and no plausible truthmakers can be found among the denizens of Worlds 1 and 2.

[13] However, Niiniluoto, together with many other philosophers, finds the existence of merely possible (uncomposed) symphonies a somewhat preposterous ontological postulation. Such a view would be too Platonistic even for a World 3 theorist. Merely possible, non-actual symphonies may be said to exist in possible worlds different from the actual one; however, the present paper does not deal with the metaphysics of modality. Of course, one may say that possible symphonies "exist" in the extremely weak sense that the musical systems required for composing them exist in World 3.

symphonies and institutions are "nothing but" complex material (and possibly mental) entities have the burden of proof: they owe us a method of reducing all the obvious truths about World 3 entities, such as the truth that the University of Helsinki was established in 1640, to truths about World 1 and World 2.

Historically, the formation of Worlds 1, 2, and 3 takes place through an *evolutionary* process. The brain is the material basis (in World 1) of the self, which is, again, "more" than the brain (Popper & Eccles 1977, pp. 16, 120). The "cultural evolution" producing World 3 entities continues the biological evolution which is restricted to World 1. A central role in the constitution of World 3 is played by the symbolic function, which makes language possible. This, in turn, is an emergent property of suitably organized matter (certain kinds of physical organisms). Once created, World 3 theories, problems, conceptual schemes, artifacts, etc., "downwardly" influence their creators, structuring human life in a human environment. Both Popper and Niiniluoto thus argue that the human self is a World 3 entity, irreducible to the body (World 1) and to the subjective states of consciousness (World 2). "As selves, as human beings, we are all products of World 3 which, in its turn, is a product of countless human minds" (ibid., p. 144). The self "emerges in interaction with the other selves and with the artefacts and other objects of his environment" (ibid., p. 49), achieving the capacity to use language. Thus, World 3 plays a decisive role in the gradual formation of human consciousness and personality (see also Popper 1978; 1980, pp. 166–167; 1992, pp. 189–193).[14]

It is not difficult to see why Popper's World 3, associated with Plato's Forms and Frege's "third realm", has not been met with sympathy among critics. Anthony O'Hear (1980, pp. 183ff.), for instance, argues that everything Popper wants to say in World 3 terms can be said in non-Platonist World 2 terms (which, for materialists, are reducible to World 1 terms). Mario Bunge (1981, chs. 7–8) strives for a materialist theory of culture as a "concrete system" and rejects World 3 as "unworldly" and Platonistic, as a "monster of

---

[14] Critics, such as O'Hear (1980, pp. 200–204), have resisted Popper's postulation of an immaterial self. I am not sure that Popper's view leads to such a Cartesian outcome; nonetheless, philosophy of mind is not my main interest in this paper. It should be noted, however, that both Niiniluoto and Popper emphasize the processual – as distinguished from substantial – nature of the mind (see, e.g., Popper & Eccles 1977, p. 146; Niiniluoto 1994).

traditional metaphysics" (p. 148).¹⁵ L. Jonathan Cohen (1985), in turn, locates conceptual problems in Popper's theory: since World 3 contains (the "contents" of) inconsistent theories, and also undiscovered logical consequences of theories, it contains any proposition whatsoever (because anything follows from a contradictory pair of propositions, and it is highly probable that there are such pairs among the innumerable denizens of World 3). Brian Ellis (1990, pp. 231–232) also argues against a reification of the contents of books, papers, inscriptions, etc., in a third world. However, these critics, correctly claiming that the step from "objective knowledge" to a separate "world" in which that knowledge is somehow stored is a *non sequitur*, seem to view World 3 as too Platonic and static. We should, rather, place it within a more dynamic, processual metaphysical scheme in which it is constantly made and remade by habitually acting human beings.

In my own reading of Popper (a reading which Popper himself would undoubtedly have rejected), such a *pragmatism* provides the framework for an ontologically tolerant, pluralist metaphysics in which World 3 entities do have their place in reality along with World 1 and World 2 ones.¹⁶ A pragmatist background metaphysics enables us to avoid, say, O'Hear's (1980, p. 182) claim that World 3 entities are "rooted in another realm altogether", different from the natural history of human beings, as well as Bunge's (1981, p. 156) claim that they are nothing but instances of reification. For a pragmatist Popperian, World 3 entities cannot be detached from worldly human *practices*.¹⁷ They do not exist completely autonomously in any Platonic heaven, independently of material, mental, and conceptual ways of human interaction. Yet, they can be seen as real, insofar as they have practical ways of affecting human life. We pragmatically *need* to postulate various kinds of World 3 entities in order to make

---

¹⁵ Bunge's (1989) naturalist conception of value and morality is, of course, meant to account for values without invoking anything like World 3.

¹⁶ See Pihlström (1996), ch. 5. Popper had a remarkably narrow-sighted conception of pragmatism, erroneously picturing the classical pragmatists as naive instrumentalists. Niiniluoto's 1978 paper, "Notes on Popper as a Follower of Whewell and Peirce" (reprinted in Niiniluoto 1984) is an informative discussion of the many Peircean elements of Popper's philosophy of science.

¹⁷ For comparison, see Rouse's (2002) account of the embeddedness of (scientific) normativity in material-historical (scientific) practices. Rouse, however, does not explicitly speak of either pragmatism or World 3.

sense of our lives and the world we live in. If we are prepared to endorse a pragmatic ontological tolerance (Pihlström 1996; 1998), there is no need to follow reductive materialists in their rejection of World 3. On the other hand, the reification charge must be taken seriously: World 3 is *not* to be construed as a separate realm, or world – all the three worlds are, for us humans, an inseparable mixture. Reality, or human reality at least, is the totality of Worlds 1, 2, and 3.

If this idea is taken seriously, then we should maintain, with both classical and more recent pragmatists (e.g., John Dewey), that there is no clear nature/culture dichotomy. Nature, for us, is inevitably culturalized, while culture is undeniably based on fully natural processes.[18] Popper's own trichotomy between the three worlds is, for a pragmatist, too sharp, but we need not follow him dogmatically here. Our world is a mixture, as I said, but for practical and heuristic purposes we may choose to emphasize the differences between Worlds 1, 2, and 3. An anti-reductionist ontology of the mind and of culture can be articulated in Popperian terms by claiming that whatever belongs to Worlds 2 and 3 is real, in addition to World 1.[19]

3. Cultural Entities

We now arrive at a more detailed discussion of the ontology of *cultural entities*, based on the concept of World 3. In particular, we should take a look at how values "exist" as such entities in World 3.

Niiniluoto (1999, p. 33) characterizes a cultural entity as a physical object with both physical and non-physical properties existing in World 3. Such an entity has, he says, "a material kernel in World 1, but enriched with its relations to World 2 and 3 entities it becomes a cultural object with some mind-

---

[18] See especially Dewey's classical naturalist work, *Experience and Nature* (Dewey 1929). Cf. Pihlström (1998), ch. 4. See also, again, McDowell (1996) on "second nature", and Pihlström (2003a), ch. 4, on McDowell's program of "naturalism of second nature".

[19] I believe I am here more or less in agreement with Popper and Niiniluoto, but my pragmatism departs from their views in the rejection of the materialist idea that there is, or even could be, purely material World 1 entities with no relation whatsoever to human culture. In this weak sense, my view is (or may come close to) "panculturalism" (cf. note 20 below).

involving relational properties" (ibid.). Cultural entities are thus "mixed", but we may "conceptually distinguish" between their World 1 and World 3 "parts" (ibid.). There is a problem here that deserves attention. Niiniluoto seems to think that a cultural entity must stand in *relations* to World 2 and World 3 entities; without such relational properties, it would be a mere physical entity. Thus all cultural entities are essentially relational in nature. But how, let us ask, was the *first* World 3 entity formed? It could not claim its status as a cultural (World 3) entity because of its relations to other (earlier, already existing) World 3 entities, because, we assume, there were none prior to it. Niiniluoto's view entails that there was no single first World 3 entity but that a whole group (at least two) World 3 entities emerged, were created (constructed), simultaneously. But could they have emerged simply on the grounds of their relations to each other, and to the World 2 entities that already existed? Somehow the very emergence of World 3 (and, of course, World 2) still remains a mystery.[20]

Moreover, one may ask what actually *is* the "it" that has a "material kernel" in World 1 but is a cultural entity because of "its" relations to Worlds 2 and 3. Can this underlying entity be ontologically individuated independently of its relations to the three worlds, or is it fundamentally a World 1 material entity? If the basic entity, the ontological individuation basis of World 2 and World 3 entities, is simply material, how does this view really differ from reductive materialism?

Let us agree with Niiniluoto that cultural entities are physical objects, with World-3-level non-physical relational properties. Then it seems that they are *particulars*. Here another question arises. Wouldn't it be more natural to treat only some World 3 objects as particulars and some others as *universals*? After all, we have already seen that the *repeatability* of such entities as literary or musical works is, for Popper and Niiniluoto, a crucial argument in favor of

---

[20] In the case of the emergence of World 2 entities, particularly conscious experience, a wildly speculative theory that would avoid the "mystery" is *panpsychism*, the view that in some sense (proto-)mentality or (primitive) experience is present in the most fundamental units of the physical world. (On the varieties of and arguments for and against panpsychism, see Clarke 2004.) Analogously, one might hold the *panculturalist* (that is, radically constructivist) position that regards World 3 as ontologically primary: everything there is is a cultural entity (in some, perhaps rudimentary, sense). I am not endorsing such a view, but my pragmatism (cf. especially section 7 below) might bear some affinities to it.

their reality *qua* World 3 entities. Persons or paintings may be unrepeatable particulars, while symphonies or novels might, rather, be *types* whose *tokens* are particular spatio-temporal entities. The "manifestation" or "embodiment" requirement of abstract World 3 entities would then be interpreted as the instantiation requirement that at least some tokens of such abstract types must be realized in Worlds 1 and 2 in order for the World 3 type (universal) to exist.[21] Neither Niiniluoto nor Popper, as far as I know, has ever made it entirely precise how the categorial division of entities in particulars and universals (and tropes) is related to the three worlds ontology.

Let us skip these difficulties, however, and return to Niiniluoto's account of cultural entities, equipped, however, with his insight that in order to be a cultural entity, a thing – whether particular or universal – must be somehow related to a human cultural context (i.e., World 3). He further says that World 3 contains, in addition to material artifacts, also "*abstract* constructions" (novels, symphonies, numbers, etc.), which do not have physical properties at all but can be "documented in World 1" and "manifested in World 2" (ibid.). Values, presumably, ought to be ontologically classified as such abstract constructions, although they may be said to supervene on facts or the factual properties of things and events, particularly on human actions, including their material (World 1) aspects as well as, say, emotional (World 2) ones. As to their ontological status, values might stand somewhere between more clearly abstract, immaterial entities such as numbers and directly embodied cultural entities such as statues. In the case of values, we cannot point to a corresponding material World 1 entity or to a mental World 2 entity (e.g., an emotion); nor, however, should we think of their abstractness on the model of the abstractness of mathematical entities. Values are always tied to concrete human circumstances and ways of doing things. If they are treated as universals (or as types rather than tokens), they should not be construed as Platonic, uninstantiated ones, but as more Aristotelean universals always requiring instantiation in the humanly experienced and manipulated world.

---

[21] A related issue is the ontological status of emergent properties in general. Are properties, usually only loosely invoked in emergence discussions, best construed as universals, as tropes, or perhaps nominalistically as classes (of *possibilia*)? What kind of properties are postulated, e.g., in the kind of discussion of mental (downward) causation that authors like Kim (1998) have engaged in? See Pihlström (2002a) for some remarks on these problems.

In order to provide an illuminating comparison, I want to mention a couple of related accounts, which a pragmatistically inclined ontologist of culture might invoke as support for the pragmatist rearticulation of Popper's theory. First, let us note that a strikingly similar, yet more complexly arranged, picture of the emergence and embodiment of cultural entities (such as artworks, but also persons and, presumably, values) has been defended for decades by Joseph Margolis, both in his earlier (Margolis 1978; 1980; 1984) and in his more recent work (Margolis 1995; 2002; 2003). According to Margolis, cultural entities are embodied yet autonomous "tokens-of-types". They need a material basis, but they cannot be adequately accounted for in any naturalized theory restricted to that basis. "Naturalizing" strategies, according to Margolis, desperately fail as theories of culture – and as theories of the mind.[22] We should be able to ascribe to cultural entities a causally relevant (and thus also explanatorily relevant) role – in this sense, they must be seen as autonomous, without sacrificing the materialist demand for a material basis of embodiment (see Margolis 1984, p. 14).[23]

Sandra Rosenthal (1986, ch. 8), in turn, discusses the emergence of value within her metaphysical system of "speculative pragmatism" synthesizing the classical pragmatists' (i.e., Peirce's, James's, Dewey's, Mead's, and Lewis's) views. Value, for the speculative pragmatist, is neither subjective nor simply "there"; value qualities of objects and situations are "as ontologically real in their emergence as the processes within which they emerge" (ibid., p. 174). Thus, value, "as a quality of events within nature, is immediately experienced and irreducible" (ibid.); valuing experiences are "traits of nature", "novel emergents" (ibid., p. 175). As Rosenthal further writes: "The occurrence of the immediate experience characterized by value is an ontological emergent within nature [...]. Values are as real as are other emergents within processive concreteness; the presence of value is an ontological condition within the context of intermeshing matrixes characterized by organic activity." (Ibid., p. 176.) There are interesting similarities between this view and Popper's: both stress process metaphysics (cf. ibid., ch. 5) and insist that the emergence of value (and

---

[22] See especially Margolis (2003) for a devastating critique of scientistic assumptions in twentieth century American philosophy.

[23] For Margolis's somewhat complex characterizations of the concept of emergence, see his (1995), pp. 219, 257; for a discussion, see El-Hani & Pihlström (2002).

of culture in general) ought to be taken seriously in an ontological sense. Yet, Rosenthal, as well as Margolis, would undoubtedly resist World 3 as an unnecessary reification – as much as they would presumably resist the construal of values or any other cultural entities as universals. For these two pragmatists, the Popperian "worlds" would be thoroughly and inseparably mixed, and any remnant of Platonism would have to be abandoned.

In any event, these two pragmatist emergentists offer examples of the pragmatic way in which I would like to reinterpret Popper's and Niiniluoto's World 3 realism. In order to defend the emergent, and constantly dynamically emerging, reality of World 2 and World 3 entities or processes, one can in the end easily dispense with the whole terminology of "worlds", while retaining the kind of tolerant, anti-reductionist ontology this terminology is intended to capture.

4. Moral Realism and Relativism

Ilkka Niiniluoto defends both the reality of values in World 3, as we saw in the previous section, and a "modest form" of *moral relativism* (Niiniluoto 1999, pp. 229–236). He correctly notes that moral relativism is not simply the factual "diversity" thesis about the relativity of moral views and principles to cultures and societies but a philosophical thesis about the truth and justification of those views (ibid., p. 231). Relativism must be distinguished from (i) moral *nihilism*, which denies the existence of moral values, (ii) moral *skepticism*, which denies that we may have moral knowledge, (iii) Mackie's *error theory*, which claims that moral statements are meaningful but false, and from (iv) the metaethical applications of semantic *anti-realism*, which deny that moral judgments have truth values (ibid.). Unlike these negative doctrines, which I cannot even start to examine in this paper, moral relativism is "a positive thesis which accepts that moral views may be true or justified in some *relativized* sense" (ibid., p. 232).[24]

---

[24] Niiniluoto's position comes close, it seems, to the well-known combination of scientific realism (culminating in the view that science aims at an "absolute conception of the world") and ethical relativism defended by Williams (1985). My criticism of Niiniluoto is in many ways parallel (and indebted) to Putnam's (1990; 2002) criticism of Williams. Indeed, as the conjunction of scientific realism and moral relativism is a rather usual one in the realism discussion today, my critical reflections should have a

The main argument Niiniluoto offers against moral realism is that there is no independently existing "moral reality" to which true moral judgments could correspond. Both Platonist and naturalist attempts to defend such a reality are implausible, while non-natural moral properties face Mackie's charge of "queerness". Even Richard Boyd's (1988) sophisticated naturalist moral realism commits, in Niiniluoto's view, G. E. Moore's traditional naturalistic fallacy. (Niiniluoto 1999, pp. 232–233.) On these grounds, Niiniluoto arrives at an articulation of moral relativism in Popperian terms as the view that moral values are constructed and, therefore, culturally relative:

> Moral realism fails to appreciate the fact that *morality is human-made* – a historically developing social construction. According to what may aptly be called *moral constructivism*, moral values as social artefacts belong to Popper's World 3.[25] Besides rejecting naturalist attempts to reduce morality to the physical World 1, this view also excludes the claim of moral *subjectivism* which takes morality to be simply a matter of personal attitude or feeling, hence an element of World 2. The constructivist view of morality as a social fact helps us to understand why a moral system has binding or coercive force (in Durkheim's sense) within the community where it has been constituted or accepted. (Ibid., p. 233.)

This, however, is precisely the problem we should be interested in: in order to be able to so much as identify a "system" as "moral", we need to acknowledge that it is meant to have "binding or coercive force" *not merely* within the community in which it has been constituted or accepted but also outside that community, in principle *universally*. This is what the "overriding" and uncompromising character of a genuine moral commitment amounts to. If I say that I am committed to the view that murder is wrong, I am *not* claiming that this holds for those who are committed in the way I am, for the

---

broader relevance independently of how well I succeed in my evaluation of Niiniluoto's position. (Moreover, it must be understood that not all scientific realists accept moral relativism. Some try to defend a form of moral realism, too, while others think that an anti-realist, typically expressivist, account of ethical language is the best choice for a scientific realist.)

[25] One of Niiniluoto's references here is Searle's (1995) theory of social facts as results of agreement. Cf. also, e.g., Tuomela (2002) and Schatzki (2002) for recent clearly un-Popperian theories in the ontology of social practices and institutions.

like-minded, or for those who share my cultural background; what I mean by uttering such words, or what I *ought to* mean (a moral "ought"!), is that this holds for everyone, even those who think otherwise. If this basic requirement of morality is not acknowledged, then I fail to see how we could even be speaking about morality or "moral systems" here.

Niiniluoto admits that it is possible to make true or false statements about morality "relative to some socially constructed system of moral ideas", such as an ethical theory or a community's "moral code". Relativized statements like "In system S, a is good/right/holy" thus have truth values, unlike their absolute cousins "a is good/right/holy". The relativized statements (e.g., "In Christian ethics, respect for your parents is good") are statements about World 3 structures, and the human sciences investigating World 3 structures may arrive at such truths about reality. We may also say that "semi-autonomous" World 3 entities make such statements true (if they are true). Thus, relativized norm propositions may be among the results of the human sciences, realistically interpreted.[26] Such relativized statements may also be formulated in terms of relative goodness or relative rightness ("a is good-for-x", "a is right-for-x"). The reason why these relativized statements are not absolute is that World 3 is, obviously, subject to change as our human lives and thinking develop. We may simply change our moral codes – and have done so. (Ibid., pp. 233–234.) In short, moral realism fails and relativism prevails because "morality is a social construction, progressively constituted by the consensus of human communities", and there is no "external criterion" for evaluating whether a consensus reached in a finite time (or even ideally "in the limit") is true or right (ibid., p. 236). In contrast, in science, or the scientific pursuit of truth, there presumably *is* such an external criterion (according to realists like Niiniluoto): theories can be empirically, if only indirectly, tested against a theory-independent reality.

However, again, I take it as obvious that moral statements, insofar as our pre-understanding of them counts here, are in an important sense *not* like the

---

[26] See also Niiniluoto's (1981) investigation of norm propositions on these grounds. In that early paper, Niiniluoto argues that language is an historical entity and that the truth conditions for statements about language can be accounted for by means of the correspondence theory of truth. He places both language and the legal order in Popper's World 3 (ibid., p. 172).

relativized ones Niiniluoto considers – or *should not* be regarded as such relativized statements. Morality is about what is good or right, period. It is not about what is good or right for someone or for some particular community. Nor is it about what would be the good or right thing to do from the point of view of an ideal community; it is about what is absolutely, yet personally, required from me right here, right now.[27] This kind of "absolute" moral commitment is *personal*, all right, but also *universal* in the sense that it requires any serious moral agent in relevantly similar circumstances (whatever this ultimately means) to commit her- or himself to the same kind of ethical duties. Regarding community consensus as something more fundamental than a moral duty would be a serious misunderstanding of what it is to have a moral duty or obligation at all.

It can now be seen that Niiniluoto problematically assumes a "God's-Eye View" from which it seems to – but only seems to – make sense to ascribe relativized truth values to evaluative statements. For us, there is no such view from nowhere; we always already find ourselves within one or another system of value commitments, which, if truly moral, makes uncompromising demands on us. This shows why my pragmatist criticism of moral relativism is closely connected with a similar criticism of metaphysical realism. It is by assuming metaphysical realism that a critic of moral realism like Niiniluoto is able to account for the relativized truth values of moral judgments at an allegedly value- and commitment-neutral meta-level. What I am urging, in contrast, is that value commitments will always have been made insofar as one so much as attempts to occupy such a level. One cannot explain away the demands of evaluative practices simply by attempting to relativize their truth claims at an allegedly more fundamental level.

Niiniluoto does, we must admit, reject (with good reasons) as internally inconsistent what he calls "radical moral relativism", i.e., the thesis that all moral systems are equally good or equally well justified (ibid., p. 235). His relativism is merely the modest view that a "relativizing" reference to a system S or an auditory x is needed in the formulation of a moral statement with a

---

[27] I develop this "absolute" yet pragmatic view of moral realism at more length in Pihlström (2005). It must be noted that the absoluteness of moral commitment by no means precludes irresolvable moral dilemmas or tragic choices. It is precisely because of this absoluteness that some of our ethical situations are tragic.

truth value, such as "In system S, a is good", or "a is good-for-x" (ibid., pp. 234–235). He even says that his modest relativism is compatible with the idea of *moral progress* (and regress). Philosophical ethics can explicate "higher-order principles", such as consistency, universalizability, agreement with moral intuition, harmony with other valuations, etc., that can be used to evaluate and compare moral systems, although there is again a clear difference to the more absolute (though also fallible) comparison between rival scientific theories in terms of their truth/falsity or truthlikeness. Factual knowledge about the consequences of the adoption of certain value systems may also be relevant in such an evaluation; also personal and collective moral experiences and critical conversation may be important in this regard. (Ibid., pp. 235–236.) Niiniluoto further claims that attempts to defend moral "objectivism" (or moral objectivity "in the long run") by referring to the critical discussion of free citizens in a democratic community (Jürgen Habermas) or to the pragmatist insight that fallibilism and the pursuit of consensus apply equally well to science and ethics (Putnam)[28] only yield relativized moral judgments, not absolute ones, because "some more basic value assumptions" (e.g., the principles of equality and democracy in Habermas's case) must be in place in order for such arguments to get off the ground (ibid., p. 236).[29]

It is worth noting that Niiniluoto – somewhat surprisingly – refers to Richard Rorty in a quite positive or at least neutral manner when he observes that "Rorty defends the idea of moral progress without Moral Truth by appealing to a kind of universalizability principle: moral progress means 'an increase in our ability to see more and more differences among people as irrelevant'" (ibid., p. 235n11; see Rorty 1998, p. 11). Thus, in this case, Niiniluoto sides with Rorty against Putnam and Habermas. For a critical scientific realist defending Enlightenment ideals, this is an odd choice. I will propose in the following sections that Niiniluoto should, rather, opt for a moderate moral realism – although he might claim, with some justification, that the

---

[28] I will return to Putnam's defense of moral objectivity below in section 6.

[29] There is no reason for the (pragmatic) moral realist to dispute this: there indeed are always prior value assumptions at work, not only in arguments for moral objectivity in the long run but also – as I shall try to argue in what follows – in allegedly purely factual arguments and truth-seeking.

difference between "moderate" moral realism and an equally "moderate" relativism is little more than a matter of words.[30]

We should perceive already here, however, that Niiniluoto's (qualified) Rortyan sympathies are by no means innocent. Arguably, from Rorty's perspective, there is, as recent critics have pointed out, "no point to the morality-prudence distinction" at all (Guignon & Hiley 2003, "Introduction", p. 26). This elimination of a basic ethical distinction may be argued to lead to downright immoralism or nihilism, even if the initial motivation was only to avoid postulating a dubious Platonic abstraction of "Moral Truth". Rorty's "ironism" makes it impossible for us to understand why people may sacrifice or risk everything, even their lives, in order to do the right thing (ibid., pp. 37–38).

This is emphasized in more detail by Jean Bethke Elshtain (2003), who argues compellingly that in some cases, such as the saving of Jews during World War II, people need *more* – something more universal (though not for that reason "foundationalist") – than mere Rortyan "final vocabularies" in order to be ethically motivated to, say, save other human beings' lives in dangerous circumstances. Not everything can just be "redescribed" in a Rortyan manner to make it look good (ibid., p. 147). There are features of human moral life that seem to require a particular kind of description in order to be identifiable as moral at all. After all, organizations such as Amnesty International see themselves as defending fundamental human dignity as an "ontological given" (ibid., p. 152) – not as a mere "redescription" of something that might be described in many other ways, too. Saving fellow human beings is a stronger ethical urge than saving contingent neighbors (as Rorty might put it); mere contingent final vocabularies which we may try to extend to others is simply not enough, if we try to understand moral commitment as a real human phenomenon. Nor is it clear what it would even mean to say that Amnesty International is defending a local, contingent set of values: its activities are precisely intended to demonstrate the universal significance of some moral values.

Rorty's problems are not, of course, the same as Niiniluoto's. What this brief digression to a recent critical rejoinder to Rorty should remind us of,

---

[30] See here also Sihvola's (2004) discussion of the possibility of defending a moderate moral realism on the basis of Aristotle's virtue ethics.

however, is that a stronger moral realism than Niiniluoto's moderate relativism may be needed for us to account for the kind of phenomena examined by Elshtain. Mere vocabularies are not sufficient in the construction of moral values; what we need are genuine, objectively binding moral values. Yet, as we learn from Popper and Niiniluoto, these values can be seen as "existing" in the humanly constructed World 3, instead of any superhuman layout of reality. We just should not consider them relative to contingent communities or frameworks, or to a consensus reached in them, in the sense in which Niiniluoto and Rorty propose.

5. Ethics and Mathematics

Let us briefly turn from the ontology of values to mathematical objects and mathematical truths for a moment. In this area of philosophy, Niiniluoto's choice is *constructive realism*, which should be distinguished from Platonist mathematical realism as well as from anti-realist (e.g., fictionalist) doctrines that deny the existence of mathematical entities altogether (see especially Niiniluoto 1992). For Niiniluoto (and for Popper), mathematical objects such as numbers are, again, World 3 entities socially constituted by the human mind and the mathematical practices we engage in, yet autonomous in the sense of transcending any individual mental states. However, in mathematics, unlike in morality, there is a place for absolutely true statements. Insofar as a mathematical system (say, the system of natural numbers and the basic laws of arithmetics) has been constituted in World 3, certain truths (e.g., 2+2=4) hold (necessarily) in it.

Should we describe this view as a mathematical relativism? After all, mathematical truths are, like the moral ones, relative to the systems in which they hold. Someone might invent an entirely different mathematics, just as different communities have held different ethical principles.[31] However, we might be tempted to claim that even if some people used mathematical

---

[31] Of course there are different systems of mathematics and logic, even ones in which certain obvious-looking classical logical truths are given up (e.g., paraconsistence logics). Examples of "exotic" mathematical practices and other relevant cases abound in Wittgenstein's philosophy of mathematics and his rule-following considerations (see Wittgenstein 1953; 1978), from which I do not draw any Kripkean skeptical conclusions, however (cf. Putnam 2001).

expressions very differently from the way we use them, certain truths we take to be true, including 2+2=4, would be true – even in a world populated entirely by such non-standard mathematicians. Turn now to the ethical case. Similarly, even if everyone in the human community were converted to Nazism, we might – that is, *we* who are now arguing about the matter – resist considering Nazism a moral truth. We might, analogously to the mathematical case, say that even if all people became Nazis, Nazism would still be absolutely morally wrong, just as 2+2=5 would be false even if everyone believed it to be true.

What, if any, is the relevant difference between these two cases? Why not adopt a similar pragmatic-cum-constructive realism in both areas? Why be a relativist about morality but not about mathematics? In both cases, we are dealing with human-made constructs, albeit quite different ones. What is the point of, or motivation for, treating them differently in metaphysics?

Morton White (1986), a somewhat neglected neopragmatist, argued, extending W. V. Quine's famous "indispensability argument" for the existence of mathematical entities, that values should be acknowledged along with numbers and that the Quinean holistic empiricist should reject the fact/value dichotomy. That is, both factual beliefs and normative commitments should be incorporated in the holistic web of belief that, in Quine's memorable phrase, meets the "tribunal of experience" as a corporate body, on its edges. (Cf. further White 2002; Pihlström 2003b.) This idea has also been developed by Putnam, who has been crucially influenced by both Quine and White (see especially Putnam 1994, pp. 153–154, 503–504). Putnam draws an analogy between mathematics and ethics: no mysterious "objects" need to be postulated in order to account for the objectivity of the truths we commit ourselves to in these two different areas of human thought, but both factual and evaluative beliefs are involved in our attempts to cope with the world and our experiences of it (cf. Putnam 2002; 2004). Our mathematical and ethical practices are such that we are equally committed to objectivity in both; this commitment does not require any queer entities as its metaphysical ground. In both areas, and everywhere else as well, we should be satisfied with "objectivity humanly speaking". This corresponds to the semi-autonomous or relatively autonomous character of World 3. No metaphysical super-autonomy or super-objectivity is needed – or possible for us.

However, our acceptance of this Putnamean point should not lead us to endorse Putnam's thesis that ontology is to be given up altogether. Indeed, one of Niiniluoto's (2007) complaints against Putnam is that Putnam speaks about the relation between facts and values in terms of judgments and terms, thus following his teacher Rudolf Carnap in employing a "formal" as distinguished from a "material" mode of speech. According to Niiniluoto, the view that values are relative to human practices is naturally analyzable as the view that values exist as World 3 entities (in the sense explained above), although there are – contrary to what Putnam and other pragmatists typically claim – also World 1 entities ontologically independent of practices and culture. While I do not share Niiniluoto's strong realism and materialism asserting the primacy of World 1 over World 3 (as should have become clear by now), I do share his defense of ontology. I also believe that Putnam goes too far in his repudiation of ontological questions about value, about mathematical structures – and about everything else (cf. Pihlström 2006). Even so, the parallel between the ethical and the mathematical cases should be taken seriously, without sacrificing the very real differences between these human practices.

6. Putnam's Moral Realism

The above reflections lead me to a more detailed comparison between Niiniluoto's and Putnam's views. I have earlier critically compared these two thinkers' positions in the general debates over realism and truth (see Pihlström 1996, ch. 4; 1998, ch. 3), and I now have a chance to continue the comparison in the field of moral realism and value theory.

Putnam rejected the fact/value dichotomy already decades ago (see Putnam 1981; 1990; 1994), but he has returned to the topic in his more recent books (Putnam 2002; 2004).[32] Putnam's argument, as I read it, is both pragmatist and transcendental (see Pihlström 1996, ch. 5; 2003a, ch. 7; 2005, ch. 2). He tells us that there can be no (empirical) world without values at all, because the notions of fact and rationality are dependent on each other, and the notion of rationality is a value-involving one. Our talk about facts involves and requires

---

[32] Another (neo)pragmatist, Sandra Rosenthal (1986, p. 176), also rejects the "fact/value distinction" – but does she really want to abandon the *distinction*, or only, as Putnam (2002) does, the corresponding unpragmatic *dichotomy*? It should be obvious that in many contexts, for good pragmatic reasons, we must make a distinction between facts and values; drawing a clear dichotomy between them is much more problematic.

criteria of relevance, which in turn involve "all our values" (Putnam 1981, p. 201). As there is no hope in trying to reduce or eliminate normative notions such as rationality and justification, or semantic notions, such as truth and reference, from our scientific picture of the world – the very picture that some philosophers skeptical of objective values, including Mackie (1977), claim to support their moral nihilism – we should simply accept, as a pragmatic postulate, the reality of both epistemic and ethical (and other) values. If we choose to give up moral values, then we should give up their "companions in the guilt", too, namely, the epistemic values we simply cannot give up without giving up the scientific conception of the world which was supposed to make values ontologically problematic in the first place. (Cf. Putnam 1981, pp. 206–211; 1990, pp. 37, 140–141.) In brief, our life within both epistemic and ethical fields of commitment is shot through with values. There is no way of going on living a human life without value commitments. Thus, facts and values are indeed inseparable and mixed up. There are no fact-independent values nor any value-independent facts. We may arrive at this position through a transcendental-cum-pragmatic consideration of the ways in which our actual (moral) practices of evaluation and deliberation are arranged, of the ways in which we actually commit ourselves to objectivity in these areas of life.[33]

---

[33] Recent studies (in addition to Putnam's) in pragmatist ethics, defending (qualified) moral objectivity, include Fesmire (2003), Lekan (2003), and Misak (2000; 2004, ch. 5; 2006). Fesmire (2003) emphasizes, among other things, "empathetic projection" in Dewey's theory of imagination and argues against scientistic readings of Dewey's notion of "dramatic rehearsal"; his discussion of the socio-historical contextuality of moral character, beliefs, reasoning, and imagination fits well the pragmatic moral realist's account of values as World 3 entities. Lekan, in turn, appeals to the Deweyan concept of habit and sees pragmatist metaethics as rejecting the theory vs. practice dichotomy, urging that "practice is primary"; this leads to a *"reconstructive revisionism* of moral beliefs" (Lekan 2003, p. 3) and to an investigation of the "contingent sources of morality in human practices" (ibid., p. 9). A qualified kind of moral objectivity can be defended on Deweyan grounds, because our norm-governed practices always transcend individual lives (ibid., pp. 137–138). Misak, in turn, has in a series of writings returned to Peirce rather than Dewey in her project of defending cognitivist metaethics, emphasizing the truth-aptness of moral discourse, insofar as the concept of truth is understood along Peircean pragmatic lines. These views are in some ways close to the pragmatic moral realism I articulate in Pihlström (2005), but there are a number of divergences I cannot take up here. For further recent discussions of pragmatism about values and morality, see, e.g., the essays collected in the proceedings volumes of the

In a recent paper (in Finnish), Niiniluoto (2007) again argues that facts and values should be kept ontologically, semantically, and epistemologically distinct, in contrast to what Putnam has suggested. As in his (1999) book, he proposes a moderate value relativism and constructivism, placing values in World 3 and relativizing value statements in the manner explicated above – resisting, however, radically constructivist attempts to relativize World 1 facts to World 3 cultural constructions. Nonetheless, his "moderate" relativization presupposes that a value statement can be divided into a factual and an evaluative element – and this is the move attacked at length by Putnam (1981; 2002; 2004). Such a division is impossible, if our concepts of fact and factuality already presuppose values. Niiniluoto (2007) further argues that facts and values are both dependent on human practices *only* if we mean by "facts" World 3 facts; if we extend the concept of a fact to cover the facts belonging to World 1, as a realist like Niiniluoto of course urges us to do, then we must admit that factual statements about World 1, unlike "absolutely" formulated (non-relativized) value statements, have an objective, non-relativized truth value – and that they are in this sense true or false objectively, independently of us or our social constructions. Now, this leads us back to the more general discussion of realism, viz., whether there is a sense in which World 1 facts and entities can exist independently of World 3. I have considered this issue at length elsewhere (e.g., Pihlström 1996; 2002b), but I will say a few more words on it below in section 7.[34]

Central European Pragmatist Forums (Ryder & Višnovský 2004, Ryder & Wilkoszewska 2004, Ryder & Wegmarshaus 2007).

[34] In his recent paper, Niiniluoto (2007) also criticizes the "absolute" views on moral duty that I have recently defended (in some Finnish writings of mine; see also Pihlström 2005). He says that my position violates the pragmatist "symmetry thesis", according to which both facts and values are in a similar manner relative to human practices (and therefore, as Putnam suggests, we should be as realistic about values as we are about facts, albeit *not* metaphysical realists in either area). Niiniluoto claims that the absoluteness of morality and "metaphysical guilt" (cf. Pihlström 2007) presuppose a "God's-Eye View", which is inconsistent with pragmatism. I do not find this criticism convincing, however. In my view, guilt is something we perceive, or *ought to* perceive, in our lives *from the point of view of our subjectively led lives themselves*, "from within" our commitment to morality, a commitment constituted by our capacity for experiencing guilt. My notion of guilt is, then, closer to, say, Kierkegaard's than to any absolute (e.g., traditional Christian) moralists'.

Meanwhile, let me note another disagreement, which is related to the way in which Niiniluoto draws the fact/value distinction *within* World 3 itself. Not only are the values existing (culture-relatively) in World 3 distinguishable from the facts of World 1; in addition, Niiniluoto (2007) tells us that while World 3 is value-laden in the sense of being constituted and maintained through human practices, its objects (e.g., artifacts) have both factual and cultural properties, which can be distinguished from each other. For example, technology is value-laden, because it is always used for certain human purposes; yet, there are purely factual questions we may ask about technological instruments (e.g., "what is the probability of a nuclear accident?"), to be answered by technological experts. In a morally neutral sense, a "good" medicine serves its purpose by curing an illness, while a "good" weapon serves *its* purpose by being an efficient killer. The technical, factual properties of these World 3 entities can – and should, according to Niiniluoto – be distinguished from the ethical evaluation of the (also World 3) purposes for which they are used.

Again, the plausibility of this position depends on the possibility of drawing a clean dichotomy between Worlds 1 and 3. If the very ontology of World 1 entities and properties is partly constituted through the conceptual classifications made in World 3, as I suggest, then the argument hardly goes through. For the kind of pragmatist (like Putnam) who opposes the fact/value dichotomy and, to the contrary, insists on the entanglement of facts and values, the analogous entanglement of Worlds 1 and 3 is a natural view to hold. The "factual" properties of technological inventions can hardly be identified entirely independently of the cultural context in which those technological constructs are designed, produced, and used for human purposes. These purposes guide the identification of the very properties that Niiniluoto describes as purely factual (and thus as distinguishable from "cultural" value-properties).

The pragmatic moral realist, moreover, can admit that the "objectivity" of moral values (and other normative structures belonging to World 3) is not to be assimilated to the objectivity of physical facts. There are, the pragmatist insists, many different kinds of objectivity, depending on the differences between our human practices. Moral values and duties are both personally binding and objective (that is, not subjective or simply "relative"), though of course they are not objective in the sense in which sticks and stones or electrons are objective. Objectivity, in brief, may be seen as a Wittgensteinian family-resemblance notion. There is no essence of objectivity – and bringing this insight to the fore may be among the most significant meta-level achievements of pragmatism and the pragmatic and pluralist approach to the moral realism discussion. In this sense, our discussion up to this point is not merely relevant to moral philosophy; the problem of values leads us to the heart of more general philosophical issues of realism, normativity, objectivity, and so on.

Yet, Niiniluoto might claim that the "identification" of entities or properties is not the main issue here (cf. Niiniluoto 1999, ch. 6). It is one thing to exist; it is quite another thing to be identifiable by human conceptual means. However, for a pragmatist even this *might* simply be a distinction without difference – although I cannot adequately argue for this view in the present essay. To exist is to play a role in humanly relevant practices, to (possibly, conceivably) make a difference to human conduct. Existence requires potential pragmatic relevance, when ontology is thoroughly pragmatized.

7. Moral Realism and Scientific Realism

My lengthy groundwork has been done, and I now finally arrive at the way I want to state the main argument of this paper. What I am proposing is that moral realism, pragmatically articulated, is a *more fundamental* form of realism than Niiniluoto's favorite version of realism, critical scientific realism, even if the latter contains ontological realism, the postulation of a mind- and language-independent world, as one of its main theses.[35]

I do not merely have in mind the relatively uncontroversial (and of course plausible) view that science itself is an ethically structured human practice and that the scientific pursuit of truth is a normative project striving for an ethical ideal, having moral and political implications. In addition to defending the irreducible normativity of truth-seeking and the various ethical virtues it includes, such as honesty or openness to dialogue, I wish to formulate the following kind of a *transcendental argument*: In order to defend scientific realism – in order to hold that science is, or aims to be, about real entities and

---

[35] For different versions of realism (and the problem of realism), including ontological, epistemological, semantic, methodological, and axiological realisms (and their alternatives), see the classifications in Niiniluoto (1999), ch. 1; Pihlström (1996), ch. 2.

processes and that we can reasonably test our scientific theories by empirical means – we have to be committed to epistemic values, such as truth, information, consistency, simplicity, and the like. Now, arguably, there is no clear way to draw a sharp dichotomy between epistemic and non-epistemic (ethical and aesthetic) values.[36] Thus, if we wish to defend scientific realism – in Niiniluoto's manner or, *mutatis mutandis*, in some other – then we ought to find certain values basic and irreducible in our normative system of commitments, more basic than the (allegedly) "pure facts" that our value-governed scientific theories claim to obtain in the world. Moreover, the facts investigated by science cannot be strictly separated from the values through which we see (view, interpret, construe) "the facts" we think our science is about. *All* facts, whether scientifically investigated or not, will inevitably be "colored" by values, for any human investigator; just as no object is colorless for the human sight, no object, event, or situation is value-indifferent for a human spectator and thinker.[37]

The last sentence might evoke fierce criticism. Of course there are value-indifferent objects and situations. Consider a stone lying deep on the bottom of an ocean. Such a stone is hardly valuable for us in any plausible sense. Yet, the very possibility of identifying a stone as a stone (and not, say, as a collection of molecules) already implies a whole system of value commitments. A comprehensive conception of humanly relevant values, of "human flourishing" – to borrow Putnam's (1981) Aristotelian way of speaking – must be in place in order for us to find it relevant to distinguish stones from non-stones and to find it relevant to place such entities in spatiotemporally described contexts

---

[36] This is not to say that no distinction at all can be drawn between these different kinds of value. As Putnam (2002), ch. 1, reminds us, to reject a dichotomy or a dualism is not to reject a corresponding (harmless) distinction. Certainly there are clear cases of epistemic values, but they are to be found (only) in a context in which all kinds of ethical (and other) values are already in place.

[37] Yet, I must, even when defending this position through a transcendental argument (as in Pihlström 2003a, ch. 7; see also Pihlström 2005; 2007), carefully distinguish my views from those of more strongly transcendental moral thinkers, such as Karl-Otto Apel and philosophers influenced by him (cf. Illies 2003). My argument is transcendental but does not claim apodictic certainty for its conclusion. Moreover, McDowell's (1996) moral realism and, more generally, his realism about the irreducibly normative "second nature" (cf. also Thornton 2004, ch. 2), is obviously close to the position I hope to defend (as noted already in Pihlström 1996, ch. 5).

(e.g., an ocean, its bottom, etc.). Of course, the particular stone I mentioned may be irrelevant to us, but it is a metaphysical prejudice to claim that it is somehow valuationally irrelevant or indifferent *per se*, that is, that the stone itself metaphysically contains no values. This is either trivially true (because all valuation requires human subjectivity and practices, i.e., because the stone, obviously, possesses no mystical, queer, value qualities that we might simply discover in it in the way we may measure its weight, for instance), or false (because when placed in a humanly relevant context of conceptualization, it does contain, or relate to, all kinds of human values without which no practice of conceptualizing would be possible for us).

Whether or not there are in some sense facts "out there" entirely independently of us, and thus independently of our "coloring" of them with values, is the more general issue of realism that I cannot hope to resolve here. For a pragmatist, it is, however, natural to arrive at a position resembling Kantian *transcendental idealism*: the factual world is, to be sure, "empirically real", but it is – at a transcendental level – constituted (and constantly reconstituted) by our habitual practices and their normative commitments, including the moral values we subscribe to when engaging in our practices. Another way to put this is to say that, at the empirical and factual level, World 1 entities and facts are, obviously, independent of our beliefs, theories, concepts, and perceptions, as any reasonable realist ought to admit, while at a transcendental level of philosophical reflection, which even a pragmatist ontologist should invoke, World 1 entities and facts (as well as, in principle, World 2) receive their ontological structure through World 3 interpretations and conceptualizations. In this sense, they do not and could not exist in the full sense of "exist" independently of their transcendental conditions – which, however, should not be sought (in Kant's own manner) in the a priori capacities of the human mind but in the evolving, worldly (and thus also World-1-involving, i.e., material) human practices of conceptualization and evaluation.[38] In this sense, and in this

---

[38] Putnam (1990) says that *if* he were a metaphysician, he would be willing to claim not that (as Berkeley or Mill would have it) to exist is to be a permanent possibility of sensation but that to exist is to be a permanent possibility of valuation. Still, the claim that anything there is in the (humanly relevant) world can, contingently, be ethically evaluated must be distinguished from the stronger (transcendental) claim that, necessarily, there is nothing in the (humanly relevant) world that could exist independently of being (possibly) evaluated.

sense only, facts – even World 1 facts – are "dependent" on values. Facts and values, to borrow one more term from Putnam, "interpenetrate", reciprocally defining and containing (and being defined by and contained in) each other.[39]

The basic structure of my argument for the priority of (pragmatic) moral realism to scientific realism can now be somewhat loosely formulated as follows:

1. Scientific realism is true/acceptable as a philosophical interpretation of scientific theories (assumption).[40]
2. In order to be a scientific realist, one must acknowledge the (historically changing, contingent) grounding of the scientific pursuit of truth and knowledge, and even of scientific facts and truths themselves, in (material, cultural) human practices – that is, in Popper's terms, World 3.[41]

---

[39] To avoid the tricky issue of how value, or normativity in general, is generated out of thoroughly natural processes devoid of value and normativity, one might once again invoke the "panculturalist" analogy to panpsychism mentioned in note 20 above. According to such a view, value, or normativity, has – from our human perspective, which is the only one we have – been there all along, because anything *we* are able to regard as a fact (or as a thing of any kind) has always already been placed in an evaluative context of potential moral relevance. Besides having affinities with pragmatism and Kantian transcendental idealism, the panculturalist position might come close to Hans Lenk's "interpretationist" constructivism, which combines the thesis that all entities (*Gegenstände*) are (quasi-transcendental) interpretational constructs with the pragmatic realist thesis stating that we need, again as an interpretational construct, commit ourselves to realism for pragmatic reasons (see Lenk 1995; 2003).

[40] What scientific realism more specifically amounts to is open to dispute (see Niiniluoto 1999), but here we may simply characterize it as the view that scientific theories have truth values and purport to describe a theory-independent world, postulating theoretical entities for explanatory purposes, and that science attempts to discover truths about such a world. This characterization does not, as such, tell us what truth is; scientific realism *can*, for instance, be combined with a broadly pragmatist understanding of truth.

[41] For a comprehensive discussion of the notion of a scientific practice relevant here, see Rouse (2002).

3. These practices are not (narrowly) scientific ones but more general human practices, to be characterized in terms of our pre-scientific "lifeworld" experiences and habits of action.[42]
4. In order to be able to engage in the practices necessary for the possibility of science as a pursuit of truth (2, 3), one must be *committed* to their normative structures or values.
5. These values are *not* purely epistemic but, as structural elements constitutive of our lifeworld, involve (in Putnam's terms) "all our values", "our whole system of value commitments".
6. In order to be a scientific realist, one must be committed to such a holistic set of lifeworld-structuring values, including moral values (4, 5).
7. This kind of commitment can only be adequately understood if it is understood realistically, as a commitment to values (and norms) independent of contingent subjective valuations or preferences.[43]
8. Therefore, a broader pragmatic realism about values and normativity in general (including, in particular, a pragmatic moral realism), is prior to, and more basic than, any scientific realism. (However, values are *not* transcendent, other-worldly entities. This anti-Platonist qualification is based on a general naturalist attitude.)
9. Therefore, pragmatic moral realism is true/acceptable (1, 8).

In short, pragmatic moral realism functions as a ground for any reasonable scientific realism, analogously to the way in which transcendental idealism, according to Kant, provides the context in which (only) empirical realism can be saved. Here, of course, I do not expect Niiniluoto or any other contemporary realist to agree with me.[44] I have, however, argued at length for a

---

[42] See Pihlström (2003a), ch. 5. We could, invoking Husserlian phenomenology, even speak of a transcendental a priori grounding of science in a pre-scientific lifeworld (*Lebenswelt*), but I will not adopt this terminology here, because I want to insist that only a pragmatic, non-foundationalist "grounding" is needed. Another natural vocabulary one might adopt here is the Wittgensteinian one of "forms of life".

[43] This premise, in turn, may require a supporting (pragmatic) theory of what moral commitment is (cf. Pihlström 2005).

[44] In particular, I am not claiming that the "truth" of (pragmatic) moral realism could itself be correspondence-theoretically analyzed. For a pragmatist, pragmatic moral realism is itself a pragmatic truth, a postulation required for our human purposes. Thus, my argument for the priority of moral realism to scientific realism is itself a pragmatic

moderate form of realism – moderate enough to be reconcilable with both pragmatism and transcendental idealism, and/or their combination – and a transcendental reading of pragmatism elsewhere (see Pihlström 1996; 1998; 2003a), and I believe my case for pragmatism (indeed, for a "transcendental pragmatism") can be used to support the case for a pragmatic realism about cultural entities, such as values, that I have developed here (as well as the case for a pragmatic realism about the kind of physical World 1 theoretical entities that natural science is concerned with).

8. Conclusion

Ilkka Niiniluoto is right to defend an ontologically anti-reductionist picture of Worlds 1, 2, and 3, although one surely need not formulate such a picture in these Popperian terms. Thus, I am happy to join him in the attack against reductive naturalists and physicalists who too easily give up the ontological status of World 3 (and even World 2) entities. I also join him in his defense of serious ontological reflection: I do believe, *pace* Putnam, that the philosophical issue of values deserves not merely semantic or epistemic but also ontological attention, though I also think that these fields of philosophical reflection are inseparable from, and mutually presuppose, each other.

However, I find Niiniluoto's position less convincing when it comes to moral realism and relativism. I have argued that it is a mistake to reject moral realism in the way he does – a way shared by several materialistically and realistically oriented contemporary thinkers who find science ontologically prior to other human practices. Niiniluoto's (and many others') view is unstable here, and a pragmatic critique such as the one I have provided may highlight this point.

Let me note, in closing, that I have not claimed that values are among the final ontological inventory of World 3. Such metaphysical claims would be contrary to my basic pragmatist point of leaving the ultimate ontology of reality in itself untouched. What I have claimed is that *we* – whoever we are – pragmatically need to postulate such cultural entities as values, and that we need

---

argument in favor of a certain kind of priority-order of some of our basic (yet fallible) philosophical postulations, not a demonstration of the metaphysical order of priority of some eternal philosophical truths.

to postulate them as equally objective as we consider scientific or common-sense facts to be. World 3, humanly created and maintained, is never finished; its population is never decided once and for all. Our philosophical disagreements themselves may continuously affect its ontological structure. This I consider a most pragmatic attitude to ontological theorizing – about values or about anything else we need to populate our world with.[45]

University of Helsinki
University of Tampere
University of Jyväskylä

References

Armstrong, D. M. 2004: *Truth and Truthmakers*, Cambridge University Press, Cambridge.
Boyd, R. 1988: "How to Be a Moral Realist", in Sayre-McCord (1988), pp. 181–228.
Bunge, M. 1981: *Scientific Materialism*, D. Reidel, Dordrecht.
Bunge, M. 1989: *Ethics: The Good and the Right*, D. Reidel, Dordrecht.
de Caro, M. & D. Macarthur (eds.) 2004: *Naturalim in Question*, Harvard University Press, Cambridge, MA & London.
Clarke, D. S. (ed.) 2004: *Panpsychism: Past and Recent Selected Readings*, SUNY Press, Albany.
Cohen, L. J. 1985: "Third World Epistemology", in G. Currie & A. Musgrave (eds.), *Popper and the Human Sciences*, Martinus Nijhoff, Dordrecht, pp. 1–12.
Dewey, J. 1929: *Experience and Nature*, 2nd ed., Open Court, La Salle, IL, 1986 (1st ed. 1925).
El-Hani, C.N. & S. Pihlström 2002: "Emergence Theories and Pragmatic Realism", *Essays in Philosophy* 3:2, www.humboldt.edu/~essays.
Ellis, B. 1990: *Truth and Objectivity*, Blackwell, Oxford.
Elshtain, J. B. 2003: "Don't Be Cruel: Reflections on Rortyian Liberalism", in Guignon & Hiley (2003), pp. 139–157.
Fesmire, S. 2003: *John Dewey and Moral Imagination: Pragmatism in Ethics*, Indiana University Press, Indianapolis.
de Gaynesford, M. 2004: *John McDowell*, Polity Press, Cambridge.
Guignon, C. & D. R. Hiley (eds.) 2003: *Richard Rorty*, Cambridge University Press, Cambridge.

---

[45] In addition to Ilkka Niiniluoto, I would like to thank the participants of the NOS-H workshop, *Reason, Emotion, and Evaluation* (Bifröst University College, Iceland, June 2005), at which this material was partly presented – especially Marika Enwald, Leila Haaparanta, Jón Olafsson, Håkan Salwén, and Dan Zahavi.

Hasker, W. 1999: *The Emergent Self*, Cornell University Press, Ithaca, NY & London.
Illies, C. 2003: *The Grounds of Ethical Judgement: New Transcendental Arguments in Moral Philosophy*, Clarendon Press, Oxford.
Kelly, E. I. 2004: "Against Naturalism in Ethics", in de Caro & Macarthur (2004), pp. 259-274.
Kim, J. 1998: *Mind in a Physical World: An Essay on Downward Causation and the Mind-Body Problem*, The MIT Press, Cambridge, MA & London.
Korsgaard, C. M. 1996: *Creating the Kingdom of Ends*, Cambridge University Press, Cambridge.
Kotkavirta, J. & M. Quante (eds.) 2004: *Moral Realism*, Acta Philosophica Fennica 76, The Philosophical Society of Finland, Helsinki.
Lekan, T. 2003: *Making Morality: Pragmatist Reconstruction in Ethical Theory*, Vanderbilt University Press, Nashville, TN.
Lenk, H. 1995: *Interpretation und Realität*, Suhrkamp, Frankfurt am Main.
Lenk, H. 2003: *Grasping Reality*, World Scientific, Singapore.
Mackie, J. L. 1977: *Ethics: Inventing Right and Wrong*, Penguin, Harmondsworth.
Margolis, J. 1978: *Persons and Minds: The Prospects of Non-Reductive Naturalism*, D. Reidel, Dordrecht.
Margolis, J. 1980: *Art & Philosophy: Conceptual Issues in Aesthetics*, The Harvester Press, Brighton.
Margolis, J. 1984: *Culture and Cultural Entities: Toward a New Unity of Science*, D. Reidel, Dordrecht.
Margolis, J. 1995: *Historied Thought, Constructed World: A Conceptual Primer for the Turn of the Millennium*, University of California Press, Berkeley.
Margolis, J. 2002: *The Rediscovery of Pragmatism: American Philosophy at the Turn of the Millennium*, Cornell University Press, Ithaca, NY & London.
Margolis, J. 2003: *The Unraveling of Scientism: American Philosophy at the Turn of the Millennium*, Cornell University Press, Ithaca, NY & London.
McDowell, J. 1996: *Mind and World*, 2nd ed., Harvard University Press, Cambridge, MA & London (1st ed. 1994).
Misak, C. 2000: *Truth, Politics, Morality: Pragmatism and Deliberation*, Routledge, London & New York.
Misak, C. 2004: *Truth and the End of Inquiry*, rev. ed., Oxford University Press, Oxford (1st ed. 1991).
Misak, C. 2006: "Truth in Science and Ethics", in H. J. Koskinen, S. Pihlström & R. Vilkko (eds.), *Science – A Challenge to Philosophy?*, Peter Lang, Frankfurt am Main, pp. 277-291.
Niiniluoto, I. 1981: "Language, Norms, and Truth", in I. Pörn (ed.), *Essays in Philosophical Analysis*, Acta Philosophica Fennica 32, The Philosophical Society of Finland, Helsinki, pp. 168-189.
Niiniluoto, I. 1984: *Is Science Progressive?*, D. Reidel, Dordrecht.
Niiniluoto, I. 1990: *Maailma, minä ja kulttuuri: Emergentin materialismin näkökulma* [World, Self, and Culture: The Point of View of Emergent Materialism], Otava, Helsinki.

Niiniluoto, I. 1992: "Reality, Truth, and Confirmation in Mathematics: Reflections on the Quasi-empiricist Programme", in J. Echeverria, A. Ibarra & T. Mormann (eds.), *Space of Mathematics*, de Gruyter, Berlin, pp. 60–78.
Niiniluoto, I. 1994: "Scientific Realism and the Problem of Consciousness", in M. Kamppinen & A. Revonsuo (eds.), *Consciousness in Philosophy and Cognitive Neuroscience*, Lawrence Erlbaum Associates, Hillsdale, pp. 33–54.
Niiniluoto, I. 1999: *Critical Scientific Realism*, Oxford University Press, Oxford & New York.
Niiniluoto, I. 2007: "Arvot ja tosiasiat – samaa vai eri paria?" [Facts and Values – Same or Different?], forthcoming in E. Kilpinen, O. Kivinen & S. Pihlström (eds.), *Pragmatismi filosofiassa ja yhteiskuntatieteissä* [Pragmatism in Philosophy and the Social Sciences], Gaudeamus, Helsinki.
O'Hear, A. 1980: *Karl Popper*, Routledge & Kegan Paul, London.
Pihlström, S. 1996: *Structuring the World: The Issue of Realism and the Nature of Ontological Problems in Classical and Contemporary Pragmatism*, Acta Philosophica Fennica 59, The Philosophical Society of Finland, Helsinki.
Pihlström, S. 1998: *Pragmatism and Philosophical Anthropology: Understanding Our Human Life in a Human World*, Peter Lang, New York.
Pihlström, S. 2002a: "The Re-Emergence of the Emergence Debate", *Principia* 6:1, 133–181 (special issue on emergence and downward causation, ed. C. N. El-Hani).
Pihlström, S. 2002b: "Contextualization, Conceptual Relativity, and Ontological Commitment", *Human Affairs* 12:1, 26–52.
Pihlström, S. 2003a: *Naturalizing the Transcendental: A Pragmatic View*, Prometheus/ Humanity Books, Amherst, NY.
Pihlström, S. 2003b: Review of White (2002), *Transactions of the Charles S. Peirce Society* 39, 305–313.
Pihlström, S. 2004: "Metaphysics without Methology? A Pragmatic Critique", *Philosophy Today* 48, 185–221.
Pihlström, S. 2005: *Pragmatic Moral Realism: A Transcendental Defense*, Rodopi, New York & Amsterdam.
Pihlström, S. 2006: "Putnam's Conception of Ontology", *Contemporary Pragmatism* 3.
Pihlström, S. 2007: "Transcendental Guilt: On an Emotional Condition of Moral Experience", *Journal of Religious Ethics* 35.
Popper, K. R. 1972: *Objective Knowledge: An Evolutionary Approach*, Clarendon Press, Oxford.
Popper, K. R. 1974: "Replies to My Critics", in Schilpp (1974), pp. 961–1197.
Popper, K. R. 1980: "Three Worlds", in S. M. McMurrin (ed.), *The Tanner Lectures on Human Values I*, University of Utah Press, Salt Lake City, and Cambridge University Press, Cambridge, pp. 141–167.
Popper, K. R. 1984: *Auf dem Suche nach einer besseren Welt*, Piper, München.
Popper, K. R. 1992: *Unended Quest: An Intellectual Autobiography*, rev. ed., Routledge, London (1st ed. in Schilpp 1974).

Popper, K. R. & J. C. Eccles 1977: *The Self and Its Brain*, Routledge & Kegan Paul, London, 1986.

Putnam, H. 1981: *Reason, Truth and History*, Cambridge University Press, Cambridge.

Putnam, H. 1990: *Realism with a Human Face*, ed. J. Conant, Harvard University Press, Cambridge, MA & London.

Putnam, H. 1994: *Words and Life*, ed. J. Conant, Harvard University Press, Cambridge, MA & London.

Putnam, H. 2001: "Was Wittgenstein *Really* an Anti-Realist in the Philosophy of Mathematics?", in T. McCarthy & S. C. Stidd (eds.), *Wittgenstein in America*, Clarendon Press, Oxford, pp. 140–194.

Putnam, H. 2002: *The Collapse of the Fact/Value Dichotomy and Other Essays*, Harvard University Press, Cambridge, MA & London.

Putnam, H. 2004: *Ethics without Ontology*, Harvard University Press, Cambridge, MA & London.

Rorty, R. 1998: *Truth and Progress*, Cambridge University Press, Cambridge.

Rosenthal, S. B. 1986: *Speculative Pragmatism*, The University of Massachusetts Press, Amherst.

Rouse, J. 2002: *How Scientific Practices Matter: Reclaiming Philosophical Naturalism*, The University of Chicago Press, Chicago & London.

Ryder, J. & E. Višnovský (eds.) 2004: *Pragmatism and Values*, Rodopi, Amsterdam & New York.

Ryder, J. & K. Wilkoszewska (eds.) 2004: *Deconstruction and Reconstruction*, Rodopi, Amsterdam & New York.

Ryder, J. & G.-R. Wegmarshaus (eds.) 2007: *Education for a Democratic Society*, Rodopi, Amsterdam & New York.

Sahavirta, H. 2006: *Karl Popper's Philosophy of Science – and the Evolution of the Popperian Worlds*, Ph.D. Diss., Philosophical Studies from the University of Helsinki, vol. 12, Department of Philosophy, University of Helsinki.

Sayre-McCord, G. (ed.) 1988: *Essays in Moral Realism*, Cornell University Press, Ithaca, NY & London.

Schatzki, T. R. 2002: *The Site of the Social*, Pennsylvania State University Press, University Park & London.

Schilpp, P. A. (ed.) 1974: *The Philosophy of Karl Popper*, vols I–II, Open Court, La Salle, IL.

Searle, J. 1995: *The Construction of Social Reality*, The Free Press, New York.

Shafer-Landau, R. 2003: *Moral Realism: A Defence*, Oxford University Press, Oxford, 2005.

Sihvola, J. 2004: "Aristotle and Modern Moral Realism", in Kotkavirta & Quante (2004), pp. 201–229.

Stephan, A. 1999: *Emergenz: Von der Unvorhersagbarkeit zur Selbstorganisation*, Dresden University Press, Dresden & München.

Thornton, T. 2004: *John McDowell*, Acumen, Chesham.

Tuomela, R. 2002: *The Philosophy of Social Practices: A Collective Acceptance View*, Cambridge University Press, Cambridge.
White, M. 1986: "Normative Ethics, Normative Epistemology, and Quine's Holism", in L. E. Hahn & P. A. Schilpp (eds.), *The Philosophy of W. V. Quine*, Open Court, La Salle, IL, pp. 649–662.
White, M. 2002: *A Philosophy of Culture: The Scope of Holistic Pragmatism*, Princeton University Press, Princeton, NJ.
Williams, B. 1985: *Ethics and the Limits of Philosophy*, Fontana, London.
Wittgenstein, L. 1953: *Philosophical Investigations*, trans. G. E. M. Anscombe, Blackwell, Oxford, 1958.
Wittgenstein, L. 1978: *Remarks on the Foundations of Mathematics*, trans. G. E. M. Anscombe, Blackwell, Oxford.

Juha Sihvola

Religion within the Limits of Rational Acceptability

The Thomist philosopher Sir Anthony Kenny has argued in his book *Faith and Reason* for a view according to which the rational acceptability of beliefs requires proper justification (1983, pp. 25-46). The way of justification depends on the nature of the beliefs in question. It is rational to accept a belief if it can be shown to basic, i.e., self-evident or fundamental, evident to the senses or to memory, or defensible by argument, inquiry, or performance, or, in the case of non-basic beliefs, if it can be articulated how they can be based on the evidence of the basic beliefs. Basic beliefs are accepted without articulated evidence, but even in this group there is a large number of beliefs in regard of which it is in principle possible to provide evidence if a defense is needed. Those beliefs in regard of which there is no such evidence have to be shown to be either self-evident (such as *a priori* propositions in logic) or fundamental in the sense that giving up them would obviously cause chaos in our noetic structure. As to the belief in the existence of God, Kenny argues that it is rationally acceptable if and only if evidential support can be provided for it through proofs in natural theology.

    Kenny's view implies that, with a few exceptions, the rational acceptability of beliefs requires evidential support, and even in regard of those exceptions, some other kind of justification. Kenny expresses a commitment to an evidentialist attitude in epistemic issues, a commitment that, I believe, is also shared by Ilkka Niiniluoto, although their conclusions on religious faith are very different. Evidentialists are suspicious about importing any elements into our belief-systems without subjecting them to a careful argumentative or otherwise intersubjective scrutiny. Evidentialist conviction can also be understood as being in the background when Niiniluoto has criticized religion and defended atheism as a philosophical position in several of his writings.[1]

Unlike many of his fellow atheists, Niiniluoto is also well-versed in modern developments in the philosophy of religion. Despite congenial conversations with theologians and philosophers with a positive or neutral attitude to religion, his sympathies have not warmed towards what he has called the trendy currents in the philosophy of religion.[2] He has used this expression to refer to views according to which religious beliefs do not need any evidentialist support or, more radically, do not even have much to do with the ordinary sense of believing, i.e., holding something as true of the reality in which we live.

Niiniluoto has also criticized those philosophers of religion who have been inspired by Ludwig Wittgenstein's ideas. They have interpreted the language of religion in anti-realistic terms and claimed that it involves no assertions about ordinary reality. In Niiniluoto's view, Wittgensteinian anti-realism in the philosophy of religion leads to radical relativism that does not allow religions to be criticized from the outside. In the following, I shall take up Niiniluoto's challenge and argue that certain versions of anti-realistic theism can be acceptable positions to a rational person who is committed to a strong form of evidentialism with respect to beliefs about reality. I shall also argue that anti-realistic theism does not necessarily make actual religions immune to external criticism.

1. The Types of Theism, Atheism, and Agnosticism

I shall first clarify the terms I use and make some restrictions concerning the scope of my study. Religion is a vague notion of which it might be impossible to provide a definition that would cover all of its instances. I shall restrict my study to theistic religions, i.e., those that use the notion of God in their practices. God is almost as difficult a notion as religion. Paradigmatically, God can be assumed to be a personal supernatural being, considered in one way or another important to life and regarded as sacred and worthy of worship in one way or another. There are also religions and theological views that do not

---

[1] Most of Niiniluoto's writings on atheism and the philosophy of religion have been only published in Finnish, e.g., Niiniluoto (2003; 2005); see also (1999).
[2] See Niiniluoto (2003).

include an idea of a personal God. However, I shall not discuss these non-theistic religions.

The theistic thesis is simply the sentence "God exists". Those who accept the theistic thesis are theists. They believe that God exists. The theistic thesis can be denied in two ways. The atheists deny the theistic thesis and accept the contradictory atheistic thesis "God does not exist". So the atheists believe that God does not exist. The agnostics join the atheists in their denial of the theistic thesis, but they do not accept the atheistic thesis either. When they deny the theistic thesis, they only deny the epistemic attitude of acceptance that is directed to it. As to the substantial contents of the theistic and atheistic theses, the agnostics are indifferent. The agnostics do not believe that God exists, but they do not believe that God does not exist either. Strictly speaking, the atheists deny the existence of God, whereas the agnostics only reject the belief in the existence of God. The atheists' concern is ontological, whereas the agnostics are more interested in epistemology.

Theistic beliefs can be accepted either on the basis of evidence or without evidence. With regard to this difference, the upholders of the theistic thesis can be divided into evidentialists and fideists. Classical evidentialist theism, such as used in medieval proofs concerning the existence of God, claims that the existence of God can be conclusively proved by deductive arguments.[3] Modern evidentialist theism does not usually go as far, but its upholders think that confirmatory empirical evidence can be brought forward to support the belief in the existence of God.[4] This evidence is thought to make the existence of God probable. Therefore, the modern evidentialist theist holds that that a belief in the existence of God is an intellectual duty for a rational person, because the evidence for the theistic thesis is stronger than the evidence for its denial. Unlike the evidentialists, the fideists consider the theistic thesis acceptable, even though neither conclusive nor even confirmatory evidence can be found to

---

[3] The classical formulations of the arguments for the existence of God, the most famous of which were the cosmological, ontological, and teleological arguments were developed in mediaeval philosophical theology. They were to a large extend based on arguments found in the writings of ancient Greek philosophers. See especially Thomas Aquinas, *Summa theologiae*, 1.2.2–11; Anselm of Canterbury *Proslogicon* 2; Duns Scotus, *Reportatio* 1.2.1.3. See also Plantinga (1974); Craig (1980); Swinburne (1979).

[4] The best known proponent of modern evidentialist theism has been Richard Swinburne (1979).

support it. According to the fideist view, the theistic thesis is rationally acceptable without any reference to evidence.⁵

The contrary opposite of theism is atheism, the upholders of which believe that God does not exist. Atheism can also be divided into three types corresponding to the above classification of theism. The strongest version of evidentialist atheism is the view that it can be conclusively proved that God does not exist. The weaker versions of evidentialist atheism claim that there is confirmatory empirical evidence for the belief that God does not exist, or that the existence of God is so improbable in the light of the available evidence that it is both rational and intellectually obligatory to believe that God does not exist. There can also be fideist versions of atheism. The fideist atheist admits that the evidence we have does not support either the theistic thesis or the atheistic thesis. He thinks that it is rationally at least as acceptable to believe that God does not exist as it is to believe that God exists. However, as an atheist, he just decides to choose the former view.⁶

The line between evidentialist and fideist atheism is not quite clear. There is a version of atheism according to which the available evidence may not add to the probability of God's non-existence, but the acceptance of the atheistic thesis may still be regarded as being based, broadly speaking, on evidentialist reasoning. Niiniluoto (2003), for example, has argued that since there is absolutely no evidence to support the belief in the existence of God, the rational person is obliged to believe that God does not exist, since the burden of proof is on the side of those who contest the atheistic thesis. Atheism is here adopted on the basis of ontological commitment to the simplest possible universe and the avoidance of unnecessary existential assertions. If the basis of any existential assertion is required to be evidentialist, it is reasonable to think that none of those entities that exist in some possible world exist in the actual world if there is no evidence to support the corresponding existential claims.

---

⁵ On fideism, see Penelhum (1997).

⁶ In the definition of atheism, I follow Niiniluoto (2003), who restricts atheism to those who positively believe that God does not exists, and excludes agnostics. Many philosophers, however, support a more extensive definition according to which atheism includes all those who do not believe that God exists. On the different possibilities, see Edwards (1967); Martin (1990); Flew (1997).

There are also several varieties of agnosticism.[7] According to a standard view, the theistic thesis is possibly true and possibly false, but there is no evidence in either direction. In this kind of situation, the agnostic, unlike the proponent of the last version of atheism introduced above, remains neutral with respect to both the belief that God exists and the belief that God does not exist. Like the ancient Skeptics, this kind of agnostic simply suspends his judgment in regard of the existence of God.[8]

Another version of agnosticism is based on ideas developed in logical empiricism.[9] The logical empiricists did not hold the theistic thesis as false but meaningless, because even in principle there is no way to verify it as either true or false. The unverifiability of God's existence is not merely contingently caused by the unavailability of appropriate evidence. God is said to be supernatural, and all existential claims concerning supernatural entities are for conceptual reasons rendered unverifiable. In the language of logical empiricism, God cannot even be a possible being – only a conceptual contradiction. This may seem to be pretty close to atheism, but it should be noticed that the atheistic thesis has the same logical properties as its theistic correspondent. Since the logical empiricists do not regard the claim that God does not exist as verifiable, perhaps it is better to count them as agnostics rather than atheists.

2. Realistic and Anti-realistic Theism

The classifications above become more complicated if we add another distinction to that between evidentialism and fideism. Positions in the philosophy of religion can also be classified depending on how we understand the function of religious language. Some versions of evidentialism and fideism assume that the theistic thesis refers to a being that exists in the same reality as the beings referred to in ordinary languages and scientific theories. The word 'believes' in the sentence "X believes that God exists" means holding the theistic thesis as true in the same sense as we hold ordinary existential asser-

---

[7] The term agnosticism was introduced by Thomas Henry Huxley in 1869 (see Huxley 1893–1895), but its roots extend to pre-Socratic philosophy. The most famous of the ancient agnostics was Protagoras, see also Sihvola (2006).

[8] On the ancient Skeptics' philosophy of religion, see Knuuttila and Sihvola (2000).

[9] See especially Ayer (1936).

tions as true or false. I shall use the term *epistemic belief* of believing in this ordinary sense. Whether or not religious beliefs are based on evidence, they are considered to be integrated to ordinary and scientific beliefs to produce a unified rational conception of reality. I shall use the term *realistic theism* of these kinds of integration theories of religion.

There are, however, also views according to which religious language makes no assertions at all about the world referred to by ordinary languages and scientific theories. Perhaps the best known among these views are those inspired by Ludwig Wittgenstein's later philosophy and his scattered remarks about religion.[10] According to the Wittgensteinians, religious language may include beliefs, but in a different sense from ordinary and scientific ones. If one can even speak of the epistemic conditions of acceptance and denial in the context of religious beliefs, they have to be understood in a special way. Religious beliefs have to be ascribed an autonomous position in relation to ordinary and scientific beliefs. They are not thought of as being among the non-evidentially accepted basic beliefs belonging to our ordinary world-view, as in some fideist positions. In the autonomy theories of religion, religious assertions have a completely different function from that of providing an account of empirical external reality. Religious beliefs are not beliefs in the sense of holding something as true of the reality around us. There is no ontological commitment involved in them. I shall use the term *non-epistemic belief* of the religious beliefs understood in the Wittgensteinian way. If we accept this distinction, we can say that the language of religion only has a meaning in the practice of worship. The function of worship is thought to be to express the existential experiences and emotions of the religious believer. Religion is an autonomous practice that cannot be criticized from the outside, at least as far as it remains inside its own limits without presenting imperialistic claims about ordinary language, science, or morality. I shall use the term *anti-realistic theism* of the autonomy theories of religion.

The distinction between realistic and anti-realistic versions of theism is often confused with the distinction between evidentialism and fideism. These are, however, two different distinctions. It often happens that evidentialists are realistic theists and fideists anti-realistic theists, but this is not a necessity. Alvin Plantinga and some others among the so-called reformed epistemologists are

---

[10] See, e.g. Phillips (1993).

fideists but also realists.¹¹ Referring to Thomas S. Kuhn, Michel Foucault and others, they hold that every rational world-view necessarily includes basic beliefs that are accepted without evidence, and moreover, that world-views that include theistic basic beliefs are at least as acceptable as those that do not include theistic beliefs. Reformed epistemologists also hold that it is a religious duty for a Christian to defend theistic alternatives in the competition between different world-views. Reformed epistemology is obviously incompatible with a strong epistemic evidentialism.

The combination of evidentialism and non-realism might at first sight seem to be a non-starter, but in fact, Thomas Aquinas subscribed to something rather close to it. According to him, the theistic thesis and some other basic religious beliefs could be conclusively proved by deductive arguments. However, Aquinas also distinguished religious faith in the proper sense of the word from those theistic beliefs that could be argumentatively proved. If his view is accepted, then language related to religious faith in the proper sense of the word should not be interpreted in the realistic sense, because it does not follow evidentialist principles. By distinguishing religious faith from its evidence-based presuppositions, Aquinas managed to combine evidentialism concerning the presuppositions of religious language with the non-realistic nature of religious language itself.[12]

3. The Minimal Presupposition Theory of Religious Faith

Aquinas' idea of the distinction between religious faith and its epistemic presuppositions has also inspired criticism against the strict form of anti-realistic theism, even though his proofs for the existence of God are more or less out of date. Simo Knuuttila, for example, has argued that the complete autonomy of religion in relation to epistemic beliefs is not plausible.[13] His argument is the following. It seems difficult to commit oneself to Christian worship, at least, if at the same time one believes that God does not exist or the notion of God is meaningless at the level of ordinary language. The religious language of Christian worship seems to presuppose the active presence of a

---

[11] On reformed epistemology, see Plantinga (1998); Koistinen (2000).
[12] On Aquinas' philosophy of religion, see Jordan (1993); cf. Kenny (1983).
[13] See Knuuttila (2004); cf. Knuuttila and Sihvola (2000).

personal God at the same level of reality as the one at which we ordinarily live. However, a Christian should be able to make a distinction between her religious and epistemic beliefs. She does not need to use religious notions or take a stand on God's existence, when she is moving at the level of her epistemic beliefs, especially when practicing science. She does not have to believe that God exists in any epistemic sense. Still, she has to believe that the existence of God is in principle possible, although there is no evidence to find out whether this possibility is actualized in the world we live in. This kind of moderate agnosticism can be combined with Christianity, if one's orientation to life is based on a religious hope that the God of Christianity exists and the teachings about him are true. There is an autonomous dimension in the religious language so that its notions and sentences do not need to meet the requirements of intersubjective acceptability and empirical verifiability. Those beliefs that are in the proper sense religious are non-epistemic. However, religious commitment also presupposes something about the believer's epistemic beliefs. The minimum condition at the epistemic level is that the person is not an atheist, and regards the notion of a personal God as meaningful. I shall call Knuuttila's view the *minimal epistemic presupposition theory of religious faith*.

Several problems follow, if we define the epistemic presuppositions of religious faith in this way. First, it is not obvious that it is rationally acceptable to regard the existence of God as possible even in principle, if the notion of God is defined with a reference to any substantial characterizations of the divine being found in the Christian tradition. It might be that the notion of God is logically inconsistent at a very deep level. If this is the case, the rejection of the strict verifiability thesis after logical empiricism is not sufficient to make the notion of God meaningful. Leaving aside the problems related to the concept of the supernatural, the concepts of omnipotence and providence are especially problematic. If the idea of an omnipotent and benevolently providential being is logically inconsistent, it is not rationally acceptable to believe that it is possible for this kind of being to exist. The problem could be perhaps solved, if we think that all logically inconsistent characterizations of divinity are religious expressions and should not be understood literally. Beliefs related to them are non-epistemic. The personal divine being, the possibility of whose existence we are supposed to accept at the epistemic level, cannot be characterized by traditional religious notions. But does this kind of divinity any longer have anything to do with Christianity or any other actual religion? Clearly, at least, our conception of rational acceptability and logical consistency provides the limits within which we can speak about possibly existing beings and the epistemic presuppositions of religious faith.

Another problem is related to the notion of religious hope. Let us suppose we have somehow solved the problem of logical inconsistency. If we are rational persons, are we yet entitled to hope that God really exists and religious teachings are true? It is obvious that the mere fact that something is possible is not sufficient also to make it rationally acceptable to hope that it exists. It would be a strange world if we could acceptably hope for the existence of whatever we regard as possible. There should be some criteria coming from outside religion according to which the acceptable objects of an acceptable religious hope should be defined. As religion is usually thought to be closely related to morality, it seems natural to think that our moral beliefs limit and possibly even define what kind of religious beliefs we are rationally entitled to hope to be true. These moral views should be independently accepted in relation to religion and at least partially based on rational reflection. The relation between religion and morality is of course complicated. There can, for example, be a danger of falling into the vice of wishful thinking. This might happen, if we think that an acceptable religion is something needed to support our motivation to act in accordance with moral beliefs, although the beliefs themselves would have been adopted on the basis of rational reflection. However, it seems clear that morality should take some sort of primacy in relation to religion in the world-view of a rational person.

Finally, do we really need any epistemic presuppositions to support our religious faith? Do we have to believe at the epistemic level that the existence of God is possible in order to commit ourselves to Christian or other forms of religious worship? The requirement of this kind of epistemic presupposition seems to me unnecessarily strong. Why should a praying person have to believe that the God he asks for help is actively present as a separate person in the external world? Perhaps one could rather give a symbolic or metaphorical interpretation to languages used in prayers, confessions and hymns of praise. If the practice of religion does not need any epistemic presuppositions, then there should be no problem, if we combine participation in religious practices with stronger forms of agnosticism or even full-blown atheism at the epistemic level.

4. Evidentialism and Theism

It seems to me that a strong evidentialist attitude to epistemic beliefs, even if it is thought to imply radical agnosticism or philosophical atheism, allows space for anti-realistic theism. It is even consistent with a view according to which the epistemic presuppositions of religious faith are defined so that the epistemic aspects of religious notions are required to be logically consistent. Both of these views are compatible with an evidentialist conviction that one should never accept any epistemic beliefs without a good reason. If the epistemic presuppositions of religious faith are defined in an extremely minimal way, they only say the following. There might be something outside the reach of our cognitive abilities and possibilities of empirical research, such that there is no evidence available or forthcoming on the basis of which we could say anything more or less probable about this thing. Neither our abilities nor the possibilities open to us are pre-ordained, and one should be convinced that critical scientific research is the one and only way to extend the limits of our knowledge. The only thing we can reasonably say about the unreachable is that although we cannot say with any probability that there is such a thing, if there is anything like it, it has to be something that can in principle be described in a logically consistent way, although this description will remain unavailable to us. Perhaps this could be reasonably acceptable even to a hard-headed evidentialist. But the defender of religion does not have to worry about these complexities, if she subscribes to anti-realistic theism. From her perspective, religious faith does not even need minimal epistemic presuppositions.

It is important to note that, *contra* Niiniluoto, neither the minimal presupposition theory nor anti-realistic theism leave religions immune to criticism coming from the outside. There might be some Wittgensteinians who think that religions are autonomous practices that can only be understood from the inside. However, this cannot be the whole story. The anti-realist can admit that scientific research determines what kinds of entities we are entitled to believe, in the epistemic sense, to exist in the world. It is irrational to believe or even hope that something would exist that would be inconsistent with the scientifically plausible.[14] Our criteria of logical consistency determine the limits

---

[14] There is an exception. It is often rational for a scholar to hope that something would emerge from his studies that would be logically and empirically inconsistent with what is regarded as true at the moment. But then his aim is to prove that the earlier

within which we are entitled to regard something as possibly existing. Sometimes it might be rationally permissible if not recommendable to hope for the existence of something, given that it is consistent with the results of science and within the limits of logical consistency. There may also be moral criteria for a rationally acceptable hope. These are the philosophical limits within which the actual religions should be allowed to move. Insofar as religions involve beliefs that cannot be interpreted within this epistemic space, they have to be understood in an anti-realistic way. Niiniluoto might be right in his suspicion that churches and the believers do not readily conform to these restrictions, but that is another problem, the solution to which does not primarily concern the philosophers.

Let us for a moment take it as given that there is an existing religion that accepts the limits of rational acceptability. What would then be the point of this kind of religion? Why should anybody practice it? I guess Niiniluoto might readily agree with a friend of mine who gave me some comments on a paper I once wrote about anti-realistic theism and the minimal presupposition theory. The cynical commentator wrote in the marginal to a passage in which I described God as a possibly existing being about the probability of whose existence we cannot say anything: "Poor God, poor priests!" However, even after these concessions to epistemic evidentialism, religion can still be an appropriate way for some people to reflect and meditate on existential problems such as the meaning of life, the experience of being a person, and the possibility of understanding and caring for others. Religion can also support moral motivation and moral education, even though a rational person is advised to understand the contents of morality independently of any religious convictions. However, it is clear that not everybody needs a religion. I agree with Niiniluoto's view that there are also purely secular ways to realize respectable moral and intellectual ideals in one's life.

Helsinki Collegium for Advanced Studies

---

scholars have been wrong and their views have to be corrected. It is, however, irrational to hope that the results of science in their perfected form would be logically inconsistent.

References

Ayer, A.J. 1936: *Language, Truth, and Logic*, Viktor Collanz, London.
Craig, W. 1980: *The Cosmological Argument from Plato to Leibniz*, Macmillan, London.
Edwards, P. 1967: "Atheism", in P. Edwards (ed.), *The Encyclopedia of Philosophy*. vol. I, Macmillan, New York, 174-189.
Flew, A. 1997: "The Presumption of Atheism", in P. L. Quinn and C. Tagliaferro (eds.), *A Companion to Philosophy of Religion*, Blackwell, Oxford, pp. 410-416.
Huxley, T. H. 1893-95: *Collected Essays*, vol. 5, Macmillan, London.
Jordan, M. D. 1993: "Theology and Philosophy", in N. Kretzmann and E. Stump (eds.), *The Cambridge Companion to Aquinas*, Cambridge University Press, Cambridge, pp. 232-251.
Kenny, A. 1983: *Faith and Reason*, Columbia University Press, New York.
Knuuttila, S. 2004: "Biblical Authority and Philosophy", in D. Z. Phillips and M. von der Ruhr (eds.), *Biblical Concepts and Our World*, Palgrave Macmillan, New York, pp. 113-127.
Knuuttila, S. & J. Sihvola 2000: "Ancient Scepticism and the Philosophy of Religion", in J. Sihvola (ed.), *Ancient Scepticism and the Sceptical Tradition*, Acta Philosophica Fennica 66, Helsinki, pp. 125-144.
Koistinen, T. 2000: *Philosophy of Religion or Religious Philosophy. A Critical Study of Contemporary Anglo-American Approaches*, Luther-Agricola Society, Helsinki.
Martin, M. 1990: *Atheism: A Philosophical Justification*, Temple University Press, Philadelphia.
Niiniluoto, I. 1999: *Critical Scientific Realism*, Clarendon, Oxford.
Niiniluoto, I. 2003: "Ateismi", in T. Helenius, T. Koistinen, and S. Pihlström (eds.), *Uskonnonfilosofia*, WSOY, Helsinki, pp. 122-148.
Niiniluoto, I. 2005: "Tämän ajan etiikka", *Vartija* 118, 167-176.
Penelhum, T. 1997: "Fideism", in P. L. Quinn and C. Tagliaferro (eds.), *A Companion to Philosophy of Religion*, Blackwell, Oxford, pp. 376-382.
Phillips, D. Z. 1993: *Wittgenstein and Religion*, Macmillan, London.
Plantinga, A. 1974: *The Nature of Necessity*, Clarendon, Oxford.
Plantinga, A. 1998: *Warranted Christian Belief*, Oxford University Press, New York.
Sihvola, J. 2006: "The Autonomy of Religion in Ancient Philosophy", in V. Hirvonen, T. Holopainen, and M. Tuominen (eds.), *Mind and Modality. Studies in the History of Philosophy*, Brill, Leiden, pp. 87-99.
Swinburne, R. 1979: *The Existence of God*, Clarendon, Oxford.

Matti Sintonen

The Social and Cultural Roots of the Scientific Method

1. Science as Product and as Process and Procedure

Ilkka Niiniluoto is a leading exponent of scientific realism, committed as he is to the view that sciences progress towards truth. Part of his realism is ontological and epistemological: there is an inquirer-independent world, and the main products of scientific inquiry, most notably theories, try and capture it in representation that are, or at least are intended to be, true. Part of his realism has to do with method and the cognitive procedures of discovering truths: sciences consist of organised inquiries, that is, gradual or stepwise and perhaps occasionally revolutionary processes of building increasingly truthlike (as well as explanatorily powerful, simple and perhaps otherwise pragmatically virtuous) theories. How this process has been characterized has varied during the times, and the various kinds of heuristics and methods of scientific discovery focus on the way scientists can order their steps in a fruitful way. Finally, science is a not just a cognitive but also a culturally mediated and essentially social enterprise. Scientists share a common language and a set of goals, their work is organized in often highly specific ways, they must market their results to their peers and, ultimately, to the community at large. Just as we can ask what the cognitive desiderata of science are and what sorts of epistemic norms and methodological canons could be developed for promoting them, we can look at the norms and values of scientific institutions as well as at the larger social setting which house these norms and values.

Niiniluoto defends a view of the method of science that derives from the long tradition of thinkers who have emphasized external permanence, coupled with debate and free criticism, rather than the inner light of the (individual or communal) reason, as the foundations of the scientific method. Charles Peirce (together with, e.g., Karl Popper) looms large behind Niiniluoto's view, not just in the general view of truth as the most important aim of inquiry, but above all in the confidence on experience and reason as determinants of the outcome. Peirce did give a concession to instinct in his attempt to explain why abduction works and why we ought to expect that reason is not the sole authority in matters vital to survival. And when addressing the scientific method in particular, he drew a parallel between the animal instinct and inevitability in which the method of science leads to truth.[1] Yet, on balance, Peirce gave an account of the epistemic credentials of the scientific method, based on experience and reasoning, that could well be considered *the* corner-stone of our scientific image.

Looking at science in its social context readily takes us into wider vistas about how science has been justified. In the concluding chapter of his *Critical Scientific Realism* Niiniluoto focuses on how realism has been coupled with a variety of non-cognitive interests. He dismisses, in his gentle way, views that hold that in science anything goes (and nothing is special), and those who have based the epistemic authority of science on the special mental or moral characteristics of scientists (intelligence or moral strength) or on such historically contingent institutions as trust. Free inquiry does not amount to all being equal in the sense that layman authority would or should rule. Science is not democratic in this sense but is rather a system of expert knowledge which owes its epistemic authority to the fact that scientists "use public, critical methods of investigation, and the reliability of their claims can be evaluated" (Niiniluoto 1999).

Acknowledging that science is social and culturally mediated is broad and innocent enough to be universally accepted. But the details are hotly debated. Does acknowledging the social and cultural roots of the method of science deprive it of its epistemic authority? Is the *differentia specifica* of the scientific enterprise in the goals of truth and other epistemic virtues or is there something unique in its method, conceived on the most general level of reasoning on the basis of experience and not just on the level of specific research

---

[1] "Taking the phenomenon as a whole, then, without considering how it is brought about, science is foredestined to reach the truth of every problem with as unerring an infallibility as the instincts of animals do their work, this latter result like the former being about some process of which we are as yet unable to give any account." (Peirce, CP, 7.77)

methods? In what sense is science and especially the scientific method, including most saliently induction, deduction and abduction, social?

This note is a largely descriptive look at how the social roots of the method have been conceived at an important juncture, the turn of the 20th Century. The reference in my title to the social and cultural roots of the method of science is intentionally ambiguous between the causal or, shall we say, socio-historical origins of the method and the normative claim that the proper method *should* be social in nature. Peirce pioneered a construal of the success of science through the self-correcting methodological procedure. But self-correctness operated on two levels, the cognitive-methodological and the social. It has been pointed out (Anderson 1997) that the method of science in Peirce relied on "a well-functioning or healthy community" and that, therefore, there was a political dimension to this method. Be that as it may, I shall suggest that Peirce combined, in a unique way, the self-corrective methods of induction, deduction and retroduction with the self-corrective manner in which a community of scholars operates. But Peirce left two important and related lacunae in his account. One concerns the roots of the "Gnostic instinct" and the will to learn necessary to the workings of the method (CP, 7.58). The other has to do with the nature of knowledge as a public enterprise, and the precise mechanisms which make the procedures of science impersonal and the pursuit of truth disinterested. The distinctly social way of filling in these lacunae comes in Robert K. Merton's ethos of science, as I suggest in sections 3 and 7. This ethos and the norms of science that describe the way the ethos is made manifest in the expansion of certified knowledge is, roughly, Peirce's method of science made explicitly sociological.

2. The Moral Dimension and the Social Impulse

Why the method of science – and what is the method? The canonised way of understanding the virtues of the scientific method comes from Charles Peirce, a pragmatist with strongly realist inclinations (and hence, a pragmaticist). Science in general, Peirce thought, distinguishes itself from everyday belief systems by two pervasive features. One is its systematic nature, and the other one is its reliance on method. Neither one of these features is news to the philosophical tradition, in that the former has been the default view amongst "modern writers", no doubt as a result of the work of Immanuel Kant and

Whewell Whewell in particular. The latter feature, though equally old or even older, is in Peirce's view more fundamental than the former since it cuts deeper. It captures the thought of Plato and Aristotle according to which mere true belief does not equal knowledge, since knowledge requires knowing the reason why (CP, 7.49).

In the popular article "The Fixation of Belief" Peirce gave an explanation of why the method of science works. According to Peirce inquiry begins when doubt creeps into mind, as when observations and others' opinions undermine one's confidence. The resultant 'irritation of doubt' is a motive (and the only immediate motive) in "the struggle to attain belief". Something must be done to eliminate this doubt, to fix belief. There are four different methods for settling opinion, the ultimate aim of inquiry. The methods of tenacity and authority, as well as the *a priori* method all have their virtues, but since opinions and authorities can conflict with each other, and since intuitions are too subjective, they are incapable of settling differences of opinion in an orderly and permanent fashion.

The key to the success of the scientific method is the idea that there is a mind-independent reality, 'secondness' which forces the inquiring mind, by help of the senses, into irritating beliefs:

> [T]here are Real things, whose characters are entirely independent of our opinions about them; those Reals affect our senses according to regular laws, and, though our sensations are as different are our relations to the objects, yet, by taking advantage of the laws of perception, we can ascertain by reasoning how things really and truly are; and any man, if he have sufficient experience and he reason enough about it, will be led to the one True conclusion". (Peirce, CP, V)

Peirce was writing at a time when various forms of idealism and neo-Kantianism dominated the image of science and knowledge, and he therefore considered his reality principle a "new conception." It is no longer a new conception, partly because of Peirce's influence. Indeed, Niiniluoto's critical scientific realism is a technically detailed defense of the position: there is a mind-independent reality of things (including events and processes), properties and relations; theoretical concepts (at least aspire) to refer; theoretical claims do have truth values; realistically conceived truth is an aim of inquiry; institutional, pragmatic, empirical and cognitive success in science is best cashed out and explained by help of increase in truthlikeness – and the method of science is the best means of approaching truth.

But the reality principle is not the only warrant of the unique status of the method of science rested. Another pillar, one that at first sight seems to undermine the first one, reaches into what Peirce identifies as the *moral* realm. In a detailed portrayal Peirce (ibid., 7.79–7.86) gave a step-by-step account of the method of science, starting with the precise identification of the problem and the developing of appropriate mathematical methods, to questions of logic and methodeutic and, finally, to testing and acceptance. But he then goes on to what he thinks is more important still than this logical way of ordering an inquiry:"The most vital factors in the method of modern science have not been the following of this or that logical prescription – although these have had their value too – but they have been moral factors." (CP., 7.87)

The first one of these moral factors has been "the genuine love of truth and the conviction that nothing else could long endure". The modern scientific era distinguishes itself from that of the middle ages in that the new founders of the science were not afraid of the consequences of this or that theory and hence did not fear giving up their favourites. Instead of the focusing on teaching and transmitting knowledge already at hand they understood that they did not yet know the truth and genuinely sought to seek it. Secondly, the "next most vital factor of the method of modern science", Peirce wrote, "is that it has been made social". This in turn has two aspects to it. One is the public nature of the facts: "what a scientific man recognizes as a fact of science must be something open to anybody to observe." The other aspect of sociality has to do with "the solidarity of its efforts": "The scientific world is like a colony of insects in that the individual strives to produce that which he himself cannot hope to enjoy." There is still a third moral factor in operation, viz., the self-confidence of the method of science in ultimate victory. Although experience and experiments have the largely negative role of telling when we are wrong, modern science builds on the presumption that truth will come out.

Peirce commented on the force of the the social or moral dimension in his account of one of the revals to the method of science, that of the method of tenacity, that of sticking to one's beliefs come what may. Although it has been highly successful in the past, it won't carry the day:

> Social impulse is against it. The man who adopts it will find that other men think differently from him, and it will be apt to occur to him, in some saner

moment, that their opinions are quite as good as his own, and this will shake his confidence in his belief. This conception, that another man's thought or sentiment may be equivalent to one's own, is a distinctly new step, and a highly important one. It arises from an impulse too strong in man to be suppressed, without danger of destroying the human species. Unless we make ourselves hermits, we shall necessarily influence each other's opinions; so that the problem becomes how to fix belief, not in the individual merely, but in the community. (CP, 5. 378).

I submit that social impulse here does not confine to the fact that fellow inquirers undermine the confidence of our beliefs but that it extends to the specific form in which social feedback is processed. In any case, Peirce suggested that the method of science is superior to the other methods precisely because it involves, at one and the same time, two kinds of irresistible force. On one hand there is this realm of the real, independent of the inquiring mind, giving rise to experience, "that which we are constrained to be conscious of by an occult force residing in an object which we contemplate" (CP, 581). On the other hand it has a social aspect to it, in the Gnostic instinct, in the public nature of the facts, in the solidarity of effort, and in the self-confidence that no experimentally accessible facts will remain secrets in the long run. In a sense it is the community and not the individual which is the subject of knowledge, not just in the generation of an idea but in its manner of justification.

Furthermore Peirce had a specific way of looking at how inquiry proceeds as a dialogue within the community. He says that in asking how we should go about fixing beliefs in a community we should also focus on the conditions of a well-functioning or healthy community (Anderson 1997, 224). The success is in "the cooperative character of scientific inquiry": ideally at least inquirers engage in "unreserved discussion with one another", "each being fully informed about the work of his neighbour, and availing himself of that neighbour's work." And Peirce resorts to the time-honoured image of the virtues of cooperation and collaboration in the production of knowledge: "in storming the stronghold of truth one mounts upon the shoulders of another who has to ordinary apprehension failed, but has in truth succeeded by virtue ... of his failure" (CP, 7.51, see also CP 7.87).

3. Realism – the Scientists' Official Philosophy

Peirce's metaphor, "one mounts upon the shoulders of another" is of course an explicit reference to Isaac Newton's words: "If I have seen further, it is by standing on the shoulders of Giants". It is a reminder both of the essentially social *and* historical nature of inquiry, of the fact that science is a cumulative affair that always builds on the work of previous traditions and communities of inquirers. And as Susan Haack remarks, Peirce here connects the scientific attitude with the social character of scientific inquiry. For Peirce search for truth, and in this sense the very notion of truth, cannot be viewed as "the work of one man's life, but as that of generation after generation, indefinitely" (CP, 5.589). And Haack takes note of the metaphor in a slightly altered form: "the idea ... is to pile the ground ... with the carcasses of his generation, and perhaps of others to come after it, until some future generation, by treading on them, can storm the citadel" (CP, 6.3).

It is of some interest to note that Robert K. Merton, the father of modern sociology of science, adopted this metaphor and used it both to locate his own place in the sociological tradition, and to highlight his view of science as a historical and social enterprise.[2] When he formulated his view of the famous ethos of science and the norms that manifested its *modus operandum* he was heavily influenced by the sociological giants, especially Durkheim and Weber. He had studied the roots of modern science and technology in his dissertation of 1935, arguing that there was a link between Protestant ethics and its implicit ethos, and the rise of experimental science and technology. Merton's general idea was that the norms and values in one realm, such as religion or politics, could provide a pattern for another, such as science.

Merton's historical proposal of the roots of science have been questioned. Similarly, it is doubtful if Merton managed to identify the ethos (if there is a unique one) of modern science in unambiguous terms. What is interesting is that there is a remarkable parallel between Merton's account of inquiry and that of Peirce. Peirce had referred to the well-functioning community as a cornerstone of scientific method and hence as a normative pattern for scientific

---

[2] Merton went into great lengths in his attempt to find the roots of the metaphor – and he managed to trace it to Bernard of Chartres in the 12th Century: "We are like dwarfs upon the shoulders of giants, and so able to see more and see further than the ancients" (see Merton 1965 and Sztompka 1986, p. 21).

rationality. Merton shared this view. But when he referred to the pragmatists' view of inquiry, and on the way science was to be organised to be efficient in its functions, the reference was to John Dewey rather than to Peirce (for reasons that I shall return in section 7 below).

Peirce had given the scientific method its current justification: it is the one we ought to follow because it alone can deal with the irresistible "social impulse". There is no way of insulating a mind against the opinions of others. The method of science depends on the willingness of the members of the scientific community to critisize and subject the views proposed to closer scrutiny. The key to the success and steady progress of science is the essentially social and self-corrective procedure that is built into the scientific method. This is line with Merton's explicitly sociological approach into science: science is special in its goal of expanding certified knowledge, but its manner of operation is not. Or more exactly, the mechanisms that guarantees cumulative progress of knowledge and the success of science as a relatively autonomous pursuit of knowledge (and utility) is the same as those in other domains, the difference being that here the standards are more exacting and regulated in detail.

Let's have a closer look at the parallel as well as where it breaks. First, just like Peirce Merton was an ontological and epistemological realist, as well as a critic of radical relativism. True, knowledge is socially (and culturally) mediated, and an individual scientist always is a member of a particular community with a distinct thought style. But although an asocial theory of knowledge, such as Descartes's, is a non-starter, a fundamental assumption of science is that, "along with the social and cognitive interactions of scientists in collectives, there is also – to use his [Ludwig Fleck's] words – the objective reality (that which is to be known)."[3] Nor is this just an abstract ontological assumption but one which is backed up by the deeper metaphysical assumptions of knowability and effability. Inquiry cannot even start but on the premise that the world – nature – comprises an intelligible order. The way Merton phrases this insight testifies to his deep interest in the historical roots of modern science: "...the very notion of experiment is ruled out without prior assumption that Nature constitutes an intelligible order, so that when appropriate questions are asked,

---

[3] Merton, R. K. 1981: "On Sociological Ways of Thinking about Thinking and Thought", Bicentennial Symposium of American Academy of Arts and Sciences (mimeographed), quoted from Sztompka 1986, p. 77).

she will answer, so to speak. Hence, this assumption is final and absolute" (Merton, 1968, p. 635). This is the voice of a realist sociologist of science, miles apart from the voice of some social constructivists who have employed the same metaphor. Just witness Karin Knorr-Cetina's view from the 1980's according to which e.g. laboratory scientists "do not 'ask questions to nature'" for the simple reason that "nowhere in the laboratory do we find the 'nature' or 'reality' which is so crucial to the descriptive interpretation of inquiry" (Knorr-Cetina 1983, 119).

Secondly, Merton, like Peirce, was a realist in his axiology and a cognitivist with respect to the aims and motives of inquiry. For Peirce the scientific attitude is all-important in that it is based on *disinterested* love of truth. As Susan Haack puts it, a Peircean scientist-philosopher, unlike a businessman or a teacher or a theologiacn, is animated by the pure truth-seeking attitude. As far as science and inquiry are concerned, cognitive virtues are more important than practical aims: Peirce thought that knowledge is not just an instrument for the achievement of practical utility, as with simple pragmatists, but has value of its own. A scientific inquirer desires to know the truth and prefers to do so regardless of its practical utility or convenience. It is not that utility could not be an eventual outcome, but that it is bracketed during inquiry. This is very much the way Merton viewed science. He adopted the standpoint of Peirce's pragmaticism rather than Dewey's (and Schiller's) pragmatism. As Sztompka (1986, p. 51) sums it all up, *"truth comes first"*.

Third, both Peirce and Merton back up their cognitivism a methodological and strategic argument. "The point of view of utility", Peirce wrote, "is always a narrow point of view" and, as such, not the likeliest path to success in the long run (CP, 1.641). The reason is that knowledge builds on knowledge, and that there is no way of knowing in advance what sort of knowledge could provide a basis for future practical innovations. Making practical utility the sole purpose of inquiry would not just be metaphysically and axiologically perverse but also likely to hamper the achievement of goals, both cognitive and practical. In very much the same spirit Merton held that making science serve the immediate needs of the society would be counterproductive. He writes that "fundamental scientific knowledge is a self-contained good and that, in any case, it will in due course lead to all manner of useful consequences serving varied interests in society". (Merton 1982, p. 214.)

But how, more precisely, is the value of knowledge cashed out? The two most fundamental values are *objectivity* and *originality*, where the former is geared towards safeguarding truth, and the latter towards enlightening, non-trivial or interesting truths. When these values are internalised they comprise virtues of the scientific mind, such as the tastes for truth and novelty, as well as the mental characteristics of rationality and open-mindedness. Merton did acknowledge that, apart from the two purely cognitive values of objectivity and originality there is a third one, the utilitarian value of *relevance*, which sanctions the pursuit of individually or socially important non-cognitive aims, indeed "almost anything other than the advancement of knowledge for its own sake" (Merton 1982, p. 271).

4. The Scientific Mind and the Gnostic Instinct

Before going into the differences between the ways Peirce and Merton cashed out the efficacy of the method of science, let's have a closer look at their views on individual motivation and the norms of science. In Merton's view scientific inquiry is governed by social norms, both moral and technical.[4] These social norms are based on communally shared values, along the lines of organized scepticism, and they give rise or support in individual scientists personality characteristics that comprise what could be called a scientific mind. The outcome, in line with Peirce's views, was that scientific inquiry is not a solitary affair of one individual engaging in causal, experimental and observational commerce with nature. Rather scientific inquiry and consequently the method of science moves in the two dimensions noted by Peirce: there is the causal commerce with the nature and consequently the technical norms which govern good scientific contact. On the other hand there are the social or moral norms which regulate the conduct of inquiry amongst members of the community.

Merton later explicated these rules by suggesting that scientific inquiry aims at widening or expanding certified knowledge. To be able to be efficient these

---

[4] For an excellent review of sociology of knowledge, see Zuckermann (1988). The account given below of Merton is in heavy debt to Zuckermann as well as to Sztompka (1986). The aim here is to point to a parallel between Merton's norms of science and Peirce's account of the scientific method in the Fixation of Belief. It would be interesting to see to what extent Merton actually was influenced by Peirce's lectures and writings.

societies or scientific communities must be organised around a number of norms which guide this process. The first norm is the norm of impersonality or universalism, the idea that truth-claims are to be subjected to pre-established impersonal criteria. Accepting or rejecting a claim is not a personal affair, and here race, nationality, religion or ethnic group should play no role (Merton 1968, p. 607). The second norm for Merton was communalism which requires that scientific knowledge itself is common property, and should be freely transmitted from one member to another. Here the idea is that no one in a scientific community is entitled to withhold information. This is a moral norm, but it is partially justified because spreading knowledge is instrumental in the success of the community on the whole. The third norm is called disinterestedness, the idea here being that the sole motive for inquiry is the motive provided by the satisfaction of finding out new truths. The fourth norm is organized scepticism or organized criticism. When an individual scientist or group proposes a result, others have an incentive to probe into the foundations of the claim – the more so the more interesting or central the claim is, that is, the more potential the results has towards expanding knowledge. The norm for organised scepticism calls for "public criticism by scientists of claimed contributions to scientific knowledge, both their own contributions and, more easily perhaps, those of others" (Merton 1980, pp. 18–19).

These norms could well have been written by Peirce. Impersonality and universalism are safeguarded by the two sides of the method science, external permanence and the social impulse, as well as the idea that the scientist-philosopher is a pure truth-seeker. Communalism is at the very core of Peirce's notion that science builds, both historically and socially, on results of joint labour, on ground piled by previous generations. Flouting this norm would be flouting the most important norm of inquiry "to be written upon every wall of the city of philosophy: Do not block the way to inquiry to" (CP 1.135). Organized scepticism is but another way of expressing the effects of the social impulse.

But how about disinterestedness as a motive of inquiry? As we saw, Peirce (like Merton) explicitly distinguished between two kinds of motives and hence between two (or more) types of research, one that aimed at pure truth and another that had a mixed goal, the epistemic goal of truth and the varieties of practical utilities. Now, it is easy to see what the motivational mechanism is in the pursuit of money and fame, but we may still wonder how the desire for truth in itself could have such a grip over scientists as to form the most pervasive feature of their character of mind.

There are several possible ways of accounting for motivation, one being through the desire to learn. Peirce formulated something he called the first rule of reason which is expressed in the following passage:

> Upon this first, and in one sense the sole, rule of reason, that in order to learn you must desire to learn, and in so desiring not be satisfied with what you already incline to think there follows one corollary which itself deserves to be written upon every wall of the city of philosophy: Do not block the way to inquiry. (CP 1.135)

As Susan Haack (1997) observes, in a perceptive paper on the first rule, Peirce maintained that the slogan "Do not block the way of inquiry" was a corollary of the first rule of reason, that is, the obligation not to block (permanently) inquiry can be derived from the desire to be learn. Now Haack notes that Peirce formulates the desire to learn in terms reminiscent of Aristotle's dictum "all men by nature desire to learn". Yet, as she continues, Peirce did not think that this attitude was the norm in society, now or before, but only characterised the scientific mind. The desire to learn in this sense is an outcome of a long historical development during which, with a lot of help from the irresistible social impulse, men have developed the appetite for truth. This appetite in turn has been honed by the irritation caused by doubts propelled by the social impulse. And when this idea is coupled with the realization that truth (in many quarters at least) is not within the reach of the inquirer there and then, the love of truth must be cashed out an appreciation that truth *eventually* will come out, at least ideally and in the long run.

In a remarkable move Haack turns this insight into a solution to a puzzle provided by Peirce's first rule and the view inquiry he outlines in the *Fixation of Belief*. If the only *immediate* motivation for inquiry is that of dispelling the irritation caused by doubt, whence comes the idea that a philosopher-scientist's motivation is that of finding truth? What paves the way from the motivation of getting rid of irritation to that of finding the truth? Haack's answer comes in terms of a distinction between a belief being fixed in a strong sense, that is, in a way that survives future experiences and future challenges from fellow inquirers, and a belief being fixed temporarily, as when there is no doubt (and hence no irritation) for the time being. The latter type of fixation is something that can be witnessed by actual inquirers. But some of these inquirers, the most sophisticated cognizers, realize that the more permanent sort of fixation of belief only materializes when the belief arrived at is true. And since a sophisticated inquirer must also realize that, in theoretical matters at least, future setbacks are always possible, infallibility is out of question. But realizing this is tantamount to acknowledging that something can be temporally settled, and in this sense fulfil all operational criteria for truth, and yet not be true. And Haack (1997, p. 247) writes: "The most sophisticated cognizers, seeking fixed belief in the stronger sense, are motivated (at the second order, so to speak), by doubt in the weaker sense." In this way the simple homeostasis of seeking relief from irritation is turned into the genuine thing, the scientific attitude or pure truth-seeking attitude.

5. The Biological Pattern: The Survival Value of Truth

The story so far seems to be this: the method of science, induction, deduction and abduction, relies on external permanence and social organization. Together they conspire to make the scientific enterprise into a self-correcting process. The beauty of this conspiracy is that this self-corrective nature is built into the method of science, indeed it is the only method for fixing belief that has this characteristic. But we may still wish to be advised on some remaining puzzles, both of them having to do with the needed link between truth as a value in itself, and the motivation for pursuing it. There are many things in the world worth desiring, money, fame and reputation being amongst them. Given that truth is the aim of science, how does it come an end for individual scientists? What accounts for the desire for truth, and hence what is the key to the formation of the scientific 'temperament' or 'attitude', in Peirce's terms? Could there be a link from truth as a value in itself to truth as a value for an individual inquirer, and hence an explanation for the strong appetite for truth. More specifically, is this appetite a character of a scientists' psychological constitution, as Peirce seems to think, or could self-correctiveness arise without this assumption?

Two possibilities for such a link, for an explanatory mechanisms, come to mind. One was offered by Moritz Schlick and vulgarized by more recent sociobiologists. The other one is by Merton – and this is, I suggest, where

Merton took a decisive step. I shall call these two patterns the biological and the socio-cultural patterns of explanation.

Schlick's reasoning is, on the face of it, armchair biology. His starting point was not the epistemic value of truth but the esthetic value of beauty. (For the modern ear the link between truth and beauty appears somewhat contrived, but till the turn of the 20th Century it was more of a commonplace). In his early treatise on aesthetics Schlick singles out the question "Why does anything whatsoever appear beautiful?" as the *Grundproblem*. The tradition, from Kant on, had taken it that aesthetic pleasure is disinterested delight, contemplation from a point of view different from utility. Beautiful things do offer us pleasure, of the disinterested variety, and one can engage in discourse over the details. But an aesthetically pleasing object is not aesthetically valuable *because* we enjoy it. Rather, we enjoy it because it has such and such aesthetically pleasing properties. However, there is no deeper explanation, of the naturalistic sort as we would now say, for the fact that we value particular properties. Indeed, Wilhelm Wundt had suggested, in consonance with Kant, that "the question of why we feel pleasure, displeasure, and so on, is thus really just as vacuous as if we sought to ask why we touch, smell, taste and so forth, and why utterly different sensations have not developed instead of these".

Here Schlick begged to disagree. According to him the *Grundproblem* has the form of a straightforwardly causal query: "The 'why' here must be taken to be asking for a real causal explanation; so it is not a matter of specifying the properties in virtue of which an object becomes a beautiful one, but rather of discovering the causes that lead to these properties actually having such an effect." Here Schlick departs from Kant, because he thinks that "the good is good not because it has "a value in itself", but because it gives joy". The next question is: How would this explanatory strategy extend to cognitive or epistemic values? Is the suggestion that the true is true not because it has "a value in itself" but because it gives us joy?

Schlick set out to bite the bullet. His example was not the cognitive value of truth but the pragmatic-cognitive value of simplicity, but the same pattern of explanation could be at work in both cases. Knowledge in general, he wrote, contributes towards "the preservation of the individual and the species". The drive for knowledge undoubtedly falls under this general principle: "In its origin, thinking is only a tool for the self-maintenance of the individual and the species, like eating and drinking, fighting and courting." The explanation of the riddle of the value of knowledge, he wrote, " will indicate the place occupied by cognition in relation to other human activities ... the solution of the problem must of necessity lie in the province of biology". Now the pragmatic-cognitive property of simplicity is instantiated when we manage to reduce one thing into another. Such reduction, Schlick said, of one thing to another affords us pleasure. Generalizing this we can say that "the value of knowledge consists quite simply in the fact that we enjoy it". But this cannot the end of the story. It is easy to imagine that possessing a theory that is simple, or true, could have survival value (and this has been emphasized by more recent armchair sociobiology) – but there is a crucial piece missing still. Why would truth (and simplicity) become motives for individual action? Agreed that we seek pleasure and joy, why is it that truth delivers these good?

6. Instinct – Gust or Gnostic?

It is to Schlick's credit that he had an answer to this question, too. There is, he thought, an evolutionary explanation that connects the survival of the species to the motivating mechanisms of its members. His explanation of how these values-in-themselves become values-for-us indeed comes from "the province of biology", because pleasure is naturalised. Whatever is useful for an individual or a species *must appear to the individual as agreeable*. Although the direct causes of pleasurable factors are, from a psychological point of view, irreducible, they are not so from an evolutionary point of view. Adaptation, as a matter of biological fact, brings about that what is useful to the survival of the species betokens pleasure for the individual. Thus although an individual acts on the hedonistic principle, the objects of pleasure have an evolutionary explanation. And Schlick notes that once an organism achieves periods when it is not caught in the struggle for existence – when all energy does not go to the satisfaction of the primary needs – all perceptual and ideational activities which once were associated with pleasurable activities in the primary sense come to produce pleasure.

With this piece of armchair biology, let us return to "The Fixation of Belief". The problem Susan Haack picked out was the seeming conflict between the *immediate* motive for inquiry, viz. doubt and irritation, and the more sophisticated but abstract motive for seeking truth. Her solution was that this conflict vanishes into mere appearance, provided we make the distinction between temporary and indefeasible fixation of belief. As she notes, this is in line with Peirce's own gloss, the year before Schlick published his evolutionary explanation, of what he means: "if Truth consists in satisfaction, it cannot be any actual satisfaction, but must be the satisfaction which would ultimately be found if the inquiry were pushed to its ultimate and indefeasible issue." (CP, 6.485)

Peirce shared the view of Schlick and many contemporaries that "the deepest characteristic of normative sciences" is to be sought in esthetics, for here, he said, we have the dualism of the normative sciences in a softened form, in the opposing poles of the attractive and the repulsive. Parallel to that there are, in logic, the opposing poles of truth and falsity, and in conduct those of wisdom and foolishness (CP, 5.551). He also puzzled over Schlick's *Grundproblem* (without referring to it in Schlick's terms) and over the distinction between the feeling of pleasure (or of pain) associated or connected to the normative sciences. Esthetic good and evil are "closely akin" to pleasure and pain. But he then distinguishes between a *feeling* of pleasure (or pain) and the pleasure (or pain) itself, suggesting that pleasure and plain are secondary feelings (or generalizations) attached to other feelings, viz. pleasurable (painful) feelings. Pleasurable (and painful) feelings in turn are distinguished in accordance with "the kind of action they stimulate". The (esthetic) good is attractive (to a mature agent at least) and the (esthetic) evil repulsive. Perhaps we could say that pleasure and pain are determinables whilst the pleasurable feelings are determinants, falling in different fields.

Peirce's account is rather engaged, but does not void attempts to deal with the *Grundproblem*. So, is the True true because finding it out is a pleasure, or does the mature agent find the truth pleasurable because it is true? And what would the determinant feelings be that associated with this pleasure? One candidate for a determinant feeling would be that which arises when curiosity is satisfied. So is satisfying curiosity what makes scientists tick?

It is here that Peirce took a stand, one which came to set him apart from both Schlick and, more importantly, the pragmatists. Schlick was not the only one to flirt with the literal use of evolutionary theory in the service of epistemology. F. C. S Schiller had published, anonymously, his *Riddles of the*

Sphinx: A Study in the Philosophy of Evolution, in 1891[5] arguing for a Protagorean relativism and humanism in which all knowledge grows out of human interests and in which man is the measure. Allied with James, Schiller had argued that indeed all thought must contribute to the survival of the organism. Human thought, to a greater degree than animal thought, is adapted to the rapidly changing environment and hence exhibits the sort of flexibility characteristic of a mind that escapes rigid instinct. But Schiller went further than this by accommodating James's views on the will-to-believe, *equating* truth with what emerges as valuable.

We know that this is precisely where Peirce drew the line and departed company with the pragmatists. No doubt, he writes, men act, "in the action of inquiry" in particular, "*as if* their sole purpose were to produce a certain state of feeling". And once that feeling is achieved in "firm belief", the effort to move further ceases, whether or not that belief is true. But this purpose of inquiry is only *as if*, precisely because observation and the social impulse may seed of doubt (CP, 5.563). Therefore, also we cannot go along with the pragmatists and think that "the True is simply that in cognition which is Satisfactory" (CP., 5.555). Similarly, although the Gnostic Instinct, curiosity, "is the cause of all theoretical inquiry", and every discovery amounts to satisfying curiosity, it still would not be the case that "pure science is or can be successfully pursued *for the sake* of gratifying this instinct." If that were the case, the motive would not be the Gnostic Instinct but the Gust-Instinct, love of pleasure (CP, 7.58, italics in the original).

7. The Social Pattern: Having the Interest of Having No Particular Interest

What marks the scientific attitude, in Peirce's view, is the persistence in which its holders refuse to yield to immediate satisfaction and go on revising their belief until truth emerges (and of course, this goal may be unattainable for individual inquirers). His account, as glossed by Haack, and perhaps with a little bit of help from armchair biology, just might explain the emergence of the appetite for truth, and hence the pure truth-seeking scientific attitude.

---

[5] The third, revised edition in 1910 has a revised title also: *Riddles of the Sphinx: A Study in the Philosophy of Humanism*, 1910).

Here we can see that the parallel between Peirce and Merton breaks down, in two related ways. One has to do with the gloss that Peirce and Merton gave for the motive of seeking truth, the other one has to do with the mechanism through which the norms of science work. Both ways hinge on what disinterestedness amounts to.

Let us start with the mechanism. Whether or not Peirce's strategy for ascent from the Gust-Interest to the Gnostic Interest works, it is clear that a crucial link in his method of science hinges on inculcating in scientists, the most sophisticated thinkers, the disinterested attitude. This is a disposition of the scientific mind, along with other virtues, such as intellectual honesty and sincerity (CP, 2.82). But is there any independent characterization of what makes for sophistication in this sense? Granted that the desire to learn is an outcome of a long historical development, what paves the way from immediate and hence concrete and mundane satisfaction to someone else's (or indeed nobody's) satisfaction in the future? For recall, "if Truth consists in satisfaction, it cannot be any actual satisfaction, but must be the satisfaction which would ultimately be found if the inquiry were pushed to its ultimate and indefeasible issue." (CP, 6.485). In any case doesn't this come at least close to the view that science, including its self-corrective method, relies on its workings and successes on scientists *as individuals* having certain virtues. The social impulse helps, by feeding in information from fellow inquirers, but the method works by relying on the features of the individual inquirers, on their desire to search true, and on their epistemic integrity.

We can now see where Merton took a step beyond Peirce. Merton's genius was in the detailed analysis of communal mechanism in which motivation and control were organized. The scientific enterprise distinguishes itself not just in its goal of expansion of certified knowledge but in the minute way in which members of the scientific community are rewarded, encouraged, and punished. To say that scientists are disinterested is to say that they are motivated through the recognition of other members to the extent they have made a valuable contribution. This recognition comes in the form of peer acknowledgement and praise, established prizes, grants. As later generations of students of science studies in particular have stressed, this recognition results in accumulated credibility, and this in turn can be cashed out in terms of further possibilities of research.

What Merton does, then, is identify the values and norms that govern this creation and distribution of credibility and recognition: "The ethos of science is that affectively toned complex of values and norms which is held to be binding on scientists. The norms are expressed in the forms of prescriptions, preferences, permissions and proscriptions. They are legitimized in terms of institutionalised values." (Merton 1982, p. 5) The force of the scientific ethos therefore is based on the idea that norms are both an efficient means of achieving the institutional goal of science, certified new knowledge, and that they are also perceived by the scientific community as morally and socially binding.

But there is a problem with this account, the vary same we already dealt with from another angle. Here we come to the other side of the coin, disinterestedness. Peirce thought that the motive for theoretical inquiry cannot be the feeling of pleasure, for this would be self-defeating hedonism. Merton also took intrinsic satisfaction to be the fundamental drive within the scientific community. But what is this satisfaction, how does he cash out disinterestedness? In the end the source of satisfaction for Merton is not the enjoyment of truth as such, the contemplation of true proposition, but peer recognition.

But seeking peer recognition, much as seeking social recognition in general, does not seem to stay within the bounds of disinterestedness. It is not the interest in truth as such but interest in being a famous and even celebrated scientist or scholar. And this is, arguably, just one variety of non-cognitive interest. Isn't recognition, getting credit and social respect for being the first to capture a fact or to formulate a doable theory, or to apply a particular method, just one variety of non-cognitive interest? It is entirely conceivable (and seems sometimes to be true) that an inquirer is motivated by the search for recognition without being motivated by the proximate or distal search for truth. In any case, for a pure truth-seeker such recognition, quite as much as money and fame, should be irrelevant!

Merton's way out was the idea that the scientific interest is the interest of having no particular interest. Seeking and finding satisfaction in peer recognition does not go against the grain of the scientific enterprise. In so far as the drive for satisfaction and recognition does not lead to deviant behaviour and pathological modes of plagiarism, falsification of data, and so forth, it is in accordance with the norms of science. And even when dysfunctions raise their heads the community need not just look in disapproval. It is well-known that those who flout the moral or social norms are reprimanded (sometimes to the degree of being excommunicated).

Although the norms of science are coupled by or associated with the psychological traits that make for the scientific mind, these traits are not the underwriters of the scientific method. Rather, it is the faceless or impersonal pattern of organization that explains why the sciences work the way they do. Indeed, the entire idea of socially enforced disinterestedness amounts, perhaps paradoxically, to disinterestedness being the interest of having no particular interest. The key is the mechanism through which incentives work. When an individual scientist or group proposes a result, her or his incentive is to be recognised as having made a contribution. Others then have an incentive to probe into the foundations of the claim - the more so the more interesting or central the claim is, that is, the more potential the results has towards expanding knowledge. The outcome is that whatever the ulterior or non-truth seeking motives of an individual are, the impersonal mechanism turns this into truth-seeking. It is thus the invisible hand of the knowledge market that guides the community towards truth.

8. Liberal Democracy: Private Vices, Public Benefits

Niiniluoto ends his book *Critical Scientific Realism* (1999) with a chapter on "Realism, Science and Society". In its first section he discusses the social reasons for realism and anti-realism, and concludes that there is no direct link between views of scientific knowledge and social values: "the same moral and political attitudes have led to very different positions about science".

It is easy to agree with 's Ilkka's concluding remarks that the liaison between any particular form of government, such as liberal democracy, and the views on science is at best contingent, although he does bring out the possibility of sociology of philosophy of a sort, viz., that of seeing what particular forms of government could be favourable to which styles of thought or philosophy. But it might still be amusing and perhaps instructive to review some possible links, however contingent.

Merton held that science does and should form a relatively autonomous area in the following sense. Although scientists are part of the larger society and therefore under the influence of socio-economic forces, they have values and norms of their own. These values and norms give them their purpose and define the limits of autonomy. The aim is the expansion of certified knowledge, and truth, and the various particularistic interests, friendships, alliances, and other social forces operative within the scientific system should be subservient to the common cause. Although scientists have this autonomy, it is relative only, for Merton nowhere accepts the idea that scientists would be free from commitments to the establishment. It is also obvious that for Merton commitment to liberal democracy and its values finds its way to his description of the scientific enterprise. He writes (STSS, p. 588):

> Science flourishes and scientists make progress in an atmosphere of free inquiry and free interchange of ideas, with the continued mutual simulation of active minds working on the same or related fields.

Peirce thought that the method of science was the correct and ultimately the most efficient means of safeguarding progress, and that pure-truth seeking attitude was the key to its success. Free inquiry and unrestrained exchange of information were crucial for the workings of the scientific community, not least because they safeguarded the social impulse (which, to wit, is impossible to thwart anyhow, in the long run). I do not know how deeply Peirce was inspired by the ideal of liberal democracy. Merton certainly was, and so was John Dewey, which gives one reason why Merton's references were more often to Dewey than to Peirce. Mertons emphasis on relative autonomy of science certainly derives from his recognition that external forces, such as a totalitarian regime, can undermine the scientific enterprise. The ethos of science is a way of securing that scientists are not sacrificed to this or that political cause, and that the scientific community is allowed to apply its own moral and technical norms.

Merton agreed with Dewey in that there was a connection, non-accidental if not quite universal, between science and liberal democracy. Yet there was a fundamental difference between the two men. Merton did not think that Dewey had got the direction of fit right. Dewey had advocated the view that has been the default view since the so-called Scientific Revolution or, at the very least, from the heyday of the Enlightenment. This view, though not universally accepted, is that science will be both a liberating force and the means of bringing about prosperity and all manner of useful innovations. Since science is synonymous with rationality, the method of science should be adopted as the model of political conduct. As Dewey put this, the future of democracy depends on the spreading and consolidation of the scientific attitude.

Merton's agreed that there was a connection, but thought that Dewey had the model and the modelled mixed up. In his grand scheme science was to be modelled on liberal-democratic society rather than the other way around. The message in Merton's reversal of procedure is precisely that the norms of science, when at their best, mirror the democratic procedure – when at its best. This is also why the democratic personality (or the scientific mind) is more of an outcome than cause of democratic (scientific) institutions. The key to scientific success is in the norms of the well-functioning community rather than the interests and moral characteristics of its members. As on a free market in general rational self-interest (and even private vices) turn into public benefits, so in science private interests work towards the good of the public goal, expansion of knowledge.

Summing up the social roots of the scientific method in this way might give the impression that scientists do not value truth as such, or that truth only emerges as an aim of inquiry through the kitchen door. Similarly, it could be thought that cooperation only arises through selfish interest and the realization of mutual benefit. This has been suggested, again and again, in recent science and technology studies. As Adam Smith said, enlightened self-interest would be enough to bring about cooperation and division of labour: "by pursuing his own interest he frequently promotes that of the society more effectually than when he really intends to promote it". For Smith the invisible hand came as a result of people's natural inclination for cooperation. Bernard Mandeville went even further, suggesting that the invisible hand would work for the public good even if the individuals lacked all manner of virtues and pursued their private vices only.

I do not wish to suggest that there is anything conceptually self-defeating in the view that people can, sometimes at least, act in order to promote someone else's good. Rather, this seems frequently to be the case. Similarly, it is entirely possible that truth for the sake of truth could be a motivation of inquiry. Merton's strategy, to model science on society and not vice versa, merely highlights that the outcome would be the same even if the scientists' individual motivations were different or, more realistically, mixed.

University of Helsinki

References

Anderson, Douglas R. 1997: "A Political Dimension of Fixing Belief", in J. Brunning and P. Foster, *The Rule of Reason. The Philosophy of Charles Sanders Peirce*, University of Toronto Press, Toronto, Buffalo, London, pp. 223–240.

Brunning and P. Foster, *The Rule of Reason. The Philosophy of Charles Sanders Peirce*, University of Toronto Press, Toronto, Buffalo, London, pp. 223–240.

Haack, Susan 1997: "The First Rule of Reason", in J. Brunning and P. Foster, *The Rule of Reason. The Philosophy of Charles Sanders Peirce*, University of Toronto Press, Toronto, Buffalo, London, pp. 241–261.

Knorr-Cetina, Karin 1983: "The Ethnographic Study of Scientific Work: Towards a Constructivist Interpretation of Science", in K. Knorr-Cetina and M. Mulkay (eds.), *Science Observed: Perspectives on the Social Study of Science*, Sage, London, Beverly Hills, New Delhi.

Merton, R. K. 1968: *Social Theory and Social Structure*, Free Press, New York. (Enlarged version, first version 1949.)

Merton R. K. 1965: *On the Shoulders of Giants: A Shandean Postscript*, Harper & Row, New York, 2nd ed.

Merton R. K. [1938] 1970: *Science, Technology, and Society in Seventeenth-Century England*, Harper & Row, New York, 2nd ed.

Merton, R. K. 1980: "On the Oral Transmission of Knowledge", in R. K. Merton and M. W. Wiley, *Sociological Traditions from Generation to Generation*, Ablex, Norwoood.

Merton, R. K. 1982: *Social Research and the Practicing Professions*, Abt Books, Cambridge, Massachusetts.

Niiniluoto, I. 1985: "Truthlikeness, Realism, and Progressive Theory-Change", in J. Pitt (ed.), *Chanage and Progress in Modern Science*, D. Reidel, Dordrecht.

Niiniluoto, I. 1987: *Truthlikeness*, D. Reidel, Dordrecht.

Niiniluoto, I. 1999: *Critical Scientific Realism*, Oxford University Press, Oxford.

Peirce, C. 1877: "The Fixation of Belief", *Popular Science Monthly* 12 (November 1877), 1–15. As reprinted in Collected Papers, 1.

Peirce, C. *The Collected Papers of Charles Sanders Peirce*, Vols. 1–8, edited by C. Hartshorne, P. Weiss and A. Burks, Belknap Press, Cambridge, Massachusetts, 1931–5 and 1958. Referred to as CP.

Peirce, C. 1992: *Reasoning and the Logic of Things*. Peirces lectures of 1898, edited by K. Ketner, Harvard University Press, Cambridge, Massachusetts.

Schiller, F. C. S. 1891: *Riddles of the Sphinx: A Study in the Philosophy of Evolution by a Troglodyte*, Swan, Sonnenschein and Co., London (3rd revised edition, *Riddles of the Sphinx: A Study in the Philosophy of Humanism*, in 1910). Reprinted by Greenwood Press, New York, 1968.

Schlick, Moritz 1909: "The Fundamental Problem of Aesthetics Seen in an Evolutionary Light", in Moritz Schlick, *Philosophical Papers*, Vol. I (1909–1922), edited by H. L. Mulder and B. F. B. van de Velde-Schlick, D. Reidel, Dordrect/Boston, 1979.

Schlick, Moritz 1974: *General Theory of Knowledge*, Second edition. Translated by Albert E. Blumberg, Springer-Verlag, Wien, New York 1974. Originally published in 1925 as *Allgemeine Erkenntnislehre*.
Schlick, Moritz 1915: "The Philosophical Significance of the Principle of Relativity", in Moritz Schlick, *Philosophical Papers*, Vol. I (1909–1922), edited by H. L. Mulder and B. F. B. van de Velde-Schlick, D. Reidel, Dordrect/Boston, 1979.
Schlick, Moritz 1962: *Problems of Ethics*, Dover Publications, New York.
Sztompka, Piotr 1986: *Robert K. Merton. An Intellectual Portrait*, St. Martain's Press, New York.

Raimo Tuomela

On the Ontological Nature of Social Groups

1. Two Ontological Views of Social Groups

It is a platitude that human beings are group beings. But how are we to think of groups in this context? I will below discuss the ontological nature of social groups taken in a wide sense appropriate to the evolutionary history of human beings. Thus, we might simply say that a collection of people who take themselves to be members of the collectivity or group form a social group in the meant wide sense. (At least this is to be taken as a necessary condition for a group.[1]) Here taking oneself to be a group member can be understood in a weak sense which does not presuppose reflective cognitive understanding of the notion of group. However, in typical cases it is still true that group members are disposed to use the first person plural "we" to refer to their group. As my discussion of the ontological nature of groups will mainly concern groups in the wide sense, it is not highly sensitive to how exactly social groups are defined as to their content. Nevertheless, I would like to make a rather sharp distinction between two kinds of social groups, viz. we-mode and I-mode groups. These two kinds of groups can be argued to exhaust social groups in the wide sense meant here (but see Section II below). A we-mode group is, roughly, one in which the members have accepted some constitutive goals, beliefs, normative standards, etc. for the group and are collectively committed to them. Because of this the resulting social group will be able to act as a group. In contrast, an I-mode group is one to which the members are only privately committed and thus can base their functioning in it merely on their personal decision without having to face normative criticism from other group members in the case of quitting (although they may be criticized e.g. for lack of sufficient rational stability, etc.). My account of the ontology of groups will

---

[1] See e.g. Brown (2000), Chapter 1.

apply to both we-mode and I-mode groups, but many of my comments and examples will relate to we-mode groups, viz. groups in the full sense.[2]

A basic question to begin with is: Do social groups exist as entities? Both a positive and a negative answer can be justified depending on what kind of existence is referred to here. My view of what is real concerns observer-

---

[2] I will here state my analysis of we-mode and I-mode groups that I have argued for elsewhere. The following analysis of a we-mode group from Tuomela (2007), Chapter 1, is cited here with the understanding that an "ethos" means the group's constitutive goals, standards, beliefs, norms, etc.:

A collective g consisting of some persons (or in the normatively structured case position-holders) is a (core) *we-mode social group*
if and only if

(1) g has accepted a certain ethos, E, for itself and is committed to it. On the level of its members this entails that a substantial number of the members of g, including its specially authorized operative members (if any), when functioning as group members have accepted E as g's (viz., their group's, "our") ethos and are collectively committed to it, with the understanding that the ethos is to function as giving authoritative reasons for thinking and acting for the group members.

(2) Every member of g ought to accept E (and accordingly to be committed to it as a group member) at least in part because the group has accepted E as its ethos.

(3) Necessarily, the members collectively accept (with collective commitment) E as g's ethos if and only if (it is correctly assertable for them that) E is g's ethos.

(4) It is a mutual belief in the group that (1), (2).

A collective g consisting of some persons is an *I-mode social group*
if and only if

(1) The members of g (privately) accept some goals, beliefs, standards, etc. as constitutive for the collective, forming the collective's ethos E, and accordingly are committed to E at least in part because the others in g (privately) accept E, and this is mutually believed in g.

(2) The members of g share the beliefs that they themselves belong to g and that the others believe that they belong to g, under suitable, perhaps collective descriptions of membership.

Note that the social condition in clause (1) that the members accept the ethos in part because the others do also holds in the we-mode case, although in that case this condition is a consequence of the fact that the members take the group's acceptance of E as their reason for accepting E in the cases where E has already been accepted as the group's ethos. This point thus concerns those who have not yet accepted E, and a central case in point is where the group has a long history and those who originally established E for the group may not any more exist as group members.

independent reality and the capability of functioning as a cause or an effect. In this paper I will deal with the duality problem of groups as entities versus as non-entities in a "nominalistically acceptable" way. I will only briefly comment on the possibility that there could be groups in a *generic* ontological sense as abstract entities – except in the mental realm as thought contents. Social groups – here understood in a broad sense ranging from informal task groups to organizations and even societies and nations – can be regarded as singular entities consisting of persons and relevant interrelations (construed e.g. as tropes) between them or, perhaps, as singular mereological entities. Accordingly, groups are taken to exist at a certain time or at several, typically consecutive times, as some kind of singular conglomerations or fusions. This I will call the (or a) *singular entity view* of groups.[3] The other view is that groups are not entities of any kind. Rather, individual persons form social groups in terms of their social relationships and (collective) activities of suitable kinds. This I will call the *non-entity view*. Even under this construal, the group members may, and arguably sometimes must, think in terms of social groups in the sense that their thoughts involve groups as entities analogously to the way one may be said to be thinking about flowers, cars, or numbers.

Irrespective of whether one adopts the entity view or the non-entity view of the ontological nature of groups, one must distinguish and take into account both the subjective and the objective aspects of social groups. By the subjective aspect I mean simply that the group members – at least in the case of we-mode groups – are required to be disposed to think (believe, intend, want, feel, etc.) in a "groupish" way (e.g. deal with group entities in their thoughts). By the objective aspect I mean that the group members are required to act and behave in a certain way exhibiting group-belongingness and they are also subject to group-related norms. (This is an "adverbial" aspect – they have to behave "groupish-ly".)

---

[3] Niiniluoto says that his favorite version of physical ontology is based on a trope view of properties (Niiniluoto, 1999, Chapter 2, pp. 21, 30). He calls this view "tropic realism". The term would be apt would it not go against prevalent linguistic practice that takes a trope view to represent a kind of nominalism (in contrast to ontological realism). In any case, it would seem that what I call the singular entity view of groups would seem to be compatible with what Niiniluoto would say about the ontology of social groups. (I have not found specific remarks by him concerning this issue.)

The view that groups are entities must of course be distinguished from the view that groups are in a literal sense persons. I take this person view to be implausible. Groups are not persons, because they have neither bodies nor minds. Furthermore, postulating such extra minds beyond the minds of group members is not only superfluous but confusing. For instance, can a group be in charge of things and exercise agency if the members do not? Surely not.[4] However, from an instrumentalistic point of view it may often be highly useful and meaningful to say that groups intend, have beliefs, and act. This is only metaphorical talk, and it must be shown to have content in terms of real agents,

---

[4] As will be seen, groups can be regarded as conglomerations of (or relational systems consisting of) their members, their interrelations, and, possibly, of relevant artifacts all extended through time. So in this sense groups can be taken to exist as entities. However, groups are not strictly speaking agents or persons. I take the conceptual framework of agents and persons to presuppose that persons have bodies and bodily sensations, as well as perceptions and feelings, and can refer to themselves by the first person pronoun "I". Groups are not this kind of entities.

Carol Rovane (1998) argues that groups are persons. She ends up proposing the following necessary and sufficient condition for the identity of a group person (pp. 164, 180):

There is a set of intentional episodes such that
(1) these episodes stand in suitable rational relations so as to afford the possibility of carrying out coordinated activities;
(2) the set includes a commitment to particular unifying projects that require coordinated activities of the very sorts that are made possible by (1);
(3) the commitment to carrying out these unifying projects brings in train a commitment to achieving overall rational unity within the set.

I take these conditions jointly to be true of we-mode groups, viz. groups that can act as groups. Whether or not they are sufficient and necessary for a we-mode group depends on how we-notions can be taken to be involved in the conditions. I cannot speculate about that here. Instead, I wish to point out that group persons in Rovane's above sense do not have to be ontological agents "in their own right". It does not follow from (1)-(3) that there are in the social world group agents in a sense going beyond (or being "over and above") joint agency. Thus, if (1)-(3) were all that group persons require this is compatible with my account.

In a recent paper Philip Pettit (2003) claims that groups are indeed persons, but his ontological view remains unclear. He says that groups may be entities supervening on their members' states and what have you. This still entails, given that supervenience is non-reductionistic, that groups have an existence *sui generis*. My account need not rely on this kind of view, which I find unclear, somewhat mystical, and unnecessary for dealing with what is to be accounted for – e.g. acting in the we-mode.

viz. human beings qua position holders who act either separately or jointly in relevant group contexts. Thus, in a nutshell, my view is that groups can, but need not, be taken as (singular) entities, and they are agents and persons only in a metaphorical sense.

According to the singular entity view, social groups – be they normatively structured or not – are suitable kinds of conglomerations of persons at each point of time at which the group exists. A group may be spatially (and temporally) dispersed in the sense that the group members are not spatially (or temporally) connected in a causal sense. The fusion of (the elements of) such a conglomeration existing at certain times over space and time yields a singular group entity. Here 'fusion' is an informal term which might, but need not, be taken in the sense of a mereological sum. We may thus speak loosely of ordered sequences of the kind $(g^1,\ldots,g^k)$ corresponding to the times $t^1,\ldots,t^k$ at which the group exists. Each g^i, $i=1,\ldots,k$, is a fusion over space of the group members existing at t^i.

I will now present a more precise statement of the singular entity view of social groups. In my presentation I will use the material mode exposition. Note, however, that later I will speak of linguistic entities as well. Below, a group name "g" will in general be taken to refer to a real group entity and a group predicate "G" will be taken to represent a specific group aspect G of some agents' activities. The central idea here is that G will include at least the kind of activities that upon analysis can be described as functioning as a group member.[5] I propose the following general elucidation of the notion of a social group (covering both I-mode and we-mode groups):

(SG) Singular entity g is a *social group* if and only if there is some (typically complex) aspect G and there are times $t^1,\ldots,t^k$ representing the history of g such that in the case of all these times t^i $(i=1,\ldots,k)$
(1) g, viewed as a fused sequence $(g^1,\ldots,g^k)$, has some persons, A^i_1, $\ldots,A^i_m$, as its members at t^i;
(2) these members $A^i_1, \ldots, A^i_m$ think and act (function) in the special G-way related to g.

---

[5] See Tuomela (2007), Chapter 1, for a detailed discussion.

We can now say that g is a social group in a diachronic sense if and only if for all times t^i, $G(A^i_1, ..., A^i_m)$, viz. clause (2), is satisfied by $A^i_1, ..., A^i_m$.

In the present analysis g is a singular rather than a generic entity. A generic entity is one which can be multiply instantiated in contrast to a singular entity. The basic argument for the singular (as opposed to the generic) entity view is that it does not postulate "unrequired" new entities in the world. The reason is simply that with singular entities one can do all that one wants to do in this connection. Thus, (a) one can account for the behavior of groups – e.g. actions performed by groups; (b) using the above "adverbial" account, which puts the main burden on the predicate "G", one can account for the continuity of groups (ontically: the continuity in a consecutive series of the existence of singular entities $g^1,...,g^k$); and (c) one can accommodate common sense locutions in this kind of context (e.g. "g performed X", "g is responsible for X", "g is cohesive", "g is egalitarian").[6]

In my account g is a singular real entity. Candidates in ontic terms are singular constellations of persons (and possibly other "hardware" such as tools of some kind) and their interrelations, conceived in terms of non-Platonistic relations. My preferred way is to think of such relations as nominalistically acceptable "tropes", viz., particularized properties and relations, which we can take to account for the internal structure of events, and states. Thus we can speak of the ontic grounds of the applicability of those predicates. This idea can be combined e.g. with Sellarsian metalinguistic nominalism that is based on a use account of meaning and "languaging".[7] Abstract entities are in that account explicated in terms of the correct uses of words, in our present kind of case words for putatively social and institutional objects and properties. No extralinguistic entities are to be postulated here. Instead, it is assumed that people can meaningfully and functionally use such language in response to "what is going on in the world", viz. various extralinguistic states of affairs and happenings. Thus e.g. "group", "organization", "government", and other similar putatively entity-expressing concepts can in principle be handled, if the program is tenable. The same goes for such explicitly normative concepts as "obligation", "right", and "responsibility". Thus groupness and normativity

---

[6] My definitions of we-mode and I-mode groups serve respectively to give general content to the G-predicate.

[7] See Sellars (1979).

would not in the first place require semantically connected referents. Only psycho-linguistic use is taken to account for the connection between language and whatever the language can in a non-semantical, contingent sense be said to be about. Language use by itself does not have ontological implications, but "ontological chips will fall where they may". When combining this Sellarsian view with a trope view of real properties and relations, it is to be kept in mind that the connections to tropes will be contingent only and, furthermore, that at least in my view there are not such things as real *normative* properties (tropes).

This kind of view is in stark contrast with the view of groups as generic abstract entities that John Searle has recently expressed. According to it, a corporation "need have no physical realization, *it may be just a set of status functions*" and there need thus be no position holders.[8] I would like to say against this kind of view that what is created when a corporation is brought into existence is rather certain meaningful normative and descriptive statements on paper, on computer hard discs, in people's minds and, especially, uses of language, etc., meaning here being understood in terms of a use account. Searle's nonphysical corporation as a mere set of status functions is not what a social group (e.g. a corporation) is – at least in this paper (see (*SG*)). Rather, a social group has social reality in that there are some G-connected people performing activities and functioning in certain ways. A group is not only a bunch of principles (or, to use my terminology below, an "ethos") without any actual members. One can debate about what ontological status propositions, concepts, statements, normative principles, etc. have, but even if they were regarded as mere abstract objects, it would not follow that groups are, or are

---

[8] See Searle (2005). He continues by saying: "The laws of incorporation in a state such as California enable a status function to be constructed, so to speak, out of thin air. Thus, by a kind of performative declaration, the corporation comes into existence, but there need be no physical object which is the corporation. The corporation has to have a mailing address and a list of officers and stock holders and so on, but it does not have to be a physical object. This is a case where following the appropriate procedures counts as a creation of a corporation and where the corporation, once created, continues to exist, but there is no person or physical object which becomes the corporation. New status functions are created among people – as officers of the corporation, stockholders, and so on. There is indeed a corporation as Y but there is no person or physical object X that counts as Y."

only, abstract entities. Searle's version of the abstract entity view is not plausible and is not rationally forced upon us.

My above nominalistically acceptable sketch of linguistic meaning arguably is not committed to abstract entities other than possibly to something like the *sameness of use* of linguistic and conceptual items.[9] The central conceptual element in my account is the property or aspect G. G stands for "groupish" thinking and acting relative to g (here regarded as a singular entity, a suitable network). Groupish thinking and acting involves the existence of (what I have called) an *ethos*, viz. some constitutive goals, values, standards, and norms and we-mode thinking and acting relative to them. In the case of a specific group g we of course are dealing with an ethos specific to g, call it g-ethos, and relevant we-mode thinking and acting relative to this g-ethos.[10] So, we have available the notion of ethos and what it ontologically involves; but, as said above, this does not allow groups to be abstract entities or at least merely abstract entities.

Social groups often are normatively organized and structured. Organizations and (formal) associations are typical examples of such groups. Consider thus a structured group, g, with positions P_i, i=1,...,m. A structured group consists of (1) an open domain, D, of position holders, (2) positions $P_1,...,P_m$, and 3) a task-right system, TR, defining and governing these positions.[11] TR typically consists of general group norms (ought-to-be, ought-to-do, may-be, and may-

---

[9] See Sellars (1979) for detailed discussion (also see Tuomela, 1985, Chapter 3); and, for the non-linguistic but conceptual aspect, see Tuomela (2002), Chapter 3.

[10] By we-mode thinking and acting I mean, roughly, that the members, while functioning as group members, think and act for the group (for its use and benefit) and are collectively committed to the contents of their thoughts. See Tuomela (2002; 2003; 2007) for a detailed account of the we-mode and the I-mode.

In general, a social group should be epistemically and doxastically adequate (recall note 2). Thus the group members should believe, in general mutually believe, that they belong to the group, and such beliefs may even be constitutive of the group. Such shared or mutual belief need not be particularized *de re* belief. The members need not know of each and every member that he is a member of the group, but they need to believe that there are other members and need to know under some, perhaps generalized description who the others are (animal lovers or stamp collectors, or whatever similar). Even such social beliefs can in some cases form the ethos of the group and can thus serve to give the group its identity.

[11] See Tuomela (1995), Chapter 1, for task-right systems.

do norms) and norms specific to the positions P_i. Thus we can say that g = <D, $P_1,\ldots,P_m$, TR>. I wish to emphasize that a position is a general entity in the sense that the class of position holders is potentially infinite. Thus, considering a position P, (x)P(x) indicates the open class of position holders. What this means is that a structured group cannot be reduced to a finite collection of group members. (However, temporary groups without structure are finite.) When one is speaking of such a group one is thus committed to more that its current position holders or any finite set of previous, current, or future position holders. We can take the group predicate "G" in principle to cover positions and task-right systems. Thus our earlier definition of a social group also applies to normatively structured groups.

Let me note that a singular group entity, g, can be connected to several ethoses and several group aspects G, G', and so on, and thus it is "intensionally aspectual". Thus if we say that g is the Research Committee of the University of Helsinki, we are saying that g is here viewed in a certain way related to university research. It might happen – power tending to concentrate – that g would also be characterizable as the Real Estate Committee of the university.[12]

I will now proceed to a view of social groups – that I have earlier defended – that does not even postulate them as entities in a strict sense. This is the non-entity view of social groups.[13] One motivation for it is that generic entity construals are ontically problematic. Nevertheless, we have the following two accounts to consider. Consider first a statement that says that an entity is a social group of a certain kind, e.g. "The University of Helsinki is a social group". Here the noun "social group" is used predicatively to apply to a proper name ("The University of Helsinki"), and it corresponds to our above predicate

---

[12] See Gilbert (1989), Chapter IV, for a similar kind of example related to the intensionality of group characterizations. Her Simmelian account of social groups is rather close to my analysis of we-mode groups.

[13] I have earlier formulated a non-entity view in in Chapter 4 of Tuomela (1995); cf. also the 1984 book for essentially the same view. We can now formulate the view in terms of the above account. The basic change that has to be made, obviously, is to remove the assumption of a real entity. Below I will try to preserve the commonsense phrase "g is a group" instead of directly using the ontologically more parsimonious locution "$A_1,\ldots,A_m$ form a group g", which was my phrasing in Tuomela (1995). My view in that book grounds a group ontically, although not conceptually, on its members' actions and interactions (see Chapters 4 and 9 of the book).

G. Consider next a statement that says that certain persons form a group or are members of a group, e.g. "g has $A_1,\ldots,A_m$ as its members" or "$A_1,\ldots,A_m$ form group g". In these cases, in contrast, "g" can be seen as a singular term referring to something like a singular nominalistically conceived network with the mentioned individuals as their members (or, perhaps, to a mereological sum with the members as its parts). However, in the present case "g" may alternatively be regarded as a non-referring singular term. Within this construal, we cannot understand the standard Tarskian truth conditions for sentences of this kind (viz. if there are no singular group entities). Accordingly, for this kind of case, I have proposed a solution relying on the "vicarious existence" of groups.[14]

Using the above machinery I will now present the vicarious view – in a stylized way – as follows:

> The sentence "g is a social group with the persons, $A_1,\ldots,A_m$, as its members" is true in the *vicarious* sense if and only if
> 1) the names "A_1",…,"A_m" in the mentioned sentence succeed in referring to some real persons $A_1,\ldots,A_m$, and
> 2) these persons $A_1,\ldots,A_m$ think and act (function) at t in a special G-way, and "g" (in the analysandum sentence) is regarded as a non-referring term serving at least implicitly to indicate and "specify" the predicate "G" (referring to the G-way).

(Note that if we in clause 2) instead take the view that "g" is a singular term referring to a singular group entity, a suitable kind of network, we do not need the last specification about g.)

We can now say this:

---

[14] Let me still mention another line of argument against the entity conception of groups. Schein (1993) argues against the existence of plural objects, claiming that postulating their existence leads to Russell's paradox. His linguistic-logical discussion is interesting, but I am not totally convinced of his approach due to the assumptions that he makes. So I still tentatively allow for the existence of plural objects, although my approach does not need them. Recall, too, that I have accounted for "we" as a predicate, not as a plural object. (I wish to thank Kirk Ludwig for the reference to Schein's work.)

The sentence "g is a social group with the persons, $A_1,...,A_m$, as its members" is true in the vicarious sense at t if and only if $A_1,...,A_m$ form a group relative to g if and only if the sentence "$G(A_1,...,A_m)$" is true.

Note that things are even simpler if we take "g" to be a singular term for singular group entities, for then we can say, given our specifications about the relationship between the group entity and its parts, that g is a group if and only if $G(A_1,...,A_m)$.

As seen, the group predicate "G" (as applicable to group members collectively) and the truth-equivalent "is a group" (as applicable to the group entity) involve the group ethos (and this may apply even to I-mode groups), which serves to define (at least partly) group identity and which is assumed to be preserved through time when the group exists at various times.

I will end this section by briefly considering group identity through time. In the literature the phrase 'collective identity' is often used. Mathiesen gives a partial, sufficient-condition analysis of collective identity in this sense.[15] My account owes to her account, although I depart from it in some respects and use my own technical terminology.

Briefly, my definition goes as follows:

g at t is (largely) the same as g' (or g is *"collectively identical"*, *c-identical*, with g') at t' if and only if

(i) the content of the ethos of g' at t' is obtained from that of g in terms of a family-resemblance transformation based on collective acceptance;
(ii) g' is a causal descendant of g;
(iii) the members of g' collectively accept, in effect, to view (i) and (ii) as true;
(iv) it is a mutual belief among the members of g' that (iii).

We say that the ethos E' of g' is obtained from the ethos E of g by means of a family-resemblance transformation if and only if for every time t_i such that lies between t and t' the ethos $E_i t_i$ has been obtained from the ethos E_{i-1} at t_{i-1} by means of the collective acceptance (here perhaps only tacit "going along with") that the transformation preserves collective identity.

---

[15] See Mathiesen (2003).

Analogously we can define causal descendance except that now we do not require collective acceptance over and above that specified by (iii) and that only partial causation need be at stake. We might require causal necessity: for all i, g_i would not be a group with ethos E_i unless the previous group g_{i-1} would not have been a group with ethos E_{i-1}. Partial causation in the present sense of causal necessity need not be transitive. Thus it need not be the case that the relation holds between g and g'. An example of dynamic sameness in the above sense is given by the Glenn Miller Orchestra. The orchestra was started by Glenn Miller in the 1930's and survived his death in 1944. Without much if any change in its ethos (the type of jazz and the kind of playing together it exemplifies) it has survived under different leaders until today, when Wil Salden is its leader.

As also Mathiesen emphasizes, a group can divide into two, viz. branching is possible. As noted, also transitivity may fail. These observations show – correct as they are – show that c-identity is not an identity notion in the standard sense that satisfies transitivity and the linearity condition that prohibits branching. Note that if and when g' and g" fuse into the sum g'+g" it may well happen – although there is no necessity – that (iii) and (iv) become satisfied and that the fusion group is c-identical with g.

2. Which View to Adopt?

It seems that, after all, there is not much to choose between the non-entity view and the singular entity view. In particular, if causality is to be the central criterion of real existence (to exist is to be capable of occurring in causal contexts as a – partial or total – cause or effect), it would seem that these two views are causally equivalent and that thus, in particular, the singular entity view does not have causal entailments that the non-entity view does not have.[16] The basic reason for this is, or so I claim, that the entity view is only a redescription

---

[16] I assume that for something to be a real entity in the meant sense, it must be capable of occurring in causal contexts, as a cause (at least a partial one) or as an effect (see Tuomela, 1985, for discussion of this common naturalistic view). This means that a group entity in the above sense really can do causal work. This causal work might even be claimed to be irreducible to the causal work done (jointly and/or separately) by the elements of the singular network that an extensional group, g, is. However, I will below argue against this view.

of the non-entity view. Groups can be viewed in both ways, it all depends on the existing collective, especially linguistic, practices and acceptances. The members of a collective can collectively accept one rather than the other usage of language. At this level there are in principle no metaphysical constraints to decide which way to go. There are suitably interrelated people involved, and whether we describe them by using the singular entity view as a system of interrelated people or as people being related in such and such ways does not make a causal difference, provided of course that we take into account all the relevant features of interaction and the ways the people in question think and act. In the case of we-mode groups it is pertinent to impose the requirement, based on the very nature of we-mode groups as involving a "togetherness-we", that the group members be disposed to think in group terms, e.g. "Our group will do X" or "Our group is better than that group at doing X". Groups can well have "intentional existence" in this sense and also have objective existence in the sense of the members interacting in some ways specific to the group in question – all this without an ontological necessity to conceptualize and describe the group even as a singular entity. However, the instrumental advantages of describing groups as entities are often considerable, and this has been widely acknowledged among group theorists, not to speak of ordinary people.

To illustrate the issue of the ontic and causal issues related to groups, let me consider briefly the view of groups ("collectivities") that Keith Graham advocates:[17] A number of individual human beings form a collectivity if and only if (i) they act in ways whose significance can be adequately captured only by an ineliminable reference to some corporate body as part of which they are acting, where (ii) what that corporate body does is distinct from anything which they as individuals do, and where (iii) the corporate body is a persisting one whose survival is relatively indifferent to the persistence of the particular individuals which compose it at any particular moment.

Graham does not aim at a non-circular analytical definition and says that he will rely on the reader's assent to the existence of collectivities by appeal to the many examples of corporate bodies such as committees, clubs, families, electorates, firms, etc. I would seem that the actions in (i) must be (to use my terminology) we-mode actions, actions performed as a member of the

---

[17] See Graham (2002), pp. 68–69.

collectivity. However, as the "phenomenon of the clique" below shows that is not invariably the case. Graham must allow that the "ineliminable reference" to the group may be only from an external describer's point of view and need not have any connection to what the members think. He also claims that we require vocabulary to speak of irreducibly collective actions and that this requires the postulation of collective entities. He even says that "an irreducibly collective action requires an irreducibly collective agent" (p. 82). However, even a simple joint action such as carrying a table or singing a duet can be taken to provide a counterexample to this claim: in such cases some agents perform the irreducible collective action, and there is no need to postulate a special entity to account for agency. If more persistence is required we can generalize the present account and speak of irreducible social practices performed by some individuals acting as group members. There is a group involved here of course, but it suffices to deal with it in an "adverbial" sense starting with the idea that some individuals form a group, viz. share some relevant properties or make a relevant "group predicate" satisfied.

Graham seems to take collective entities to exist as corporate bodies on the ground that they are part of the best descriptions of the social world, as he ends the section by this: "And if the criterion for an entity's existing is that ineliminable reference to it occurs in our best *descriptions* of the world, then collectivities exist." (p. 83) However, it is far from clear that this conditional statement, together with the assertion of the antecedent, actually entails the consequent. This is because we do not presently know what the best descriptions of the world are. Some correction surely is needed to the common sense framework in order to avoid e.g. taking the word 'witch' to refer to an existing entity. The same caution in linking language to ontology is needed in the case of collectivities.

As to (ii), Graham takes the distinction to be there on contingent grounds. (The existence of the distinction is of course a well-known fact.) The persistence requirement (iii) seems to me unnecessary, for there are certainly temporary groups formed for some particular tasks or goals. Why exclude them? On the other hand, Graham's account of collectivities is very comprehensive. It includes e.g. groups (collectivities) which act unintentionally and whose members do not know that they form a group.[18]

Graham takes intentional or "transparent" accounts of collectivities to be too narrow. An intentional account takes the members' intentional states to be somehow constitutive of collectivity. The "phenomenon of the clique" is meant to give an argument against such a view. In this example, Graham claims that there could be "a number of individuals who know each other well and tend to engage in exchanges which presuppose a great deal of prior acquaintance with their interests, their sense of humour and their ways of relating to one another. Such a group of individuals may collectively exclude other people from their social exchanges ...." He argues that the group of individuals forming the clique would constitute a collectivity. According to the present definition of a clique, a collectivity need not even amount to an I-mode group in my aforementioned sense, because the members do not have to take themselves to be members of the same group or collectivity.[19] It is crucial, according to Graham, that the members of the clique might function in the mentioned way quite unwittingly and unintentionally. They might not be aware that they form a collectivity at all or one of this particular kind; or they might be aware of all of this, but unaware that they had collectively performed some particular action on a particular occasion.

Let us consider the weakest of the aforementioned cases, which may be empirically possible. It involves 1) unintentional and 2) unconscious (unaware) action, and, furthermore, 3) the members are not aware that they form a collectivity. Suppose thus that the collectivity unintentionally freezes out a person (viz. cause this freezing-out event by their unintentional action). Thus the members do not purport (individually or collectively) to freeze out the person. Secondly, they are unaware of their action, i.e. they (individually and, of course, collectively) lack the belief that they are freezing out the person in question. As to 3), we can suppose that the members only deal with each other as individuals ("I interact with you and him, and so on").

---

[18] I do not as such object to this idea, and have even myself discussed resembling cases, such as spy rings, in my work — see e.g. in Tuomela (1984), pp. 131–132, (1995), Chapter 5. However, in those cases, the group members still have at least a vague idea that they are acting as group members.

[19] This is contrary to the rather basic "Simmelian" requirement of groupness accepted also in my treatment above.

In my view, the phenomenon of the clique does not after all require the postulation of a corporate body. All we need to say here is that there are a number of individuals who unconsciously and unintentionally and interconnectedly (or jointly in a wide sense) behave here and jointly cause the freezing in question. To simplify, suppose we are initially dealing with three persons A, B, and C in the same face-to-face situation. A and B are engaged in intense discussion with each other and seem not to pay attention to the presence of C. Eventually, C, not getting any attention, leaves the others and "the situation". Here A and B unintentionally jointly caused this consequence by their interconnected behavior, their behaviors being at least jointly sufficient and perhaps separately necessary for the consequence. While we may say that the constellation or "corporate body" consisting of the agents A and B and their relevant physical and mental interactions (not involving, however, anything like their intention to freeze out C) caused the freezing out phenomenon, we can equally well just say that A and B were the agents of the situation who unintentionally and unconsciously came to form a clique with certain causal powers involving that they by their interconnected activities (behaviors) caused the freezing out to take place. And, to account for the persistence requirement, it can be suggested that A and B (the members of the clique) may have become engaged in a social practice – recurrent social activities – of the kind resulting, as an unintended consequence, in continuous freezing out of such members as C.

The phenomenon of the clique case is an interesting case in that it demonstrates the (possible) existence of a kind of groups of the non-transparent kind – while, if I am right, it does not require the postulation of an ontological entity as a "corporate body" (not even in the sense of the singular entity view).[20] Nevertheless, as said, this kind of clique hardly deserves to be regarded as a social group and at least it fails to be even an I-mode group in my sense.

If social entities are treated as above either in the singular entity sense or in the vicarious sense, naturalism – one without spooky entities – about the social

---

[20] Let me add that, indeed, the voluminous literature on "unintended consequences" in the social sciences is in part concerned precisely with the present kind of phenomenon of non-intentional causation.

realm gains support.[21] The general point is that alleged social entities such as groups, organizations, communities, and societies will basically be analyzable in terms of individuals recurrently acting jointly, having joint attitudes, joint power, and so on. In short, we-mode attitudes and practices are in a crucial position here.[22] As far as ontological naturalism – ontological similarity with the subject matter of standard natural science – in the social sciences is concerned, the we-mode jointness level is needed and although it might be taken conceptually to involve holistic social entities it is ontologically "inter-relationistic" and compatible with naturalistic scientific theorizing.

University of Helsinki

References

Brown, R. 2000: *Group Processes*, 2nd ed., Blackwell Publishers, Oxford.
Gilbert, M. 1989: *On Social Facts*, Routledge, London.
Graham, K. 2002: *Practical Reasoning in a Social World: How We Act Together*, Cambridge University Press, Cambridge.
Mathiesen, K. 2003: "On Collective Identity", *Protosociology* 18–19, 66–86.
Niiniluoto, I. 1999: *Critical Scientific Realism*, Oxford University Press, Oxford.
Pettit, P. 2003: "Groups with Minds of Their Own", in F. Schmitt (ed.), *Socializing Metaphysics: The Nature of Social Reality*, Rowman and Littlefield, Lanham, MD, pp. 167–193.
Rovane, C. 1998: *The Bounds of Agency*, Princeton University Press, Princeton, N.J.
Searle, J. 2005: "What is an Institution?", *Journal of Institutional Economics* 1, 1–22.
Sellars, W. 1979: *Naturalism and Ontology*, Ridgeview Publishing Company, Reseda, CA.
Schein, B. 1993: *Plurals and Events*, The MIT Press, Cambridge, Mass.
Tuomela, R. 1984: *A Theory of Social Action*, Synthese Library, Reidel Publishing Company, Dordrecht and Boston.
Tuomela, R. 1985: *Science, Action, and Reality*, Reidel Publishing Company, Dordrecht.
Tuomela, R. 1992: "Group Beliefs", *Synthese* 91, 285–318.
Tuomela, R. 1995: *The Importance of Us: A Philosophical Study of Basic Social Notions*, Stanford University Press, Stanford, CA.

---

[21] See Tuomela (1995), Chapters 4 and 5.

[22] See Tuomela (2002) for social practices based on shared "we-attitudes". They also include unintentional practices ("collective pattern-governed behaviors" in the sense of Chapters 3 and 4 of the book) and arguable can provide the dynamic ontological underpinnings of social groups and structures.

Tuomela, R. 2002: *The Philosophy of Social Practices: A Collective Acceptance View*, Cambridge University Press, Cambridge.
Tuomela, R. 2003: "The We-mode and the I-mode", in F. Schmitt (ed.), *Socializing Metaphysics: The Nature of Social Reality*, Rowman and Littlefield, Lanham, MD, pp. 93–127.
Tuomela, R. 2007: *The Philosophy of Sociality: The Shared Point of View*, Oxford University Press, New York.

Ilkka Niiniluoto

Replies

The best birthday present that a scholar can imagine is a *Festschrift*, a collection of essays by friends, distinguished colleagues and former students. I am most grateful to the editors and authors for the excellent articles on important topics that cover the most central areas of my philosophical interests. It has been a stimulating and refreshing pleasure for me to read these papers.

I am also happy that the editors have given me the opportunity to write short replies to the essays. In my comments, I shall slightly depart from the order in which the articles appear in the book. Some personal remarks on my relations to the authors are also given.

Hintikka on Hilbert

It is natural for me to start from Jaakko Hintikka's paper. I have learnt from Hintikka more than any other philosopher about logical and analytical methods. I wrote my doctoral dissertation in 1971–73, when I served as Hintikka's research assistant in the Academy of Finland. I spent the spring term 1972 at Stanford with Hintikka. We wrote together three papers on surface semantics, Ramsey-eliminability of theoretical concepts, and the axiomatization of inductive logic. My main work on induction and truth-likeness has relied on techniques that I learnt from Hintikka's abundant tool box.

Game-theoretical semantics and the IF logic are important areas of Hintikka's studies where I myself have not done scholarly work. It is indeed a remarkable feature of the IF logic that, within a logic involving only first-order quantification, one can achieve some interesting things that are known to be possible within second-order logic (e.g., the definition of a truth predicate for a language in the language itself).

Hintikka's statements about Hilbert seem convincing to me. Hilbert was not a formalist, since he understood that axiomatic theories are applied to their interpretations. As Hintikka has argued elsewhere, Hilbert supported the calculus view of the language of mathematics. It is evident that Hilbert's school of metamathematics influenced Tarski, Gödel, and Carnap who developed the main ideas and results of logical semantics in the 1930s.

In fact, the routine classification of the main approaches to the foundations of mathematics – logicism, intuitionism, and formalism – is derived from a symposium in the journal *Erkenntnis* in 1931. The main papers were translated in the influential collection edited by Benacerraf and Putnam (1964). But they date back to a time, when the model-theoretical approach to the foundations of mathematical theories had not yet been developed, and the interplay between syntax and semantics was still poorly understood.

Haaparanta on Frege

Leila Haaparanta defended her doctoral dissertation on Frege in 1985. Her expertise on Frege can again be seen in her essay. I don't have objections to Haaparanta's careful formulations, but I shall just add two comments.

First, when Popper introduced his doctrine of World 3, he expressed his admiration of Frege's anti-psychologism. In my own reading, I have been convinced that Frege was a Platonist in his conception of "the Third Realm", but an interesting case can be made for a Kantian transcendentalist interpretation of Frege's project. If it turns out that Frege's views should be understood in epistemological rather than ontological terms, a similar ambiguity has been sought in Popper's position.

Secondly, when Frege's method of discovering logical categories is compared to the ancient geometrical analysis, it would be interesting know how much Frege himself knew about the applications of analysis in Greek geometry. We remember that, in opposition to analytically true arithmetical propositions, the truths of geometry were synthetic for Frege.

Von Plato on Geometry

As a young professor in 1980, I was the opponent when Jan von Plato defended his Ph.D. Thesis on probability. But, obviously with good reasons, he was too independent to be really worried about my objections. He has always had an inclination towards intuitionism and constructivism in mathematics.

Later he has become (with his wife Sara Negri) a leading proof theorist. I am particularly delighted that von Plato has been able to highlight the important achievements of my predecessor Oiva Ketonen in the Genzen-style proof theory in the 1940s.

As many contemporary intuitionists tend to operate with sophisticated proof-theoretical techniques, they might deserve the title "formalist" rather than Hilbert. Von Plato's essay on the foundations of geometry is different in an interesting way. His dialogue appeals to what can and what cannot be seen in geometrical figures drawn on a blackboard, i.e., to "intuition" in the old German sense of *Anschauung*. Commitment to intuitionistic doctrines is colored by the acknowledgment that disjunction and existence might behave in a manner which differs from the traditional view of intuitionistic logic.

Raatikainen on Tarski

Panu Raatikainen, a Ph.D. of 1998, successfully continues the Helsinki tradition in philosophical logic. I have observed with satisfaction that in many important respects our views about logic and scientific realism are close to each other.

Raatikainen gives an excellent compact summary of recent debates about Tarski's theory of truth. He correctly notes my defense of the thesis that (*pace* Hodges) Tarski employed a model-theoretical account of truth already in his early work in the 1930s. I have only one reservation to Raatikainen's conclusion: "Tarski did not yet have a *wholly* general notion of *truth-in-a-model* (as e.g. Niiniluoto (1994, 2004) seems to suggest)." Raatikainen appeals here to the point that Tarski assumed "a fixed-domain conception" of models. But this has been my claim all along. I have argued that the early Tarski represented a middle position (one world, many languages) between the universal medium view (one language, one world) and the calculus view (many languages, many worlds). A similar limitation to the one domain assumption was adopted in Carnap's semantics. A full-blown move to the calculus view was achieved only in the 1950s in the mature model theory in mathematics and the possible worlds semantics in philosophy.

Following the Lvov–Warsaw tradition, Tarski assumed that a language L has to be interpreted. For Tarski, such an interpretation guaranteed that the sentences of L have truth values and "meanings". In my view, this means that he smuggled in a semantic notion of "interpretation" – even though he officially demanded that truth should be defined without making use of semantic notions. One way of making Tarski's hidden assumption explicit is the interpretation function I of later formulations of model theory. For example, when M is a one-place predicate in a formal (now uninterpreted) language L, the function I maps M to a subset I(M) of the domain of a structure W. Applied to a natural language L (e.g. German) and its interpretation in the actual world W, we have e.g. for the predicate 'rot':

(1) The predicate 'rot' denotes the class of red things.

Raatikainen suggests instead that the notion of translation, understood as a purely syntactic or non-semantic mapping between two languages, would do the job. This can be motivated by Tarski's T-schema

(2) Sentence x is true in L iff p,

where p is the translation of x in the metalanguage ML. Written more explicitly,

(3) Assuming that p in ML is the translation of x in L, sentence x is true in L iff p.

However, it seems that (1) and (3) are two sides of the same semantical medal. This can be seen from Carnap's treatment of semantics (see Niiniluoto, 2003a): already in 1939, Carnap introduced notation for the designation function des which interprets the terms of a language L. For example,

(4) Predicate 'rot' designates the property of being red.

Thus, des(rot) = the property red. Properties determine extensions by the principle:

(5) The extension of predicate 'rot' = the class of red things
     = { x | x has the property red }.

Hence, the extensionalist semantics (1) and the property semantics (4) lead to the same truth condition:

(6) The German sentence 'a ist rot' is true iff a is red.

Carnap also showed how to define the function des by recursion for arbitrary sentences h of L: des(h) is then the proposition designated by h. In Carnap's *Introduction to Semantics* (1942), the schema corresponding to Tarski's (3) is

(7) Sentence h is true in L $=_{df}$ there is a proposition p such that des(h) = p and p.

The designation function des thereby serves to define the notion of translation between the object language and the metalanguage.

I conclude that Tarski's hidden idea was to define truth in terms of reference (cf. (1)) or designation (cf. (4)), and this proposal gives a correspondence theory of truth by schemas (6) and (7). This agrees with Raatikainen's concluding remark about what Tarski should have done to achieve a substantial theory of truth. But it was Carnap in 1942, rather than Field in 1972, who saw the light in this issue.

Sandu on Read

After repeated frustrated attempts from my side, the University of Helsinki decided to establish a new associate professorship in Theoretical Philosophy in 1997. Having served as the only professor in this subject for 20 years, I was of course very pleased about this solution. The person who was appointed, and soon upgraded to full professorship, was Gabriel Sandu, a Rumanian who did his studies in philosophy in Helsinki. With a Ph.D. degree in 1993, he became an expert in logic and philosophy of language, with important results especially in Hintikka's game-theoretical semantics and the IF logic.

Sandu's essay on the Liar Paradox gives a convincing criticism of Read's attempted solution. It is a fine example of his rigorous style and careful argumentation.

Manninen on Carnap

Juha Manninen is an old friend from the student years. When the students of philosophy in Helsinki founded their own association Dilemma in 1967, we were both members of the first governing board. Manninen became professor of General History of Ideas and Learning at Oulu already in 1977, but we have ever since had many common projects especially in the field of the history of philosophy in Finland.

One of our common interests is Eino Kaila. When I edited Kaila's collected works in Finnish for his centennial anniversary in 1990, I was struck by the scarcity of Kaila's correspondence. "My mother burnt letters after my father's death", Eino Kaila's son Olli told me. This fact partly explains why we still lack a biography of Kaila's life and work. I have myself used Carnap's letters to Kaila as an interesting material – these letters are in G. H. von Wright's collection in Helsinki, but not in the Carnap archive in Pittsburgh. As an excellent scholar, Manninen has been able to recover from various sources quite a lot of new material related to Kaila and his association with the Viennese philosophers. As a nice example in his essay, one can mention Feigl's letter telling to Schlick how Kaila in 1929 defended his probabilistic account of reality in the meeting of the Vienna Circle. I hope that Manninen will soon be able to complete his planned book-length exposition of Kaila.

Manninen's exciting paper focuses on the development of Carnap's thinking from the preliminary versions of the *Aufbau* to the acceptance of the project of physicalism. Carnap's own motivations, and his relations to Neurath, Wittgenstein and Kaila, receive a new illuminating account. I think this picture of Carnap could still be completed by considering the neo-Kantian ingredients of his early work.

After reading Manninen's analysis, one may again wonder about Carnap's 1961 Preface to the English translation of his *Aufbau* (Carnap, 1967). Even though the physicalist turn in 1931–34 is mentioned, Carnap wrote that he still agrees with "the philosophical orientation which stands behind the book". As the basic elements he would chose "something similar to Mach's elements, e.g., concrete sense data". Had he forgotten all the criticism (raised by Kaila and others) against the outdated psychology of his account of the autopsychological basis?

Aho on Epistemic Logic

Tuomo Aho's 1994 Ph. Thesis is an expert work on propositional attitudes. He has subtle knowledge about the philosophy of logic and its history. With me Aho has written a joint study which applies the basic ideas of Hintikka's account of propositional attitudes to memory (see Aho and Niiniluoto 1990).

Now Aho rejects a Hintikka-style truth-definition for attitude sentences as "completely and fundamentally mistaken". This is somewhat exaggerated, since Aho in fact accepts half of the definition (see his (2)), but he rejects that half which leads to the famous problem of logical omniscience. I don't wish to deny the seriousness of the issue about logical omniscience (see also Rantala's essay). But Aho's grounds for his negative thesis can be contested. He starts from the medieval view that the basic entities in logic are acts of judgment of the form a: p, where a is a subject and p is a mental or spoken sentence. On this analysis, there are hardly any logical relations between propositions. For example, a: p does not entail a: p ∨ q. If acts of judgments are located in time, then even a, t: p fails to entail a, t' : p, when t and t' are different points of time. Getting rid of such bonds to mental propositions was the device which allowed Bolzano and Frege to build up logic upon a non-psychologist foundation. By the same token, it is not necessary to think that attitude sentences like "a believes that p" or "a remembers that p" have to refer to the attitudinal mental states of agent a. The logic of propositional attitudes is not an attempt to describe the psychology of such attitudes.

Aho adds interesting remarks on the problem of intentional identity - the topic of Esa Saarinen's 1978 doctoral dissertation. He also sketches a reinterpretation of Hintikka's perspectival quantifiers. I would not be ready to drop the causality requirement concerning the links of perspectival perceptual world-lines to the actual world, since they are useful in formalizing the distinction between statements like "a sees b" (without a success condition) and "a looks at b". Other applications include the formalization of seeing as -statements. But Aho is right that the causality condition behaves differently when we move to other attitudes, like memory or imagination. Further critical work in this area is certainly welcome: in his recent autobiography, Hintikka complains that the duality of cross-identification methods is "probably the most important neglected idea in contemporary analytic philosophy" (Hintikka 2006, p. 32).

Rantala on Perception

Veikko "Vexi" Rantala was in Hintikka's group in the 1970s. The younger members of the team, Esa Saarinen and Lauri Carlson, called him "the creator of urn models". Rantala's works range from definability and infinitary logic to scientific theories, artificial intelligence and artistic representation.

Rantala's essay starts from Marr's problem of how a viewer-centered visual representation is transformed to an object-centered one. Hintikka's distinction between perspectival and physical cross-identification is again relevant here. Kaila believed that the tendency of interpreting our perceptions as being about three-dimensional physical objects is an inborn disposition of our psychological apparatus. Rantala links this issue with the ability of connectionist systems to approximate ideal symbolic systems. Using his skills as a logician, Rantala then formulates a logic of contextual vision (agent sees in the context α that φ), where noncontextual "pure" seeing is a special case with a tautological context, and shows how the problem of logical omniscience can be handled by allowing nonnormal possible worlds. It would be interesting to see how Rantala could add quantifiers to his propositional logic of vision.

Kakkuri-Knuuttila and Knuuttila on Meaning Finitism

Marja-Liisa Kakkuri-Knuuttila and Simo Knuuttila are also old friends from the youthful years of Dilemma. Marja-Liisa and I studied mathematics and logic in the same groups. Simo studied theology. In his doctoral dissertation in 1976, he applied Hintikka's accounts of Aristotelean modality to the interpretation of medieval logic. Marja-Liisa's dissertation was on Aristotle's dialectics, but she is also an expert on the methodology of science.

Meaning finitism is a philosophical doctrine that the Edinburgh school has used to argue that social factors influence science through the conventional character of language (cf. my reply to Kusch below). Kakkuri-Knuuttila and Knuuttila note that the background orientation of this doctrine is resemblance nominalism: the application of a classificatory term to the next case depends on its similarities to earlier ostensively learned instances. They also observe that this kind of nominalism need not be antirealist with respect to particulars and their resemblances. I agree that nominalists usually are ontological realists who assume the existence of mind-independent particulars (*Critical Scientific Realism*, p. 28). But the main problem is whether nominalism, which wishes to operate without assuming transcendent or immanent universals or properties, can explain *why* our predicates are correctly applicable to some objects and not to others.

A predicate nominalist simply explains properties by means of predication: an object is red, because we apply the predicate 'red' to it. Resemblance nominalism is slightly more sophisticated: we first apply the predicate to some paradigm instances, and then to all similar cases. If r is a paradigm instance of the predicate 'red', the extension of 'red' can be defined in the following way:

(8) The extension of 'red' = { x | x is similar to r }.

If similarity is understood as a well-defined relation, the extension defined by (8) might be taken to include all red objects – for example, all the unexamined wild strawberries that are in this summer ripe in Finnish forests and meadows. Then (8) would justify the truth condition of type (6) for the sentence 'This strawberry is red'. But the main problem with the nominalist formulation (8) lies in the notion of similarity: if the nominalist's objects are propertyless particulars, it does not make sense speak about their similarity; but if these objects have some qualities or attributes which serve as grounds of their similarity, then the notion of property is needed after all.

Recall that in Carnap's semantics predicates designate properties and properties in turn determine extensions (5) and truth-conditions (6). But it is important to note that Carnap's treatment does not presuppose properties as universals. They can as well be tropes or quality-instances (like the redness-of-this strawberry). Such tropes have in a natural way similarity relations which we may be able to recognize and discriminate by our senses. Physical objects can be understood as bundles of co-existent tropes, and properties as classes of similar tropes. Then an object is red if and only if it contains a trope which belongs the class of reddish tropes. I have called this view "tropic realism" (see *Critical Scientific Realism*, p. 30, and Niiniluoto, forthcoming).

Meaning finitists have another line of objection to (8). Barnes and Bloor argue that there are no rules which distinguish "in advance" all the things to which a term will be correctly applicable. When we encounter the next possible instance, "ostensive learning offers us *nothing* by way of guidance in the classification". In each future case, the community of language users is free to consider which similarities are relevant or irrelevant and thereby to decide whether the term is applicable or not. Meanings have to be "constructed as we go along", "as we move from case to case". According to this radically nominalist view, the meaning of a term is the class of its actual applications so far.

Kakkuri-Knuuttila and Knuuttila point out convincingly that this kind of theory leads to odd conclusions: "In our empirical world, the meaning of the term 'duck' does not change every summer when it is realized that new ducks are born or when others die." Exaggeration of formal considerations leads to an antirealist position, whereas the Edinburgh authors elsewhere explicitly acknowledge the relevance of both the "the world" and "the tradition" to classification and emphasize the empirical constraints and dependencies between successive applications of a term or applications of different terms.

Kusch on Finitism

Martin Kusch is German philosopher who has worked at Edinburgh and Cambridge. He defended his doctoral dissertation in Oulu in 1989, with a penetrating discussion of the distinction of language as calculus and language as universal medium in relation to Husserl, Heidegger, and Gadamer. I have always appreciated his unusual and breathtaking expertise of all areas and styles of philosophy. In particular, I have found Kusch's criticism of my scientific realism most helpful. As Kusch's Finnish is excellent, he has also widely read my texts published in Finland. This has allowed a friendly but uncompromising debate between us on a broad range of issues.

In *Critical Scientific Realism* (1999), I give an assessment of the Edinburgh school. Given Bloor's recent clarifications about his empiricist and naturalist intentions and his opposition to the "idealist" accounts of science by some constructivist sociologists, I conclude that much of his work is compatible with scientific realism. Concerning meaning finitism, my strategy is similar. I try give at least four different interpretations of the degree of determination of the future applications of a term: (F1) full determination, (F2) vagueness, (F3) partial determination, and (F4) full undetermination. To these alternatives we can add the case where the definition of a concept is changed (F0). I argue that F0, F1, F2, and F3 are acceptable to a realist, while F4 conflicts with the basic principle needed in semantic realism:

(9)   We choose the language L, and the world chooses which sentences of L are true.

I then argue that some of Bloor's formulations seem to support F4, but that this position is not tenable.

Principle (9) reflects my conviction that languages (both their syntactical rules and semantical meanings) are conventional social institutions. For me, meanings are not "given", as "meaning determinists" allegedly claim (Kusch

2002, p. 200), but created and supported by the linguistic community. In this institutional approach I agree with the finitists – with the proviso, accepted by Kusch, that genuine discoveries about institutions should be possible. I have applied this account to languages, legal norms, and scientific theories (see Niiniluoto 1981; 2003b). But with inspiration from Popper's World 3, I argue that this approach is acceptable to a scientific realist.

Until now I have received no responses from the advocates of "the strong programme". Kusch has given his own formulation of finitism in his *Knowledge by Agreement* (2002) without referring to my criticism. Therefore, I am now happy to see what kind of counter-arguments one of the best experts on finitism can offer.

Kusch denies that Bloor holds the extreme position F4. But still he maintains that the realist principle (9) is refuted. As he puts it elsewhere, meanings are not sufficiently "stable" to determine extensions, and it does not "make sense to assume that terms are true of their extensions" (*ibid.*, p. 201). So let us go through his five main points.

First, even though many terms of natural language are vague, I am not claiming that all concepts are vague, and I am not assuming that this is the position of meaning finitism (F2). Moreover, Kusch uses the notion of vagueness as referring to a situation where "the communal practices of evaluation allow for different speakers to apply both C and non-C to a given entity" (see also *ibid.*, p. 208). In my usage, vagueness concerns situations where neither C nor non-C applies to problematic borderline cases.

Secondly, I agree that ostension ("This [pointing to an object] is red!") is not essential, but rather the idea of meaning specifications by paradigms (Kusch's "finite learning sets") and similarity judgments (see (8)).

Thirdly, it is true that I have not emphasized Bloor's strategy of reducing meanings and rules to "our instincts, our biological nature, our sense experience, our interaction with other people" etc. This is indeed a central point, noted also by Kakkuri-Knuuttila and Knuuttila (see above). Thus, the next case seems to be at least partly determined (F3), so that the realist principle (9) might be valid after all. But here I think Bloor's suggested reduction is incomplete. A theory of meaning for a naturalist should explain how we are able to communicate with each other about the external world and what contribution nature provides to this achievement. For example, in biological taxonomy, it is not *our* biological nature that is relevant but the biological nature of plants and animals. With some exceptions ("color blind" persons who cannot visually detect a difference between red and green), normal human sensory apparatus is able to discriminate between colors. For a naturalist, the similarities and differences between color perceptions cannot be figments of our mind or imagination, but they must correspond to relations between external objects. Kusch himself admits that in classifying cats he attends primarily "to the animal", its fur, sounds, and behavior (Kusch 2002, p. 209). If this sort of realism is acknowledged, then meaning specifications may appeal to qualities, tropes, or properties, as shown above in (4) and (5), and the whole nominalist idea of building up application sets by adding individual instances one by one becomes implausible.

In my view, meaning finitism has credibility at best as a "childish" theory: it may tell how children learn to apply concepts of everyday language. When a child sees the first instances of the predicate 'pig', she will venture to new applications, but seeks the approval of her parents. But soon a child will learn also semantically relevant generalizations ("Pigs have a snout and a tail"). (For inductive generalizations in concept learning by machines, see Niiniluoto 2005a.) When I go alone to a meadow to eat strawberries, I don't need any more reconfirmation from my community to pick out and correctly identify these sweet red delicacies. When a botanist defines the plant *Fragaria vesca* by some species-specific traits, and when a physicist defines the color red (as the property of a surface which is able to reflect light rays of wavelength between 647 and 700 nm), we have moved to a level of rules and definitions which don't rely on the method of learning instances. The same holds of Kusch's (2002) account of "truth finitism": a child may learn to apply the predicate 'true' from exemplars, but a philosopher (like Tarski) employs the Carnapian strategy of explication to make this concept precise. These remarks do not change the fact that the adoption of the meanings of such terms as 'strawberry', 'red', and 'true' are social conventions open to debate and change in the linguistic community.

Fourthly, mathematics gives nice examples of full determination of meaning by means of finite specifications (F1). The infinite extension of the predicate 'odd' within the class of natural numbers is uniquely defined by the arithmetic predicate 'not divisible by 2'. The sum function is defined by recursion equations which need only two lines on paper (or in an individual mind). Unfortunately Kusch accepts Kripke's uncritical footnote where this objection is dismissed. The Wittgensteinian worries about the continuation of mathematical series concern situations where the participants already understand the sequence of natural numbers. It is not interesting to contemplate situations where participants have some contingent defects or limitations. For example, a child may be able to count only to 999, so that it is not possible to explain her sums which are larger than 1000. Most of us have learnt the concept of sum by the rule of adding numbers digit by digit, but this is no objection to "the Algorithmic Response". The possibility that the recursion equations involve only quantification up to some finite limit h has no relevance, either, since the restricted quantifier ($\forall x < h$) is of course as acceptable to a finitist as the unrestricted one.

Fifthly, Kusch points out that Bloor's meaning finitism is not the same as Kripke's meaning scepticism. It is true that Bloor proposes a "communal" solution to Kripke's worries – and Kusch has written a new book as a defense of Kripke (see Kusch 2006). So it is the case that Bloor's whole approach, especially the problem of the next case, is motivated by the Wittgensteinian problems. If I am right about the Algorithmic Response (the fourth point above), one wonders whether Kripke's Wittgenstein is investigating a pseudo-problem.

Wolenski on Induction

Jan Wolenski is an old friend of Finland. The co-operation between Finnish and Polish logicians and philosophers has been fruitful ever since the 1960s.

I wrote my master thesis in mathematics in 1968 for Gustav Elfving – the same professor who taught game theory to Jaakko Hintikka (see Hintikka 2006, p. 12). The topic was Bayesian statistical tests. The philosophical problem of induction I learnt from Georg Henrik von Wright's books, and the techniques for treating inductive generalization from Hintikka's system of inductive logic. Due to this background, I have found Popper's rejection of induction as misguided – just as Wolenski in his essay.

Wolenski links the problem of induction to Lindenbaum's theorem about maximal consistent extensions of consistent sets. A finite and consistent set C of true singular sentences confirms a generalization T if T entails the elements of C. Larger sets C of this kind give better confirmation of T than smaller sets. This is essentially the hypothetico-deductive (HD) model of confirmation.

Metalogically, Lindenbaum's theorem guarantees that C has a maximally consistent extension, but it is not decidable whether T belongs to it or not. Full justification of T cannot be achieved, since that would require the construction of an infinite set of confirmations.

One complication to Wolenski's treatment arises in the case of universal conditionals of the form $T = \forall x(Ax \to Bx)$. Here a positive instance $Aa\&Ba$ is not a deductive consequence of T, and one may debate whether other instances of the form $\sim Aa\&Bx$ and $\sim Aa\&\sim Bx$ can also have confirming power with respect to T. Perhaps Wolenski's tools of metalogic are applicable to this situation as well?

Kuipers on the Hypothetico-Probabilistic Method

I met Theo Kuipers for the first time in Warsaw in 1974. Ever since our interests in formal methodology and philosophy of science have been close and parallel: inductive logic, analogical inference, truthlikeness or truth approximation, scientific progress, and more recently abduction. I am grateful to Kuipers for this long friendship and philosophical co-operation.

I also wish to thank Kuipers for the kind words about my old book *Theoretical Concepts and Hypothetico-Inductive Inference*. Partly written together with Raimo Tuomela, it was a part of my doctoral dissertation which I defended in Helsinki on December 5, 1973. Its programme of "non-inductivist inductive logic" is still something that I would endorse, even though my best Popperian friend David Miller has never acknowledged any virtues of anything with a whiff of induction.

The main idea of the book was to study inductive probabilities of the form $P(g/e\&T)$, where g is a generalization and e is a singular evidence statement in an observational language L and T is a theory in a theoretical language which extends L by new theoretical concepts. If it turns out that g is inducible from e and T, but not from e alone, theory achieves inductive systematization. As the theoretical terms in T may be indispensable for this achievement, Hempel's important argument for scientific realism is verified. Here T is typically a hypothetical theory which is hoped to gain support from its ability to achieve inductive systematization, so that the term "hypothetico-inductive inference" seemed to me appropriate.

Kuipers correctly notes that the book also briefly suggests a hypothetico-inductive method of testing hypotheses. In the traditional HD-method, theory T entails a test statement E, and E confirms H (by the positive relevance criterion) and ~E falsifies T (by modus tollens). In the HI-method, T is positively relevant to E, so that (by symmetry) E again confirms T and ~E disconfirms T. I have noted instances of this method in some of my writings on confirmation and abduction: for example, if T inductively explains E, then E confirms T. But, as Kuipers observes, no systematic account of the HI-method has been developed so far. His formulation, where the HI-method is a concretization of the HD-method (i.e., includes HD as its extreme special case), is a beautiful piece of work.

I have recently explored some of the relations of my estimated verisimilitude function ver and Kuipers' account of empirical progress and truth approximation (see Niiniluoto, 2005b). He challenges me to investigate whether the HI-method is functional for truth approximation. In my sense, this would mean that success in HI-tests increases the estimated verisimilitude of a theory. My initial reply is positive, but the tricky details and qualifications have to be left for another occasion.

Levi on Truthlikeness

As anyone can see in the Preface of my book *Truthlikeness* (1987), Isaac Levi is – together with Carnap, Popper, and Hintikka – one of its four heroes. Levi's treatment of induction in terms of cognitive decision theory in his *Gambling with Truth* (1967) has been a rich source of inspiration for me, ever since I bought my own copy of it in 1971.

In my first paper on truthlikeness in 1975, I transformed Risto Hilpinen's Lewis-type possible worlds framework into an ultimate partition B in Levi's sense, where the complete answers are constituents of a monadic first-order language L. It is known that each generalization in L can be represented as a disjunction of such constituents. My idea – which came to my mind in listening to Hilpinen's lecture in the 1974 Warsaw congress – was to first define the distance between constituents and then to define the distance of a disjunction from the true constituent. (See also Festa's paper.) For the latter purpose, inspired by Hilpinen, I proposed the combination of the minimum and maximum distances, which seemed more satisfactory than the average function proposed by Tichý (and later by Oddie). Such a distance can then be used as a loss function of a cognitive decision problem (in Levi's and Hintikka's sense), which gives us a probabilistic method of estimating distances from unknown truth. This allows us to conceptualize scientific reasoning as the maximization of expected verisimilitude.

I was unhappy about the minimum-maximum measure, since it lost information about the disjuncts between the extreme values. It was only in my paper for PSA 1984 when I found a more adequate measure which combines, as a weighted average, the minimum distance with a normalized sum of all distances. Intuitively, it is good if the best guess of the theory is close to the truth, but the theory has to pay a penalty for all of its mistaken guesses, and each penalty depends on the seriousness of that particular mistake. To my surprise, I observed that this min-sum measure essentially reduces to Levi's definition of epistemic utility in *Gambling with Truth*, if the distance measure between complete answers is trivial, i.e., all false complete answers are equally distant from the truth. In this case, the minimum criterion reduces to truth value, and the sum criterion to a content measure.

For me, this observation was an unintended instance of the principle of "concretization" that a good philosophical theory should contain its predecessors as special cases (cf. the essay by Kuipers). When I was eager to explain this result in *Truthlikeness*, I failed to pay attention to the fact that Levi had changed his definition of content soon after *Gambling with Truth*. In his essay, he thus correctly points out that in his later treatment all false complete answers need not have the same epistemic utility.

Still, in Levi's new approach it is the case that logically stronger false answers have a larger epistemic utility than weaker ones. In my view, this is not acceptable as a general principle. Another issue between us concerns the question whether some false answers can be better than some true answers. Having learnt from Levi that a tautology as the weakest true answer corresponds to the trivial answer "I don't know" to the original cognitive problem, I have argued that some falsities may be cognitively and practically better than such complete ignorance. For example, if I need to know the population of Finland, then the almost true point estimate 5.2 million is better than the disjunction of all natural numbers. Here, I think, a miss is not as good as a mile. Of course, the best answer is provided by the true complete answer, so that in that sense truth is preferred to falsity. But, more generally, in most contexts it seems to me appropriate to require that the best false complete answers are more truthlike than a tautology, while the worst of such answers are misleading and less truthlike than a tautology. This gives a systematic way of restricting the weights of the two factors of the min-sum measure (see Niiniluoto, 2003c).

Levi thinks otherwise. In his diagnosis of the situation, he gives emphasis to his new distinction between fallibilism and corrigibilism. Levi is contemplating the epistemological value of the prospective acceptance of rival hypotheses, and his version of fallibilism requires that an accepted hypothesis will serve as a "standard of serious possibility" also in future inquiries. When a hypothesis h is accepted as evidence in this strong sense, all probabilities are conditionalized on h, so that h itself is "absolutely certain". As Levi clearly explains, this situation has to be distinguished from "suppositional reasoning", where inquiry proceeds on premises (e.g. idealizations) which are known or believed to be false.

Recall that my truthlikeness measure may be taken simply as a way of ordering hypotheses generated by an ultimate partition, and no question about acceptance is yet raised at this stage. One of the applications of these measures is a retrospective comparison and assessment of the progress of past theories, where we have no inclination to accept such already discarded theories. Thus, we can now see that Levi is opposing my measure as a basis of acceptance in his strong epistemological sense. Of course, my measure of truthlikeness as such does not define an acceptance rule, and when its expected value is chosen as the expected utility to be maximized, the acceptability of various kinds of hypotheses crucially depends on the probability distribution over the ultimate partition. But even there Levi's standards of acceptance are clearly stricter than mine. My account appeals to the idea, prominent in the history of fallibilism, that the approach to truth may progress via tentatively accepted false theories. (Kuipers calls this an "instrumentalistic" methodology in truth approximation.) So sometimes a false but close-to-the-truth theory is better than a safe-but-trivial truth. This is also in harmony with Levi's earlier recommendation that we must "gamble with truth" in order to gain information. To give a dynamic probabilistic representation of such progressive steps in science one should allow conditionalization on uncertain hypotheses, and there are many open issues in such a project (for some preliminary remarks, see Niiniluoto, 1997).

Festa on Statistical Verisimilitude

When Roberto Festa came to Helsinki for the academic year 1981–82 as a visiting scholar, he experienced a cultural shock with shy and quiet Finns. But our encounter was fruitful. I was intensively working on truthlikeness, and Festa had himself independently discovered some similar ideas. So I was able to instruct him in his work on verisimilitude as an epistemic utility. When Festa defended his Ph. D. Thesis in Groningen in 1993, I was one of his six opponents.

I am happy to see from Festa's essay that it is still possible to carry forward the research programme of verisimilitude. He starts by an excellent summary of some earlier work on the verisimilitude of qualitative generalizations, and then extends it in a novel and natural way to statistical generalizations. (Note that in his figures of monadic constituents Festa shades empty cells, while I have used shading for non-empty cells.) Festa makes a distinction between KW-strategies, which treat a generalization as a disjunction of constituents, and KI-stategies, which directly analyze the content of a generalization in terms of its claims about the cells or Q-predicates. As he notes, it is possible to derive from my KW-strategy results about his and Tuomela's KI-strategy. While the KI-strategy turns out to have the advantage that it can be nicely applied to the case of statistical verisimilitude, my guess is that there is a probabilistic KW-version from which these new KI-measures can be derived.

Oddie on the Value of Truthlikeness

Graham Oddie participated in 1977 in the congress "The Logic and Epistemology of Scientific Change" that I organized in Helsinki with Raimo Tuomela (see Niiniluoto and Tuomela, 1979). We soon started to send letters between Finland and New Zealand with putative counter-examples to our rival explications of verisimilitude. Later Oddie came to Helsinki in the spring of 1986 as a visiting scholar. He was just finishing his *Likeness to Truth* (1986), which was based upon his Ph. D. Thesis at the London School of Economics, while I was completing my *Truthlikeness* (1987).

Oddie expresses his dissatisfaction with the divide of theoretical and practical philosophy. The blame should be put on Aristotle who introduced the distinction. It has been preserved not only in Scandinavia, but in Scotland as well. I have been told that in some Scottish universities there were no doors between the Department of Logic and Metaphysics and the Department of Moral Philosophy. I agree with Oddie that such barriers should be abolished.

In particular, as his essay proves, there are many interconnections already in the field of truthlikeness. Firstly, we should understand in what sense truthlikeness and the tracking of truthlike propositions is valuable. Here Oddie's argument appeals to Platonism with respect to propositions and works of art, but a similar story could be given by accepting instead the Popperian constructivist ontology of World 3 (see below). Secondly, Oddie gives interesting novel illustrations how techniques in value theory and the theory of measurement may help to develop the theory of truthlikeness - even to bring new light to Miller's famous argument about language dependence. He appeals to Dana Scott's theorem on additive representations. I knew this theorem well, when I wrote my Licentiate Thesis in mathematics in 1971 on the representation theorems of qualitative probability, but it has never occurred to me to apply these results in the context of verisimilitude.

Sintonen on Science

Matti Sintonen complemented his Helsinki education in Oxford. With his insights about scientific explanation, he has been the right person to develop Hintikka's question-theoretical approach to scientific inquiry. I am pleased that Sintonen is acting in my position in theoretical philosophy, while I am in the office of the rector of the University of Helsinki.

Sintonen's essay gives an enlightened account of the scientific method in the sense of Charles S. Peirce. I first learned the Peircean account of the self-corrective method of science from the lectures of my professor Oiva Ketonen, whose own thinking was inspired by Kaila, Dewey and C. I. Lewis. When I started my studies in theoretical philosophy in 1965, Ketonen still included Cohen's and Nagel's *An Introduction to Logic and Scientific Method* in the curriculum. Later many Finnish philosophers have been interested in Peirce (among them, Hintikka, Hilpinen, myself, and Haaparanta), and in the younger generation we have several excellent Peirce scholars (Erkki Kilpinen, Sami Pihlström, Mats Bergman, Ahti Pietarinen).

In my view, Sintonen is right in emphasizing the realist assumptions underlying Peirce's account of method (cf. *Is Science Progressive?*). It is also correct to see that for Peirce the subject of knowledge is the scientific community (cf. Niiniluoto, 2003b). As Peirce put it in "On the Doctrine of Chances", scientific reasoning presupposes that we identify our interests with those of an unlimited community: "Logic is rooted in the social principle" (*CP* 2.654). But if this community is assumed to be an ideal one, as the transcendental pragmatics of Karl-Otto Apel suggests, there is hardly any way of relating it to the sociological study of actual human collectives. Sintonen, on the other hand, provides a sustained argument to the effect that Robert Merton's ethos of science can be understood as showing how impersonal and disinterested pursuit of truth is possible in the actual community of scientists. In this sense, Merton makes Peirce's method explicitly sociological. Even though I have some reservations to the Mertonian perspective that open expert community of investigators itself operates "in the democratic order", I certainly approve the importance of liberal democracy as a condition for the autonomy of research (*Critical Scientific Realism*, pp. 296–301).

In the collection of Merton's main essays (Merton, 1973), Peirce is not mentioned at all. However, it has been shown by Erkki Kilpinen (2000) in his dissertation that Peirce in fact had important indirect influence on American sociology. The happy similarity between Peirce's and Merton's ideas serves to indicate that pragmatism and the sociology of science, when developed in a right critical spirit, need not lead to any anti-realist conclusions.

Tuomela on Social Groups

Co-operation with Raimo Tuomela was important for my early orientation to scientific realism. Raimo is some years older than me, and he had made his Ph. D. on scientific theories in 1968. He was interested in the deductive gains due to theoretical or auxiliary concepts, to be discussed in his *Theoretical Concepts* (1973), while I wished to study their inductive gains. Our joint book in 1973 was then a part of my doctoral dissertation. Later we have given joint postgraduate seminars on scientific realism, but our views differ on the question of truth: Tuomela has opted for a Sellarsian intralinguistic notion of truth as assertability, so that he has also sympathies towards Putnam's internal realism, while my critical realism is based upon a Tarskian correspondence theory of truth.

Tuomela is a leading expert on social groups. His general aim is to analyze the social realm "without spooky entities". His essay gives a penetrating analysis of social groups in two senses: as singular entities and as non-entities existing only in a vicarious sense. It seems quite useful to represent the structure of social groups by the tuple <D, P_1, ..., P_m, TR>, where D is an open class of position holders, P_1, ..., P_m are positions, and TR the task-right system about the positions. One can agree with Tuomela's criticism against Searle's thesis that a corporation can be just a set of status functions without position holders. But there seems to be an important difference between social groups, where membership is primary to positions (e.g., a family), and social institutions, where the task-right system is primary to historically variable position holders (e.g., the University of Helsinki, the Philosophical Society of Finland). The latter type of institutions are, in my view, typical entities in the Popperian World 3.

Tuomela argues that social groups are not persons, because "they have neither bodies nor minds". If persons are required to have bodies and perceptions, it is no doubt correct to conclude with Tuomela that groups can be person only in a metaphorical sense. However, one should also recognize here that social institutions may be so-called "legal persons". For example, corporations and association are legally defined entities which carry their rights and duties, can own their own property, and make legal contracts and transactions. Similarly, as I have argued elsewhere in an essay on business ethics (see Niiniluoto, 2005c), corporations have ethical responsibilities and commitments. These social aspects of "legal persons" cannot be distributed individually to their current position holders (e.g., the debt of a limited company is not a personal debt of its owners), so that they are social agents in a genuine sense.

Mäki on Social Reality

Uskali Mäki belongs to the Helsinki school of scientific realism. In his position at Rotterdam, he has become an established authority in the philosophy of economics. Background in economics gives him an excellent perspective on broader issues of social reality.

Mäki explores the viability of realism about social objects and social sciences. He points out that the mind-dependence of social objects is "regularly neglected in the writings of realism in general". I have tried to account for this feature by locating social entities in World 3 and discussing that domain of reality separately from physical World 1 entities. I have argued against Putnam that a critical realist can combine conceptual pluralism with a non-epistemic account of truth. Mäki evaluates Putnam's distinction between metaphysical and internal realism from the viewpoint of social reality. He notes that social objects satisfy some of Putnam's aspects of internal realism (mind-dependence, social constitution of its objects), but also some aspect of metaphysical realism (the social world is pre-conceptualized by social actors, many of its objects are self-identifying). The well-argued conclusion is that Putnam's two realism are "overly burdened artificial conglomerates".

Pihlström on Values

Sami Pihlström defended his doctoral dissertation in 1996. He contrasted Putnam's internal realism and my critical realism, and with some help from classical pragmatists concluded with a synthesis, where ontology is relative to human practices. Later he has attempted a bold synthesis of pragmatism and Kantianism by "naturalizing the transcendental". I have been happy to see how Pihlström has rapidly grown to an independent thinker who seeks and defends his own ideas via metaphilosophical reflections. He has amazing knowledge about almost everything within contemporary philosophy, and in particular he knows my work perhaps better than anyone else. During a decade he has impressively published more than many philosophers in their lifetime.

In his dissertation, Pihlström followed me in accepting the basic idea of Popper's World 3, the domain of human-made artefacts and institutions (cf. Niiniluoto, 2006a). However, in the spirit of the pragmatist strategy of abolishing philosophical distinctions, he argues that "a clean dichotomy" cannot be drawn between World 1 and World 3. For my realism this is a crucial issue, since I am strongly opposed to such imperialist views of World 3 (e.g., Hegelian objective idealism and social constructivism) which locate *all* entities in World 3. Some of Pihlström's formulations of his "pragmatized ontology" seem to place him in this camp. Whereas I defend the historical and ontological primacy of World 1 over World 3, he rejects "the materialist idea that there is, or even could be, purely material World 1 entities with no relation whatsoever to human culture". But if "to exist is to play a role in humanly relevant practices", what about those billions of years when the universe had yet no human beings? Is Pihlström making the sweeping claim that *anything* in the world and its history *might* be relevant to human conduct? This is implausible, since only a omnitemporal God would have access to all those past facts that have disappeared in the history without leaving traces to us. Or is he asserting that e.g. the Big Bang came to existence when cosmologists started to write about it? Does the human language magically bring about past objects and events just by naming them? These traditional arguments against idealism can be directed at the positions of Putnam and Pihlström.

The same problem arises when Pihlström denies my distinction between factual and cultural value-properties of technological artefacts. For example, nuclear power plants produce electricity. Such institutions have been built and are in operation, because human beings wish to have energy for their practical purposes. One can agree with Pihlström that the "identification" of such factual properties of power plants happens in relation to the cultural context of human energy production and consumption. But why should that imply that the truth value of a statement like "The net production of the Loviisa nuclear power plant in 2005 was 21 804 Gwh" is in some way dependent on our values? Similarly, the fact that today children love dinosaur exhibitions in science centers does not make the truth of the sentence "Dinosaurs lived on earth 100 million years ago" value-laden. It is this kind of value-independence that is relevant to the debate about the fact-value distinction.

Pihlström offers a sharp challenge of my moral constructivism and ethical relativism. I am sure that our fruitful debate will continue in other forums as well, but here I try to give a reply to four points.

First, Pihlström's comparison of ethics and mathematics is novel - and delightful, since it in fact supports my ethical relativism. Mathematical statements (like '2 + 2 = 4') are indeed relative: either they are derivable from axiom systems (like Peano arithmetic) or they are true in mathematical structures (like the natural numbers). This is analogous to my claim about the relativity of moral claims (like 'stealing is wrong') to socially constructed systems or auditories.

Secondly, Pihlström compares my position to the "absolute conception of the world" of Bernard Williams. Given my conceptual pluralism, this is misleading. He goes on to argue that the truth values of even my relativized moral statements (e.g., 'In Christian ethics, respect for your parents is good') problematically assume a "God's-Eye View". This is somewhat surprising, as Pihlström has just himself defended the absolute and universal character of morality. It seems to me that it is his position, not mine, which needs the God's Eye View. The same issue arises with Pihlström's discussion of "metaphysical guilt", which clearly has religious connotations to Augustine's and Luther's doctrine of original sin. These absolutist tendencies – philosophically derived from Kant, Kierkegaard, and Wittgenstein – appear to be in conflict with Pihlström's own pragmatism. One gets the impression that the Kantian distinction between transcendental and empirical (by the way, why is *this* distinction saved?) is a way of reintroducing something like God's Eye into a philosophical system in the guise of a "transcendental subject".

Thirdly, it is important to distinguish between the validity of a moral statement and the demands that moral assertions make. A moral realist may claim that certain value statements are universally valid – they are "right, period", as Pihlström puts it. A modest ethical relativist may admit that often moral principles are intended by their supporters to be universally applicable (if I think that killing is wrong, then killing is wrong for anybody always and everywhere in all circumstances), but denies that there are grounds for assigning truth values to such general demands. I have argued that there still are meta-principles by which intercultural criticism of moral systems is possible. But tolerance of different positions is also needed, as Edward Westermarck well argued in his ethical relativism. Pihlström, in his support of absolutist ethics, does not stop to ask whether it is morally desirable that people have unconditional ethical commitments. Strong forms of such ethical and religious commitments are advocated by fundamentalists, who have brought about wars and misery throughout the human history. Today these fights continue as "a war against terrorism", where both parties are absolutely committed to their own ideals (cf. Niiniluoto, 2006b).

Fourthly, scientific realism acknowledges that the axiology of scientific practice includes such epistemic values as truth and information. When the scientists pursue the goals of inquiry, *they* are committed to these values. But to add that such a commitment has be "realistic" (see step 7 of Pihlström's argument) does not presuppose that such commitments have to be universally valid, i.e., valid in the sense of pragmatic moral realism. The scientific values are not shared by all groups in contemporary society. For the success of science, it is sufficient if the scientific community acts in accordance to these values – and there is enough social support to guarantee the funding of research. In practice, there is no value consensus even on the level of the

Sihvola on Religion

As a historian of ancient philosophy, Juha Sihvola is a student of Simo Knuuttila and Martha Nussbaum. His broad interests include also contemporary ethics and political philosophy. With a group of scholars that I have chaired for a decade, Sihvola has actively participated in the production of books and radio programs on the present state and future of Finland, Europe, and the global world order.

Philosophy of religion has not been my main field of interest, since I settled my relation to religion already as a teenager. My first published writing in philosophy in 1966 was a review of a Finnish translation of Lucretius. I praised his enlightened way of combining the peace of mind with awareness of human mortality. Later I have defended a science-based world view, where knowledge claims are fallible and revisable. In spite of its interaction with the dynamic knowledge base, ethics involves personal commitments that are not derivable from science. Religion, as a doctrine that involves assertions about a transcendent word, has no place in this secular world view. I am not even attracted by the idea that one should hope that God exists. But many moral principles in world religions are in harmony with my own humanist ethics.

Sihvola is thus right in treating me as a kind evidentialist. All publicly preached and practiced doctrines about the world should be open to external criticism. Hence, from my perspective, fideism and realism is the worst combination in the philosophy of religion. Concerning anti-realistic theism, I tend to agree with Knuuttila that it is difficult to see the point of practices like prayers and confessions unless one's religious faith satisfies some minimal epistemic presuppositions. But, as Sihvola observes, it is not primarily a concern of philosophers what kinds of assumptions churches and believers satisfy. As long as religious practices are not harmful to their participants or others, I am willing to accept them on the basis of the principle of freedom of thought. I have no objections to Sihvola's weak form of religious commitment, as a way of reflecting the meaning of life, as long as its practitioners are aware of the dangers of deception and the lures of self-deception in this area. In the same manner, as Sihvola acknowledges, supporters to religion should tolerate completely secular world views.

References

Aho, T. and I. Niiniluoto, 1990: "On the Logic of Memory", in L. Haaparanta, M. Kusch, and I. Niiniluoto (eds.), *Language, Knowledge, and Intentionality: Perspectives on the Philosophy of Jaakko Hintikka*, Acta Philosophica Fennica 49, The Philosophical Society of Finland, Helsinki, pp. 408-429.

Benacerraf, P. and H. Putnam (eds.) 1964: *Philosophy of Mathematics: Selected Readings*, Prentice-Hall, Englewood Cliffs, NJ.

Carnap, R. 1942: *Introduction to Semantics*, Harvard University Press, Cambridge, MS.

Carnap, R. 1967: *The Logical Structure of the World and Pseudo-Problems in Philosophy*, University of California Press, Berkeley and Los Angeles.

Hintikka, J. 2006: "Intellectual Autobiography", in R. E. Auxier and L. E. Hahn (eds.), *The Philosophy of Jaakko Hintikka*, Open Court, Chicago and La Salle, pp. 3-84.

Kilpinen, E. 2000: *The Enormous Fly-Wheel of Society: Pragmatism's Habitual Conception of Action and Social Theory*, University of Helsinki, Helsinki.

Kusch, M. 2002: *Knowledge by Agreement: The Programme of Communitarian Epistemology*, Oxford University Press, Oxford.

Kusch, M. 2006: *A Sceptical Guide to Meaning and Rules: Defending Kripke's Wittgenstein*, Acumen, Chesham.

Merton, R. K. 1973: *The Sociology of Science: Theoretical and Empirical Investigations*, The University of Chicago Press, Chicago.

Niiniluoto, I. 1981: "Language, Norms, and Truth", in I. Pörn (ed.), *Essays in Philosophical Analysis*, Acta Philosophica Fennica 32, The Philosophical Society of Finland, Helsinki, pp. 168-189.

Niiniluoto, I. 1997: "Inductive Logic, Atomism, and Observational Error", in M. Sintonen (ed.), *Knowledge and Inquiry: Essays on Jaakko Hintikka's Epistemology and Philosophy of Science*, Rodopi, Amsterdam, pp. 117-131.

Niiniluoto, I. 2003a: "Carnap on Truth", in T. Bonk (ed.), *Language, Truth, and Knowledge: Contributions to the Philosophy of Rudolf Carnap*, Kluwer, Dordrecht, pp. 1-25.

Niiniluoto, I. 2003b: "Science as Collective Knowledge", in M. Sintonen, P. Ylikoski, and K. Miller (eds.), *Realism in Action*, Kluwer, Dordrecht, pp. 269-278.

Niiniluoto, I. 2003c: "Content and Likeness Definitions of Truthlikeness", in J. Hintikka, T. Czarnecki, K. Kijania-Placek, T. Placek, and A. Rojszcak (eds.), *Philosophy and Logic: In Search of the Polish Tradition*, Kluwer, Dordrecht, pp. 27-35.

Niiniluoto, I. 2005a: "Inductive Logic, Verisimilitude, and Machine Learning", in P. Hajek, L. Valdés-Villanueva, and D. Westerståhl (eds.), *Logic, Methodology and Philosophy of Science: Proceedings of the Twelfth International Congress*, King's College Publications, London, pp. 295-314.

Niiniluoto, I. 2005b: "Abduction and Truthlikeness", in R. Festa, A. Aliseda, and J. Peijnenburg (eds.), *Confirmation, Empirical Progress, and Truth Approximation: Essays in Debate with Theo Kuipers*, vol 1, Rodopi, Amsterdam, pp. 255-275.

Niiniluoto, I. 2005c: "Onko yrityksillä moraalista vastuuta?", in I. Niiniluoto and J. Sihvola (eds.), *Nykyajan etiikka*, Gaudeamus, Helsinki, pp. 23-60.

Niiniluoto, I. 2006a: "World 3: A Critical Defence", in I. Jarvie, K. Milford, and D. Miller (eds.), *Karl Popper: A Centenary Assessment*, vol II, Ashgate, London, pp. 59-69.

Niiniluoto, I. 2006b: "The Open Society and Its New Enemies: Critical Reflections on Democracy and Market Economy", *The Tampere Club Series*, vol. 2, Tampere.

Niiniluoto, I. forthcoming: "On Tropic Realism", in L. Haaparanta and H. J. Koskinen (eds.), *Categories of Being*.

Niiniluoto, I. and Tuomela, R. (eds.) 1979: *The Logic and Epistemology of Scientific Change*, Acta Philosophica Fennica 30, Amsterdam, Horth-Holland.